The Age of Migration

Stephen Castles is Professor of Migration and Refugee Studies and Director of the Refugee Studies Center at the University of Oxford, UK.

Mark J. Miller is Professor of Political Science and International Relations at the University of Delaware.

The Age of Migration

Third Edition

Stephen Castles
and
Mark J. Miller

THE GUILFORD PRESS
New York London

Published in the United States of America by
The Guilford Press
A Division of Guilford Publications, Inc.
72 Spring Street, New York, NY 10012
www.guilford.com

Published in North America under license from
Palgrave Macmillan

Last digit is print number: 9 8 7 6 5 4 3 2 1

Library of Congress Cataloging-in-Publication Data
Castles, Stephen
The age of migration: international population movements in the modern world / Stephen Castles, Mark J. Miller.—3rd ed.
p. cm.
Includes bibliographical references and index.
ISBN 1-57230-900-8 (paper)—ISBN 1-57230-904-0 (cloth)
1. Emigration and immigration. I. Miller, Mark J. II. Title.
JV6032.C37 2003
304.8′2—dc21 2003001727

This book is printed on paper suitable for recycling and made from fully managed and sustained forest sources.

Printed and bound in Great Britain by
Creative Print & Design (Wales), Ebbw Vale

Contents

List of Illustrations, Tables, Boxes and Maps

Illustrations

Tables

Boxes

Maps

Preface to the Third Edition

The Age of Migration was originally published in 1993, with the aim of providing an accessible introduction to the study of global migrations and their consequences for society. It was designed to combine basic theoretical knowledge with up-to-date information on migration flows and their consequences for society. International migration has become a major theme of public debate, and *The Age of Migration* is widely-used by policy-makers, scholars and journalists. It is recommended as a textbook in politics and social science courses all over the world. As with the Second Edition, the Third Edition is essentially a new book. It has been thoroughly revised and updated. A new Chapter 5 has been added to examine growing concerns about the significance of migration for national security and sovereignty. This pays particular attention to the growing importance of undocumented migration and asylum for migration policies.

Much has changed in the world in the decade since publication of the first edition, yet the book's central argument remains the same. International population movements are reforging states and societies around the world in ways that affect bilateral and regional relations, security, national identity and sovereignty. As a key dynamic within globalization, international migration is contributing to a fundamental transformation of the international political order. However, what sovereign states do in the realm of migration policies continues to matter a great deal. The notion of open borders remains elusive even within regional integration frameworks, except for European citizens circulating within the European Union.

Important changes over the last decade include the greatly increased political significance of migration, which has become a major theme of domestic and international politics. Other new trends include rapid growth in labour migration to the new industrial economies of the developing world, the increase in asylum-seeker movements to developed countries, and the growth of racist violence linked to migration and diversity. At the international level, there is a strong tendency towards intergovernmental collaboration to improve control of migration.

At the beginning of the new millennium, a single event appeared to reshape public perceptions of international migration: the terrorist attacks of 11 September 2001. However, we argue that this event has not brought about fundamental changes in the complex processes which define the contemporary age of migration. Indeed, 11 September testified to an imperative to understand how international population mobility has transformed the security dilemmas of the world's most powerful state.

Governments around the world struggle to adjust to altered circumstances. Outmoded security concepts bear mute testimony to the importance of understanding the epochal transformations that characterize this period of globalization and increasing population mobility.

Other new features of the Third Edition include analysis of the proliferation of dual citizenship and of states which are simultaneously lands of emigration and of immigration. Human trafficking is increasing, and combating it has become a more important concern of states and of regional and international organizations. Similarly, recent years have witnessed heightened debate over connections between demographic change and international migration, particularly in the European context. More and more societies see themselves as culturally diverse – a trend symbolized by significant changes in citizenship rules in many places. However, citizens of many countries have become sceptical about the benefits of multiculturalism, leading to significant changes in integration policies in several places.

The authors thank the following for help in preparing and editing the manuscripts of the various editions: Gloria Parisi, Debjani Bagchi, Aaron C. Miller and Mary McGlynn in Delaware; Colleen Mitchell, Kim McCall and Lyndal Manton in Wollongong; and Margaret Hauser in Oxford. Mark Miller thanks the clerical staff of the Department of Political Science and International Relations at the University of Delaware and that of the Center for Migration Studies in Staten Island, New York, for their unflagging assistance. The maps were drawn by David Martin of Cadmart Drafting, Wollongong.

We would like to thank our publisher, Steven Kennedy, for his advice and encouragement, as well as his patience throughout the project. We are indebted to John Solomos, Fred Halliday, Ellie Vasta and Jock Collins for their constructive and helpful comments. The authors wish to acknowledge the many valuable criticisms of earlier editions from reviewers and colleagues, although it has not been possible to respond to all of them.

STEPHEN CASTLES
MARK J. MILLER

List of Abbreviations

AAE	Amicale des Algériens en Europe
ABS	Australian Bureau of Statistics
AFL–CIO	American Federation of Labor–Congress of Industrial Organizations
ALP	Australian Labor Party
ANC	African National Congress
APEC	Asia-Pacific Economic Cooperation
AU	African Union
BfA	Bundesanstalt für Arbeit (Federal Labour Office)
BMET	Bureau of Manpower, Employment and Training
CDU	Christian Democratic Union
CGT	Confédération Générale du Travail
CIA	Central Intelligence Agency
CSIMCED	Commission for the Study of International Migration and Cooperative Economic Development
CRE	Commission for Racial Equality
DIMA	Department of Immigration and Multicultural Affairs
DRC	Democratic Republic of the Congo
EC	European Community
ECOWAS	Economic Community of West African States
ECSC	European Coal and Steel Community
ESB	English-speaking background
EU	European Union
EVW	European Voluntary Worker
FAS	Fonds d'Action Sociale (Social Action Fund)
FN	Front National (National Front)
FRG	Federal Republic of Germany
FRY	Federal Republic of Yugoslavia
GATT	General Agreement on Tariffs and Trade
GCC	Gulf Cooperation Council
GDP	Gross Domestic Product
GDR	German Democratic Republic
HCI	Haut Conseil à l'Integration (High Council for Integration)
HLMs	*habitations à loyers modestes* (public housing societies)
ICRC	International Committee of the Red Cross

IDP	Internally displaced person
IIRIRA	Illegal Immigration Reform and Immigrant Responsibility Act
ILO	International Labor Organization
INS	Immigration and Naturalization Service
IOM	International Organization for Migration
IRC	International Rescue Committee
IRCA	Immigration Reform and Control Act
IT	information technology
KDP	Kurdish Democratic Party
MERCOSUR	Latin American Southern Common Market
MSF	Médecins Sans Frontières
NAFTA	North American Free Trade Agreement
NESB	non-English-speaking background
NGO	non-governmental organization
NIC	newly-industrializing country
NRC	National Research Council
OAU	Organization for African Unity (now AU)
OCW	overseas contract worker
OECD	Organization for Economic Cooperation and Development
ONI	Office National d'Immigration (National Immigration Office)
OPEC	Organization of Petroleum Exporting Countries
OWWA	Overseas Workers' Welfare Administration
PKK	Kurdish Workers Party
POEA	Philippine Overseas Employment Administration
PUK	Patriotic Union of Kurdistan
RSA	Republic of South Africa
SADC	South African Development Community
SCIRP	Select Commission on Immigration and Refugee Policy
SEA	Single European Act
SGI	Société Générale d'Immigration
TEU	Treaty on European Union
TPV	Temporary Protection Visa
UAE	United Arab Emirates
UN	United Nations
UNHCR	United Nations High Commissioner for Refugees
UNICEF	United Nations Children's Fund (formerly United Nations International Children's Emergency Fund)
UNPD	United Nations Population Division
WASP	White Anglo-Saxon Protestant
WFP	World Food Programme
WTO	World Trade Organization

Chapter 1

Introduction

The post-Cold War period began with a rush of optimism about democracy, capitalism and the prospect for humanity. Many viewed globalization as irreversible. Political scientists, sociologists and economists wrote about the demise of national states, the necessity of adaptation to market forces and the imperative of global democratization. Then, in 2001, 19 terrorists flew three fuel-laden, hijacked planes into the World Trade Centre in Manhattan and the Pentagon in Washington and the world changed forever. Or did it?

One of the defining features of the post-Cold War era has been the growing saliency of international migration in all areas of the world. International population movements constitute a key dynamic within globalization – a complex process which intensified from the mid-1970s onward. The most striking features of globalization are the growth of cross-border flows of various kinds, including investment, trade, cultural products, ideas and people; and the proliferation of transnational networks with nodes of control in multiple locations (Castells, 1996; Held *et al.*, 1999). At its core, globalization results in increased transnationalism: behaviour or institutions which simultaneously affect more than one state. The terrorist attacks on 11 September 2001 in fact constituted transnational political behaviour, as those perpetrating it were aliens engaging in violence against mainly civilian targets in another state in order to achieve political goals. Al-Qaida can be seen as an extremely effective transnational network, with multiple nodes of control.

One of the key analytical questions to be asked about '9/11' (as US observers have come to label the 2001 episode) is: how did it affect international migration? That one should feel compelled to ask such a question suggests the significance of the inquiry that informs this book. It is only recently that international population movements have been viewed as so significant that they warrant high-level scrutiny. In the post-9/11 environment, the central arguments informing this inquiry take on new urgency.

After the initial wave of euphoria over the end of the Cold War, the new era became marked by enormous change and uncertainty. Several states imploded and the very nature of warfare changed from violence waged between states to fighting within the boundaries of a state (UNHCR, 2000b: 277; Kaldor, 2001). About 90 per cent of conflicts in the post-Cold

War era have not involved classic conventional warfare between states and many of these have created large numbers of internally displaced persons or IDPs. Entire regions in Africa, Europe, Latin America and Central Asia verged on anarchy and ruin. Yet, at the same time, democratic institutions, liberal economic strategies and regional integration, although still challenged, have now become globally ascendant. The ambivalent nature of the post-Cold War period can be seen in the juxtaposition of global human rights norms with episodes of horrific savagery involving mass killings and expulsions of entire populations.

For some observers, the world at the dawn of the twenty-first century is one in the throes of systemic transformation. The global order based on sovereign national states is giving way to something new. However, the contours of the emerging new order are unclear. Hope and optimism coexist with gloom and despair. Other observers doubt that fundamental change can or will occur. The nation-state system still endures despite the growth in the power of global markets, multilateralism and regional integration. National states command the loyalties of most human beings and millions have fought and died for them in recent memory.

These contradictory trends and notions comprise the backdrop to the unfolding drama that has captured the attention of peoples and leaders: the emergence of international migration as a force for social transformation. While movements of people across borders have shaped states and societies since time immemorial, what is distinctive in recent years is their global scope, their centrality to domestic and international politics and their enormous economic and social consequences. Migration processes may become so entrenched and resistant to governmental control that new political forms will emerge. This would not necessarily entail the disappearance of national states; indeed that prospect appears remote. However, novel forms of interdependence, transnational societies and bilateral and regional cooperation are rapidly transforming the lives of millions of people and inextricably weaving together the fate of state and society. Major determinants of historical change are rarely profoundly changed by any single event. Rather, singular events like 9/11 reflect the major dynamics and determinants of their time. It is scarcely coincidental that migration figured so centrally in the chain of events leading up to the terrorist attacks.

For the most part the growth of transnational society and politics, of which international migration is a dynamic, is a beneficial process. But it is neither inevitably nor inherently so. Indeed, international migration is frequently a cause and effect of various forms of conflict. Major events underscore why this is so and why 9/11 represented a culmination of trends and patterns rather than a new departure. Two cases, which are treated in greater detail later, suffice to illustrate.

Strife in Algeria pitting Islamists against a military-controlled government spilled over to France in the mid-1990s. Islamic radicals bombed subways and trains, and one unit commandeered a plane and threatened to fly it into a major public building because the French government was aiding the Algerian government in its counter-insurgency campaign. The menace posed by Islamic rebels infiltrating into France or mobilizing support among the large population of Algerians living in France or from French citizens of Algerian Muslim background, clearly ranked as France's central national security issue by 1995.

Likewise, in Germany, the Kurdish insurgency against the Turkish government spilled over to German soil in the 1990s. The Kurdish Workers Party declared that it was waging a two-front war against both Turkey and Germany, because the German government was siding with the Turks. German security analysts estimated that there were thousands of Kurdish Workers Party members among the 2-million-plus Turkish citizens living in Germany. By the mid-1990s, political violence involving Kurds became the central national security preoccupation of the German government. Meanwhile, Turkish political scientists sympathetic to the Turkish government regarded Turkish Islamist activities on German soil as constituting a grave threat to the Turkish state.

Such events were linked to growing international migration and to the problems of living together in one society for culturally and socially diverse ethnic groups. These developments in turn were related to fundamental economic, social and political transformations that shape the post-Cold War period. Millions of people are seeking work, a new home or simply a safe place to live outside their countries of birth. For many less-developed countries, emigration is one aspect of the social crisis which accompanies integration into the world market and modernization. Population growth and the 'green revolution' in rural areas lead to massive surplus populations. People move to burgeoning cities, where employment opportunities are inadequate and social conditions miserable. Massive urbanization outstrips the creation of jobs in the early stages of industrialization. Some of the previous rural–urban migrants embark on a second migration, seeking to improve their lives by moving to newly-industrializing countries in the South or to highly-developed countries in the North.

The movements take many forms: people migrate as manual workers, highly-qualified specialists, entrepreneurs, refugees or as family members of previous migrants. Whether the initial intention is temporary or permanent movement, many migrants become settlers. Migratory networks develop, linking areas of origin and destination, and helping to bring about major changes in both. Migrations can change demographic, economic and social structures, and bring a new cultural diversity, which often brings into question national identity.

This book is about contemporary international migrations, and the way they are changing societies. The perspective is international: large-scale movements of people arise from the accelerating process of global integration. Migrations are not an isolated phenomenon: movements of commodities and capital almost always give rise to movements of people. Global cultural interchange, facilitated by improved transport and the proliferation of print and electronic media, also leads to migration. International migration is not an invention of the late twentieth century, nor even of modernity in its twin guises of capitalism and colonialism. Migrations have been part of human history from the earliest times. However, international migration has grown in volume and significance since 1945 and most particularly since the mid-1980s. Migration ranks as one of the most important factors in global change.

There are several reasons to expect what we term the age of migration to endure: growing inequalities in wealth between the North and the South are likely to impel increasing numbers of people to move in search of better living standards; political, ecological and demographic pressures may force many people to seek refuge outside their own countries; increasing political or ethnic conflict in a number of regions could lead to future mass flights; and the creation of new free trade areas will cause movements of labour, whether or not this is intended by the governments concerned. States around the world will be increasingly affected by international migration, either as receiving societies, lands of emigration, or both.

No one knows exactly how many international migrants there are. A report by the International Organization for Migration (IOM) claimed that the number of migrants in the world had doubled between 1965 and 2000, from 75 million to 150 million (IOM, 2000b). By 2002, the United Nations Population Division (UNPD) estimated that 185 million people had lived outside their country of birth for at least 12 months – just over 2 per cent of the world's population (Crossette, 2002b). Previous epochs have also been characterized by massive migrations. Between 1846 and 1939, some 59 million people left Europe, mainly for major areas of settlement in North and South America, Australia, New Zealand and South Africa (Stalker, 2000: 4). Comparison of data on pre-First World War international migration with statistics on contemporary population movements suggests remarkable continuity in volume between the two periods (Zlotnik, 1999).

However, there are great unknowns, such as the number of illegal immigrants. UN statistics on contemporary international migration reflect statistics compiled by member states concerning legal migration. Yet credible statistics are lacking in many areas of the world. Moreover, there are many reasons to believe that illegal migration has increased sharply in recent decades. Hence, the contention that the late modern world has not experienced a remarkable upsurge in international migration, based upon comparison of statistics for the two periods, must be rejected. Much contemporary international migration is simply unrecorded and not reflected in official statistics. In fiscal year 1998, 660 477 persons were recorded by the Immigration and Naturalization Service (INS) as having legally immigrated to the USA (Kramer, 1999: 1). However, analysis of the 2000 census strongly suggested that some 9 million aliens lived illegally in the USA, with between 200 000 and 300 000 new arrivals each year. Similarly, between 250 000 and 300 000 illegal entrants are estimated to arrive in Northern Europe each year (Widgren, 1994).

There were 15 million refugees and asylum seekers in need of protection and assistance in 2001 (USCR: 2002). This total can be compared to 16 million in 1993, suggesting that international population movements are neither inexorable nor unidirectional. Successful repatriation policies and the end of conflict in certain areas resulted in a decrease in overall numbers between 1993 and 2001. However, concurrently, the number of persons who were in a refugee-like situation, but who were not officially recognized as refugees or asylum seekers grew rapidly after 1990, as did the number of IDPs. The number of persons applying for asylum in Western Europe, Australia, Canada and the USA combined rose from 90 000 in 1983 to a peak of 829 000 in 1992. Following restrictive measures, asylum-seeker applications fell to 480 000 in 1995, but then began to grow, reaching 535 000 in 2000. Other types of forced migrants, who remain within their country of origin, include large numbers displaced by development projects (such as dams, airports and industrial areas), but inadequately resettled. An estimated 10 million people are displaced each year in this way, and many of them may move on to become international migrants (Cernea and McDowell, 2000).

The vast majority of human beings reside in their countries of birth. Voluntarily taking up residence abroad or becoming a victim of expulsion is the exception not the rule. Yet the impact of international migration flows is frequently much greater than is suggested by figures such as the IOM estimates. People tend to move not individually, but in groups. Their departure may have considerable consequences for social and economic relationships in the area of origin. Remittances (money sent home) by migrants may improve living standards and encourage economic development. In the country of immigration, settlement is closely linked to employment opportunity and is almost always concentrated in industrial and urban areas, where the impact on receiving communities is considerable. Migration thus affects not only the migrants themselves, but the sending and receiving societies as a whole. There can be few people in either industrial or less-developed countries today who do not have personal experience of migration and its effects; this universal experience has become a hallmark of the age of migration.

Map 1.1 *Global migratory movements from 1973*

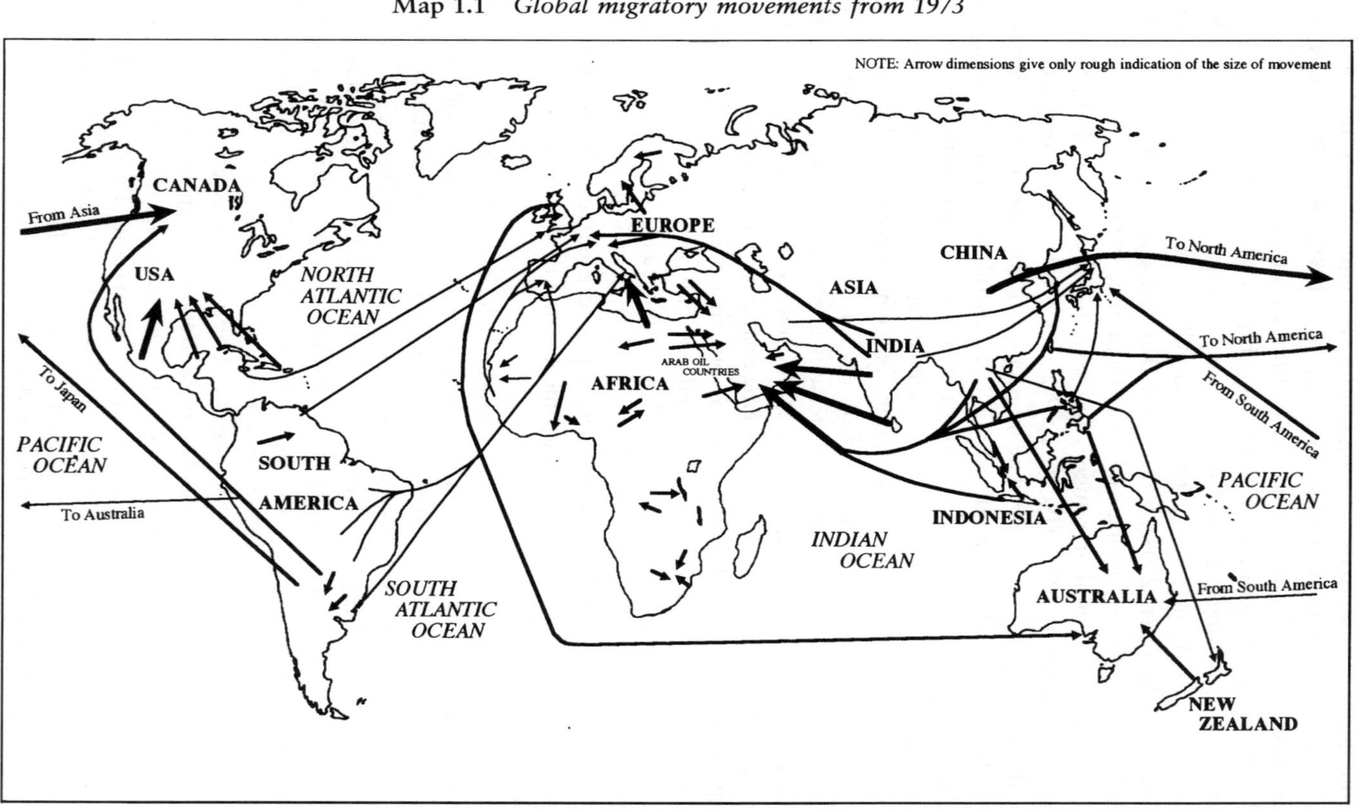

Contemporary migrations: general trends

International migration is part of a transnational revolution that is reshaping societies and politics around the globe. The differing ways in which this has affected the worlds' regions is a major theme throughout this book. Areas such as the USA, Canada, Australia, New Zealand or Argentina, are considered 'classical countries of immigration'. Their current people are the result of histories of large-scale immigration – often to the detriment of indigenous populations. Today, migration continues in new forms. Virtually all of Northern and Western Europe became areas of labour immigration and subsequent settlement after 1945. Since the 1980s, Southern European states like Greece, Italy and Spain, which for a long time were zones of emigration, have become immigration areas. Today Central and Eastern European states, particularly Hungary, Poland and the Czech Republic, are becoming immigration lands.

The Arab region and the Middle East are affected by complex population movements. Some countries, like Turkey, Jordan and Morocco, are major sources of migrant labour. The Gulf oil states experience mass temporary inflows of workers. Political turmoil in the region has led to mass flows of refugees. In recent years, Afghanistan has been the world's main source of refugees, while Iran and Pakistan have been the main receiving countries. In Africa, colonialism and white settlement led to the establishment of migrant labour systems for plantations and mines. Decolonization since the 1950s has sustained old migratory patterns – such as the flow of mineworkers to South Africa – and started new ones, such as movements to Libya, Gabon, and Nigeria. Africa has more refugees and IDPs relative to population than any other region of the world. The picture is similar elsewhere. Asia and Latin America have complicated migratory patterns within the region, as well as increasing flows to the rest of the world. Three examples of recent developments are discussed in Boxes 1.1, 1.2 and 1.3 to give an idea of the complex ramifications of migratory movements.

Throughout the world, long-standing migratory patterns are persisting in new forms, while new flows are developing in response to economic change, political struggles and violent conflicts. Yet, despite the diversity, it is possible to identify certain general tendencies which are likely to play a major role:

1. The *globalization of migration*: the tendency for more and more countries to be crucially affected by migratory movements at the same time. Moreover, the diversity of the areas of origin is also increasing, so that most countries of immigration have entrants from a broad spectrum of economic, social and cultural backgrounds.

2. The *acceleration of migration*: international movements of people are growing in volume in all major regions at the present time. This quantitative growth increases both the urgency and the difficulties of government policies. However, as indicated by the decrease in the global refugee total since 1993, international migration is not an inexorable process. Governmental policies can prevent or reduce international migration and repatriation is a possibility.
3. The *differentiation of migration*: most countries do not simply have one type of immigration, such as labour migration, refugees or permanent settlement, but a whole range of types at once. Typically, migratory chains which start with one type of movement often continue with other forms, despite (or often just because of)

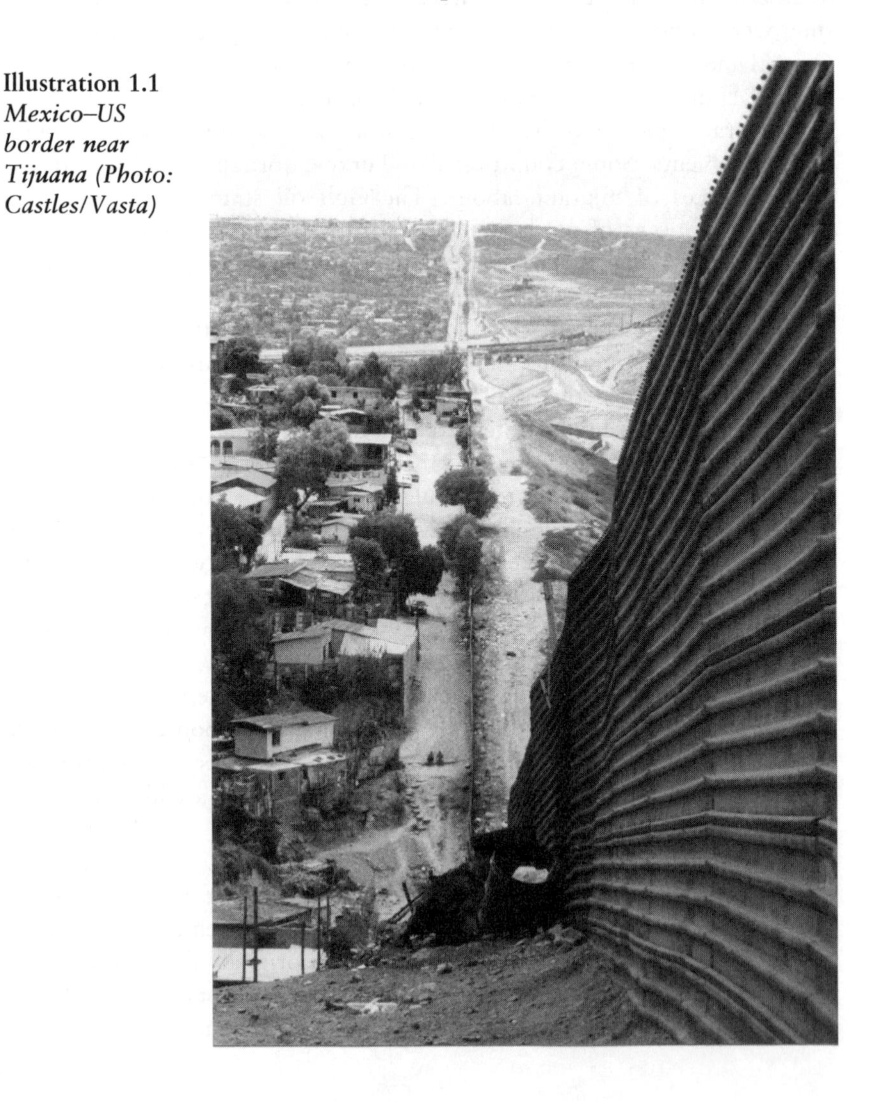

Illustration 1.1 *Mexico–US border near Tijuana (Photo: Castles/Vasta)*

government efforts to stop or control the movement. This differentiation presents a major obstacle to national and international policy measures.

4. The *feminization of migration*: women play a significant role in all regions and in most (though not all) types of migration. In the past, most labour migrations and many refugee movements were male dominated, and women were often dealt with under the category of family reunion. Since the 1960s, women have played a major role in labour migration. Today women workers form the majority in movements as diverse as those of Cape Verdians to Italy, Filipinos to the Middle East and Thais to Japan. Some refugee movements, including those from the former Yugoslavia, contain a significant majority of women as do certain networks of trafficked persons. Gender variables have always been significant in global migration history, but awareness of the specificity of women in contemporary migrations has grown.
5. The *growing politicization of migration*: domestic politics, bilateral and regional relationships and national security policies of states around the world are increasingly affected by international migration.

Illustration 1.2 *Mexico–US border near Tijuana: Each cross represents a migrant who has died crossing the border (Photo: Castles/Vasta)*

Box 1.1 Eastern Europe and the collapse of the Berlin Wall

Migration played an important part in the political transformation of Central and Eastern Europe. The Hungarian government, under the pressure of a wave of would-be emigrants to the west, dismantled the border barriers with Austria in late 1989. This destroyed a major symbol of the Cold War and created the first opportunity for emigration for East Germans since the construction of the Berlin Wall in 1961. Tens of thousands rushed to depart. The steady haemorrhage to the West helped create a political crisis in the German Democratic Republic, forcing a change in leadership. In a final gambit to maintain control, the new government opened the Wall, enabling East Germans to travel freely to West Germany. The communist regime quickly collapsed and Germany was reunited in 1990. Large-scale migration continued: at least 1 million East Germans moved west from the opening of the Wall to the end of 1991.

The collapse of East Germany had a 'domino effect' upon other communist regimes. The political transformation of the region enabled hundreds of thousands to emigrate. During 1989 alone, some 1.2 million people left the former Warsaw Pact area. Most were ethnic minorities welcomed as citizens elsewhere: ethnic Germans who had the right to enter Germany, ethnic Greeks going to Greece, or Jews who automatically become citizens according to the Israeli Law of Return. The mass arrival of Soviet Jews in Israel was viewed with alarm by Arabs who feared that one result would be further dispossession of the Palestinians.

The spectre of uncontrolled mass emigration from Eastern Europe became a public issue in the West. Before long, Italy deployed troops to prevent an influx of Albanian asylum seekers, while Austria used its army to keep out Romanian Gypsies. For Western European leaders, the initial euphoria prompted by the destruction of the barriers to movement was quickly succeeded by a nostalgia for the ease of migration control of an earlier epoch.

The disintegration of the USSR led to the creation of a plethora of successor states. Some of the 25 million or so ethnic Russians living outside the Russian Republic suddenly confronted the possibility of losing their citizenship. Economic crisis and the potential for inter ethnic violence, attendant on the reshaping of the former Warsaw Bloc area, made emigration a preferred option for many. But the great mass of Eastern Europeans did not see the welcome mat rolled out for them. Even in Germany and Israel, there was resentment over the massive arrival of newcomers from the ex-USSR and Warsaw Bloc states.

Box 1.2 The US–Mexico 'immigration honeymoon'

The elections of George W. Bush, Jr and Vincente Fox in 2000 appeared to augur well for major changes in US–Mexico relations. Both presidents wanted to improve relations and spoke of reforms of migration-related policies.

Under President Clinton, the emphasis had been on prevention of illegal migration. 'Operation Gatekeeper' was introduced in 1994 in an attempt to tighten security along the US–Mexico border. The US Immigration and Naturalization Service introduced double steel fences, helicopters, high-intensity searchlights and high-tech equipment. The number of agents enforcing the border doubled. To fund this surveillance programme, the budget tripled from 1994 to 2000, reaching $5.5 billion. However, there was no decline in the number of illegal border crossings; indeed official figures suggested an increase, and Californian agriculture experienced no shortage of migrant labour. The number of people dying as they attempted to cross the border also increased as people took ever greater risks: in 2000, 499 people died trying to cross, compared with 23 in 1994. The cause of death also altered as people moved towards increasingly remote areas in their attempts to cross the border. By 2000, they were dying mainly from dehydration, hypothermia or sunstroke trying to cross the Arizona desert, or drowning as they attempted to swim the All American Canal. The average cost of hiring 'coyotes' – who smuggle people across the border – rose from $143 to $1500 in six years (Cornelius, 2001).

President Bush had supported expanded admissions of Mexican temporary workers while governor of Texas. The Mexican president backed an amnesty or legalization programme for illegally resident Mexicans in the USA – estimated to number between 4 and 5 million. President Bush's first foreign visit was to President Fox's ranch and the US–Mexico migration initiative topped the agenda. The formation of a high-level group of US and Mexican officials was announced They were to meet regularly to shape the content of the initiative. Matters under discussion ranged from the modalities of legalization to intergovernmental cooperation on prevention of illegal migration and human trafficking. In early September 2001, President Fox made a triumphal tour of the USA to tout the initiative which culminated in an address to a joint session of the US Congress. However, it was clear that elements of the plan, the details of which were never fully disclosed, would encounter resistance in the US Congress, for example among Republicans opposed to legalization policy.

The terrorist attacks on 9/11 put the initiative on the backburner. Many US officials were rankled by the Mexican government's response to 9/11, and the attacks dramatically changed the political environment in the USA, virtually ensuring that broad legalization would not soon be authorized by the US Congress. The Fox Administration began to complain about its frustration with its neighbour to the North. A year after President Fox's address to a joint session of the US Congress, little of substance had changed in the enormously complex US–Mexico immigration relationship.

Box 1.3 Ethnic cleansing and conflict in Central Africa

Events in the former Yugoslavia and Central Africa in the 1990s made ethnic cleansing – violence directed against distinct civilian populations to drive them from a given territory – a principal problem of post-Cold War world order. In densely populated Rwanda, periodic strife between the Hutu majority and the Tutsi minority had created a Rwandan Tutsi diaspora. Tutsi exiles launched a campaign to overthrow the Hutu-dominated Rwandan government from Ugandan territory with the support of the Ugandan government. Advances by the rebels led to negotiations but then in 1994 the Rwandan president's aircraft was destroyed by a rocket, killing him and the fragile accord. The killing of the Hutu president served as the pretext for a campaign of violence by a Hutu-dominated faction targeted against the Tutsi minority and moderate Hutus. Hundreds of thousands were hacked to death in a frenzy of violence as the Tutsi-dominated rebel forces advanced. The collapse of governmental forces and their Hutu extremist allies led millions of Rwandan Hutus to flee to Tanzania and Zaire. Many of the perpetrators of the mass killings fled with them.

The governments of Tanzania and Zaire and international relief agencies scrambled to cope with the influxes. Eventually, Tanzania would force many of the Rwandan refugees to repatriate. Some returning refugees were killed. In Zaire, the government threatened to expel the Rwandan refugees en masse but may have lacked the capacity to do so. International agencies and governments around the world pleaded against the forcible return of Rwandan refugees. By 1996, military intervention by Western states, including the USA and France in coordination with the UN, was contemplated. However, an ethnic Tutsi militia from eastern Zaire and other anti-government insurgents moved against the Rwandan Hutu militants entrenched in refugee camps. This enabled many of the Rwandan Hutus to

→

International migration in global governance

Globalization has challenged the authority of national governments from above and below. The growth of transnational society has given rise to novel issues and problems and has blurred formerly distinctive spheres of authority and decision-making. As a result, authoritative decision-making for polities is increasingly conceptualized as global governance (Rosenau, 1997). The complexity and fragmentation of power and authority that have resulted from globalization typically require various levels of government to interact with other organizations and institutions, both public and private, foreign and domestic, to achieve desired goals. An important manifestation of global governance is the significant expansion of regional consultative processes focusing on international migration. An IOM report on the 11 non-binding fora that had emerged by 2001 cited

→

repatriate to an uncertain future while tens of thousands of other Hutu refugees fled deeper into Zaire to elude the insurgents.

In 1997, the beleaguered Zairian government began to arm remaining Rwandan Hutu refugees as part of a broader effort, involving the use of foreign mercenaries, to quell the anti-governmental insurgency in eastern Zaire. However, the Mobutu dictatorship in Zaire, which had long been supported by Western governments, soon collapsed. The insurgency was backed by Uganda and the new Tutsi-dominated government of Rwanda. There were also signs of spillover of Hutu-Tutsi strife to Burundi, where Hutus and Tutsis also comprise the bulk of the population.

By 1998, Rwanda and Uganda ended their support for the successor government to the Mobutu dictatorship, led by their ex-ally and former rebel, Laurent Kabila, who had been installed as president of the new Democratic Republic of the Congo. Fighting erupted between former allies. Altogether six national armies and numerous non-state groups were involved and nearly 1 million persons were displaced by the end of 1999. Subsequently Kabila was assassinated, and succeeded by his son, while 5000 UN troops attempted to monitor a tenuous peace agreement between the protagonists. Fighting flared anew in 2002 and the toll of human lives lost due to the fighting and attendant hunger and disease between 1998 and 2002 stood at 2 million with no end in sight.

The Central African crisis of the 1990s was as emblematic of world affairs in the post-Cold War period as the North American Free Trade Agreement (NAFTA) or the Asia-Pacific Economic Cooperation Forum (APEC). Ethnic violence led to mass movements of people and an emigrant-led insurgency toppled two governments and threatened several others. Mass refugee flows destabilized an entire region and the UN and major Western powers contemplated further use of military force to protect refugees and prevent further escalation of the violence.

four reasons for their inception: the post-Cold War increase in irregular migration, including human trafficking; the increase in the number of states and areas affected by international migration; the regional scope of most international migration; and the informal, non-binding nature of the consultative processes themselves (Klekowski Von Koppenfels, 2001).

Until recently, international migration had not generally been seen by governments as a central political issue. Rather, migrants were divided up into categories, such as permanent settlers, foreign workers or refugees, and dealt with by a variety of special agencies, such as immigration departments, labour offices, aliens police, welfare authorities and education ministries. It was only in the late 1980s that international migration began to be accorded high-level and systematic attention. For example, as the EU countries removed their internal boundaries, they became increasingly concerned about strengthening external boundaries in order to prevent an influx from the South and the East. By the 1990s, the successful mobilization of extreme right-wing groups over immigration and supposed threats to national identity helped bring these issues to the centre of the political stage.

Ethnic diversity, racism and multiculturalism

Regulation of international migration is one of the two central issues arising from the mass population movements of the current epoch. The other is the effect of growing ethnic diversity on the societies of immigration countries. Settlers are often distinct from the receiving populations: they may come from different types of societies (for example, agrarian-rural rather than urban-industrial) with different traditions, religions and political institutions. They often speak a different language and follow different cultural practices. They may be visibly different, through physical appearance (skin colour, features and hair type) or style of dress. Some migrant groups become concentrated in certain types of work (generally of low social status) and live segregated lives in low-income residential areas. The position of immigrants is often marked by a specific legal status: that of the foreigner or non-citizen. The differences are frequently summed up in the concepts of 'ethnicity' or 'race'. In many cases, immigration complicates existing conflicts or divisions in societies with long-standing ethnic minorities.

The social meaning of ethnic diversity depends to a large extent on the significance attached to it by the populations and states of the receiving countries. The classic immigration countries have generally seen immigrants as permanent settlers who were to be assimilated or integrated. However, not all potential immigrants have been seen as suitable: the USA, Canada and Australia all had policies to keep out non-Europeans and even some categories of Europeans until the 1960s. Countries which emphasized temporary labour recruitment – Western European countries in the 1960s and early 1970s, more recently the Gulf oil states and some of the fast-growing Asian economies – have tried to prevent family reunion and permanent settlement. Despite the emergence of permanent settler populations, such countries have declared themselves not to be countries of immigration, and have denied citizenship and other rights to settlers. Between these two extremes is a wealth of variations, which will be discussed in later chapters.

Culturally-distinct settler groups almost always maintain their languages and some elements of their homeland cultures, at least for a few generations. Where governments have recognized permanent settlement, there has been a tendency to move from policies of individual assimilation to acceptance of some degree of long-term cultural difference. The result has been the granting of minority cultural and political rights, as embodied in the policies of multiculturalism introduced in Canada, Australia and Sweden since the 1970s. Governments which reject the idea of permanent settlement also oppose pluralism, which they see as a threat to national unity and identity. In such cases, immigrants tend to turn into marginalized ethnic minorities. In other cases (France, for example), governments may accept the reality of settlement, but demand individual cultural assimilation as the price for granting of rights and citizenship.

Whatever the policies of the governments, immigration may lead to strong reactions from some sections of the population. Immigration often takes place at the same time as economic restructuring and far-reaching social change. People whose conditions of life are already changing in an unpredictable way often see the newcomers as the cause of insecurity. One of the dominant images in the highly-developed countries today is that of masses of people flowing in from the poor South and the turbulent East, taking away jobs, pushing up housing prices and overloading social services. Similarly, in immigration countries of the South, such as Malaysia and South Africa, immigrants are blamed for crime, disease and unemployment. Extreme-right parties have grown and flourished through anti-immigrant campaigns. Racism is a threat, not only to immigrants themselves, but also to democratic institutions and social order. Analysis of the causes and effects of racism must therefore take a central place in any discussion of international migration and its effects on society.

International migration does not always create diversity. Some migrants, such as Britons in Australia or Austrians in Germany, are virtually indistinguishable from the receiving population. Other groups, like Western Europeans in North America, are quickly assimilated. 'Professional transients' – that is, highly-skilled personnel who move temporarily within specialized labour markets – are rarely seen as presenting an integration problem. But these are the exceptions; in most instances, international migration increases diversity within a society. This presents a number of problems for the state. The most obvious concerns social policy: social services and education may have to be planned and delivered in new ways to correspond to different life situations and cultural practices.

More serious is the challenge to national identity. The nation-state, as it has developed since the eighteenth century, is premised on the idea of cultural as well as political unity. In many countries, ethnic homogeneity, defined in terms of common language, culture, traditions and history, has been seen as the basis of the nation-state. This unity has often been fictitious – a construction of the ruling elite – but it has provided powerful national myths. Immigration and ethnic diversity threaten such ideas of the nation, because they create a people without common ethnic origins. The classical countries of immigration have been able to cope with this situation most easily, since absorption of immigrants has been part of their myth of nation building. But countries which place common culture at the heart of their nation-building process have found it very difficult to resolve the contradiction. Movements against immigration have also become movements against multiculturalism, and such popular sentiments have led to a retreat from multicultural policies in many places, including Australia and the Netherlands.

One of the central ways in which the link between the people and the state is expressed is through the rules governing citizenship and naturalization. States which readily grant citizenship to immigrants, without requiring common ethnicity or cultural assimilation, seem most able to cope with ethnic diversity. On the other hand, states which link citizenship to cultural belonging tend to have exclusionary policies which marginalize and disadvantage immigrants. It is one of the central themes of this book that continuing international population movements will increase the ethnic diversity of more and more countries. This has already called into question prevailing notions of the nation-state and citizenship. Debates over new approaches to diversity will shape the politics of many countries in coming decades.

Aims and structure of the book

The first goal of this book is to describe and explain contemporary international migration. We set out to show the enormous complexity of the phenomenon, and to communicate both the variations and the common factors in international population movements as they affect more and more parts of the world.

The second goal is to explain how migrant settlement is bringing about increased ethnic diversity in many societies, and how this is related to broader social, cultural and political developments. Understanding these changes is the precondition for political action to deal with problems and conflicts linked to migration and ethnic diversity.

The third goal is to link the two discourses, by showing the complex interaction between migration and growing ethnic diversity. There are large bodies of empirical and theoretical work on both themes. However, the two are often inadequately linked. There is a tendency towards specialization both in academic circles and among policy-makers. Many of the research institutes which deal with migration are distinct from those concerned with ethnic relations. For instance, the International Sociological Association has separate research committees for 'ethnic, race and minority relations' and for 'sociology of migration'. Similarly, many governments have one ministry or agency to deal with immigration, and another to deal with ethnic or race relations. There is still no international regime to regulate migration, and efforts to establish a UN agency with responsibility in this area have made little progress.

Immigration and ethnic relations are closely interrelated in a variety of ways. The linkages can best be understood by analysing the migratory process in its totality. It is an ambitious (some would say elusive) undertaking to try to do this on a global level in one short book. Hence accounts of the various migratory movements must inevitably be concise, but a global view of international migration is the precondition for understanding each specific flow. The central aim of this book is therefore to provide an introduction to the subject of international migration and the emergence of multicultural societies, which will help readers to put more detailed accounts of specific migratory processes in context.

The book is structured as follows: Chapter 2 examines some of the theories and concepts used to explain migration and formation of ethnic minorities, and emphasizes the need to study the migratory process as a whole. Chapter 3 describes the history of international migration up to 1945. There is some discussion of the role of migration in the emergence of European nation-states, but the main focus is the migrations brought about by capitalism and colonialism, in the process of creating a world market.

Chapter 4 is concerned with migration to industrial countries since 1945, including new migrations to Southern and Eastern Europe. It shows the patterns of labour migration which developed during the post-war boom and discusses the differences and similarities between permanent, post-colonial and guestworker migration systems. The major changes in migratory patterns after the oil shock of 1973 are examined. Finally, the increasing volume and complexity of migrations since the late 1980s are discussed. Chapter 5 assesses the capacity of industrial states to regulate international migration. It examines illegal migration, human trafficking and policies designed to curb them. It also compares the significance of regional integration frameworks for control of migration and responses to 9/11.

Chapters 6 and 7 look at some of the new areas of migration, showing how major political, social and economic changes are leading to mass population movements. Chapter 6 is concerned with the Middle East, Africa and Latin America, while Chapter 7 deals with Asia. These areas are major sources of migrants to highly-developed countries, and it is from here that the 'next waves' are likely to come. But movements within these regions are of growing importance, particularly where the emergence of new industrial countries is leading to economic and demographic imbalances.

Chapter 8 considers the economic position of immigrants in highly-developed countries, looking at labour market segmentation, the role played by immigrants in economic crisis and why employment of migrants can continue despite high unemployment. Chapter 9 presents a comparative study of the migratory process in two countries which appear at first sight to have had almost diametrically opposed experiences of immigration: Australia and Germany. The aim is to show both parallels and differences, and to discuss the factors which determine them. Chapter 10 goes on to examine the position of immigrants within the societies of some of the other highly-developed immigration countries, looking at such factors as legal status, social policy, formation of ethnic communities, racism, citizenship and national identity.

Chapter 11 examines some of the key political effects of increasing ethnic diversity, looking both at the involvement of minorities in politics and at the way mainstream politics are changing in reaction to migrant settlement. Perspectives for the emergence of multicultural societies are discussed. Chapter 12 sums up the arguments of the book and presents some conclusions on the future of international migration, and what it is likely to mean for individual societies and for the global community as a whole.

Guide to further reading

Important information on all aspects of international migration is provided by a large number of specialized journals, of which only a few can be mentioned here. *International Migration Review* (New York: Center for Migration Studies) was established in 1964 and provides excellent comparative information. *International Migration* (Geneva: IOM) is also a valuable comparative source. *Social Identities* (Abingdon: Carfax) started publication in 1995 and is concerned with the 'study of race, nation and culture'. *Diaspora* (Toronto: University of Toronto Press) is dedicated to 'transnational studies'. Another journal with a transnational focus is *Global Networks* (Oxford: Blackwell). Journals with a European focus include the *Journal of Ethnic and Migration Studies* (Brighton: Sussex Centre for Migration Research, University of Sussex), *Migration* (Berlin: Edition Parabolis, in English and German) and the *Revue Européenne des Migrations Internationales* (Poitiers: University of Poitiers, in French and English). Britain has several journals including *Race and Class* (London: Institute for Race Relations) and *Ethnic and Racial Studies* (London and New York: Routledge). In Australia there is the *Journal of Intercultural Studies* (Melbourne: Monash University). The *Asian and Pacific Migration Journal* (Quezon City, Philippines: Scalabrini Migration Center) provides information and analyses movements in the world's most populous region. *Migraciones Internacionales* (Mexico: El Colegio de la Frontera Norte) includes articles in Spanish and English.

There are also many publications with a 'magazine' format which provide up-to-date information and shorter commentaries, such as *Asian Migrant* (Quezon City, Philippines: Scalabrini Migration Center) and *Hommes et Migrations* (Paris). A very valuable resource is the monthly *Migration News* (Davis: University of California) which is available as hard copy, e-mail or on the Internet.

Several international organizations provide comparative information on migrations. The most useful is the OECD's annual *Trends in International Migration* (Paris: OECD), which until 1991 was known as OECD SOPEMI, *Continuous Reporting System on Migration.* This publication provides comprehensive statistics on most OECD countries of immigration, as well as some data on countries of emigration. The IOM published its *World Migration Report* for the first time in 2000, and further versions are planned. Cohen's massive *Survey of World Migration* (1995) is a valuable reference work with contributions on all aspects of our theme.

There are many Internet sites concerned with issues of migration and ethnic diversity. A few of the most significant are listed here. Since they are hyperlinked with many others, this list should provide a starting-point for further exploration:

Asia-Pacific Migration Research Network (APMRN):
http://www.capstrans.edu.au/apmrn/
Center for Migration Studies, New York: http://www.cmsny.org/
Centre for European Migration and Ethnic Studies:
http://www.cemes.org/
European Council on Refugees and Exiles:
http://www.ecre.org/
European Migration Information Network (EMIN):
http://www.emin.geog.ucl.ac.uk/
European Research Centre on Migration and Ethnic Relations (ERCOMER): http://www.ercomer.org/
Federation of Centers for Migration Studies, G. B. Scalabrini:
http://www.scalabrini.org/fcms/index.html
Forced Migration Online: http://www.forcedmigration.org/
Immigration History Research Center, Minnesota:
http://www1.umn.edu/ihrc/
Institute for Migration and Ethnic Studies (IMES), Amsterdam:
http://www.pscw.uva.nl
Institute for Migration Research and Intercultural Studies (IMIS), Osnabrück: http://www.imis.uni-osnabrueck.de/english/index.htm
International Centre for Migration and Health: http://www.icmh.ch/
International Metropolis Project:
http://www.international.metropolis.net/
International Organization for Migration: http://www.iom.int/

Inter-University Committee on International Migration:
http://web.mit.edu/cis/www/migration/
Migration-Ethnicity-Racism-Refugees, Amsterdam:
http://www.pscw.uva.nl/sociosite/TOPICS/Ethnic.html
Migration Information Source:
http://www.migrationinformation.org/index.cfm/
Migration News: http://www.migration.ucdavis.edu/
Migration Policy Institute, Washington DC:
http://www.migrationinformation.org/
Multicultural Skyscraper: http://www.multicultural.net/
Refugee Net: EU Networks on the Integration of Refugees:
http://www.refugeenet.org/
Refugee Studies Centre, University of Oxford: http://www.rsc.ox.ac.uk/
Southern African Migration Project:
http://www.queensu.ca/samp/
Swiss Forum for Migration and Population Studies:
http://www.unine.ch/fsm/libri/publications.htm
UNHCR: http://www.unhcr.ch/
US Committee for Refugees: http://www.refugees.org/

Chapter 2

The Migratory Process and the Formation of Ethnic Minorities

International migration is hardly ever a simple individual action in which a person decides to move in search of better life-chances, pulls up his or her roots in the place of origin and quickly becomes assimilated in the new country. Much more often migration and settlement is a long-drawn-out process, which will be played out for the rest of the migrant's life, and affect subsequent generations too. (Migration can even transcend death: members of some migrant groups have been known to arrange for their bodies to be taken back for burial in their native soil: see Tribalat, 1995: 109–11.) Migration is a collective action, arising out of social change and affecting the whole society in both sending and receiving areas. Moreover, the experience of migration and of living in another country often leads to modification of the original plans, so that migrants' intentions at the time of departure are poor predictors of actual behaviour. Similarly, no government has ever set out to build an ethnically diverse society through immigration, yet labour recruitment policies often lead to the formation of ethnic minorities, with far-reaching consequences for social relations, public policies, national identity and international relations.

The aim of the chapter is to link two bodies of theory which are often dealt with separately: theories on migration and settlement, and theories on ethnic minorities and their position in society. This chapter provides a theoretical framework for understanding the more descriptive accounts of migration, settlement and minority formation in later chapters. However, the reader may prefer to read those first and come back to the theory later.

Explaining the migratory process

The concept of the *migratory process* sums up the complex sets of factors and interactions which lead to international migration and influence its course. Migration is a process which affects every dimension of social existence, and which develops its own complex dynamics.

Research on migration is therefore intrinsically interdisciplinary: sociology, political science, history, economics, geography, demography, psychology and law are all relevant (Brettell and Hollifield, 2000). These disciplines look at different aspects of population mobility, and a full understanding requires contributions from all of them. Within each social-scientific discipline there is a variety of approaches, based on differences in theory and methods. For instance, researchers who base their work on quantitative analysis of large data-sets (such as censuses or representative surveys) will ask different questions and get different results from those who do qualitative studies of small groups. Those who examine the role of migrant labour within the world economy using historical and institutional approaches will again get different findings. All these methods have their place, as long as they lay no claim to be the only correct one. A detailed survey of migration theory is not possible here (see Massey *et al.*, 1993, 1994, 1998), but a useful distinction may be made between three of the main approaches used in contemporary debates: economic theory, the historical-structural approach and migration systems theory (Hugo, 1993: 7–12).

Economic theories of migration

The neo-classical economic perspective has its antecedents in the earliest systematic theory on migration: that of the nineteenth-century geographer Ravenstein, who formulated statistical laws of migration (Ravenstein, 1885, 1889). These were general statements unconnected with any actual migratory movement (Cohen, 1987: 34–5; Zolberg, Suhrke and Aguao, 1989: 403–5). This tradition remains alive in the work of many demographers, geographers and economists. Such 'general theories' emphasize tendencies of people to move from densely to sparsely populated areas, or from low- to high-income areas, or link migrations to fluctuations in the business cycle. These approaches are often known as 'push–pull' theories, because they perceive the causes of migration to lie in a combination of 'push factors', impelling people to leave the areas of origin, and 'pull factors', attracting them to certain receiving countries. 'Push factors' include demographic growth, low living standards, lack of economic opportunities and political repression, while 'pull factors' are demand for labour, availability of land, good economic opportunities and political freedoms.

This model is mainly found in neo-classical economics, although it has also been influential in sociology, social demography and other disciplines. It is individualistic and ahistorical. It emphasizes the individual decision to migrate, based on rational comparison of the relative costs and benefits of remaining in the area of origin or moving to various alternative destinations. Constraining factors, such as government restrictions on emigration or immigration, are mainly dealt with as distortions of the rational market, which should be removed. Its central concept is 'human capital': people decide to invest in migration, in the same way as they might invest in education or vocational training, because it raises their human capital and brings potential future gains in earnings. People will migrate if the expected rate of return from higher wages in the destination country is greater than the costs incurred through migrating (Chiswick, 2000). Borjas puts forward the model of an immigration market:

> Neo-classical theory assumes that individuals maximize utility: individuals 'search' for the country of residence that maximizes their well-being ... The search is constrained by the individual's financial resources, by the immigration regulations imposed by competing host countries and by the emigration regulations of the source country. In the immigration market the various pieces of information are exchanged and the various options are compared. In a sense, competing host countries make 'migration offers' from which individuals compare and choose. The information gathered in this marketplace leads many individuals to conclude that it is 'profitable' to remain in their birthplace ... Conversely, other individuals conclude that they are better off in some other country. The immigration market nonrandomly sorts these individuals across host countries. (Borjas, 1989: 461)

Borjas claims that 'this approach leads to a clear – and empirically testable – categorization of the types of immigrant flows that arise in a world where individuals search for the "best" country' (Borjas, 1989: 461). On this basis, the mere existence of economic disparities between various areas should be sufficient to generate migrant flows. In the long run, such flows should help to equalize wages and conditions in underdeveloped and developed regions, leading towards economic equilibrium. Borjas has argued that this may lead to negative effects for immigration countries, notably the decline of average skill levels (Borjas, 1990). However, this finding is not uncontested within neo-classical research: Chiswick claims that migrants are positively self-selected, in the sense that the higher skilled are more likely to move because they obtain a higher return on their human capital investment in mobility. This has negative effects for countries of origin, by causing a 'brain drain' (Chiswick, 2000).

Empirical studies cast doubt on the value of neo-classical theory. It is rarely the poorest people from the least-developed countries who move to the richest countries; more frequently the migrants are people of intermediate social status from areas which are undergoing economic and social change. Similarly the push–pull model predicts movements from densely populated areas to more sparsely peopled regions, yet in fact countries of immigration like the Netherlands and Germany are among the world's more densely populated. Finally the push–pull model cannot explain why a certain group of migrants goes to one country rather than another: for example, why have most Algerians migrated to France and not Germany, while the opposite applies to Turks?

Neo-classical migration theories have therefore been criticized as simplistic and incapable of explaining actual movements or predicting future ones (see Sassen, 1988; Boyd, 1989; Portes and Rumbaut, 1996: 271–8). It seems absurd to treat migrants as individual market-players who have full information on their options and freedom to make rational choices. Historians, anthropologists, sociologists and geographers have shown that migrants' behaviour is strongly influenced by historical experiences as well as by family and community dynamics (Portes and Böröcz, 1989). Moreover migrants have limited and often contradictory information, and are subject to a range of constraints (especially lack of power in the face of employers and governments). Migrants compensate through developing cultural capital (collective knowledge of their situation and strategies for dealing with it) and social capital (the social networks which organize migration and community formation processes).

It therefore seems essential to introduce a wider range of factors into economic research. One attempt to do this was 'dual labour market theory', which showed the importance of institutional factors as well as race and gender in bringing about labour market segmentation (Piore, 1979). The 'new economics of labour migration' approach emerged in the 1980s (Stark, 1991; Taylor, 1987). It argued that markets rarely function in the ideal way suggested by the neo-classicists. Migration needs to be explained not only by income differences between two countries, but also by such factors as the chance of secure employment, availability of investment capital, and the need to manage risk over long periods. For instance, as Massey *et al.* (1987) point out, Mexican farmers may migrate to the USA because, even though they have sufficient land, they lack the capital to make it productive. Similarly, the role of remittances in migration cannot be understood simply by studying the behaviour of migrants themselves. Rather it is necessary to examine the long-term effects of remittances on investment, work and social relationships right across the community (Taylor, 1999).

The neo-classical model tends to treat the role of the state as an aberration which disrupts the 'normal' functioning of the market. Borjas, for instance, suggests that the US government should 'deregulate the immigration market' by selling visas to the highest bidder (Borjas, 1990: 225–8). But examination of historical and contemporary migrations (see Chapters 3–7 below) shows that states (particularly receiving countries) play a major role in initiating, shaping and controlling movements. The most common reason to permit entry is the need for workers – with states sometimes taking on the role of labour recruiter on behalf of employers –

but demographic or humanitarian considerations may also be important. Immigration as part of nation building has played a major role in new world countries such as the USA, Canada, Argentina, Brazil and Australia. Policies on refugees and asylum seekers are major determinants of contemporary population movements.

Thus the idea of individual migrants who make free choices which not only 'maximize their well-being' but also lead to an 'equilibrium in the marketplace' (Borjas, 1989: 482) is so far from historical reality that it has little explanatory value. It seems better, as Zolberg suggests, to analyse labour migration 'as a movement of workers propelled by the dynamics of the transnational capitalist economy, which simultaneously determines both the "push" and the "pull"' (Zolberg, Suhrke and Aguao, 1989: 407). This implies that migrations are collective phenomena, which should be examined as subsystems of an increasingly global economic and political system.

The historical-structural approach

An alternative explanation of international migration was provided from the 1970s by what came to be called the *historical-structural approach*. This had its intellectual roots in Marxist political economy and in world systems theory. This approach stressed the unequal distribution of economic and political power in the world economy. Migration was seen mainly as a way of mobilizing cheap labour for capital. It perpetuated uneven development, exploiting the resources of poor countries to make the rich even richer (Castles and Kosack, 1985; Cohen, 1987; Sassen, 1988). While the 'push–pull' theories tended to focus on mainly voluntary migrations of individuals, like that from Europe to the USA before 1914, historical-structural accounts looked at mass recruitment of labour by capital, whether for the factories of Germany, for the agribusiness of California or for infrastructure projects like Australia's Snowy Mountain Hydroelectric Scheme. The availability of labour was both a legacy of colonialism and the result of war and regional inequalities within Europe. For world systems theories, labour migration was one of the main ways in which links of domination were forged between the core economies of capitalism and its underdeveloped periphery. Migration was as important as military hegemony and control of world trade and investment in keeping the Third World dependent on the First.

But the historical-structural approach was in turn criticized by many migration scholars: if the logic of capital and the interests of Western states were so dominant, how could the frequent breakdown of migration policies be explained, such as the unplanned shift from labour migration to permanent settlement in certain countries? Both the neo-classical perspective and the historical-structural approach seemed too one-sided to analyse adequately the great complexity of contemporary migrations. The neo-classical approach neglected historical causes of movements, and downplayed the role of the state, while the historical-functional approach often saw the interests of capital as all-determining, and paid inadequate attention to the motivations and actions of the individuals and groups involved.

Migration systems theory and the trend to a new interdisciplinary approach

Out of such critiques emerged a new approach, *migration systems theory*, which attempts to include a wide range of disciplines, and to cover all dimensions of the migration experience. A migration system is constituted by two or more countries which exchange migrants with each other. The tendency is to analyse regional migration systems, such as the South Pacific, West Africa or the Southern Cone of Latin America (Kritz *et al.*, 1992). However, distant regions may be interlinked, such as the migration system embracing the Caribbean, Western Europe and North America; or that linking North and West Africa with France. The migration systems approach means examining both ends of the flow and studying all the linkages between the places concerned. These linkages can be categorized as 'state-to-state relations and comparisons, mass culture connections and family and social networks' (Fawcett and Arnold, 1987: 456–7).

Migration systems theory suggests that migratory movements generally arise from the existence of prior links between sending and receiving countries based on colonization, political influence, trade, investment or cultural ties. Thus migration from Mexico to the USA originated in the southwestward expansion of the USA in the nineteenth century and the deliberate recruitment of Mexican workers by US employers in the twentieth century (Portes and Rumbaut, 1996: 272–6). The migration from the Dominican Republic to the USA was initiated by the US military occupation of the 1960s. Similarly, both the Korean and the Vietnamese migrations to America were the long-term consequence of US military involvement (Sassen, 1988: 6–9). The migrations from India, Pakistan and Bangladesh to Britain are linked to the British colonial presence on the Indian subcontinent. Similarly, Caribbean migrants have tended to move to their respective former colonial power: for example, from Jamaica to Britain, Martinique to France and Surinam to the Netherlands. The Algerian migration to France (and not to Germany) is explained by the French colonial presence in Algeria, while the Turkish presence in Germany is the result of direct labour recruitment by Germany in the 1960s and early 1970s.

The migration systems approach is part of a trend towards a more inclusive and interdisciplinary understanding, which is emerging as a new mainstream of migration theory – at least outside the domain of neo-classical orthodoxy. The basic principle is that any migratory movement can be seen as the result of interacting macro- and micro-structures. Macro-structures refer to large-scale institutional factors, while micro-structures embrace the networks, practices and beliefs of the migrants themselves. These two levels are linked by a number of intermediate mechanisms, which are often referred to as 'meso-structures'.

The macro-structures include the political economy of the world market, interstate relationships, and the laws, structures and practices established by the states of sending and receiving countries to control migration settlement. The evolution of production, distribution and exchange within an increasingly integrated world economy over the last five centuries has clearly been a major determinant of migrations. The role of international relations and of the states of both sending and receiving areas in organizing or facilitating movements is also significant (Dohse, 1981; Böhning, 1984; Cohen, 1987; Mitchell, 1989; Hollifield, 2000).

The micro-structures are the informal social networks developed by the migrants themselves, in order to cope with migration and settlement. Earlier literature used the concept of 'chain migration' in this context (Price, 1963: 108–10). Research on Mexican migrants in the 1970s showed that 90 per cent of those surveyed had obtained legal residence in the USA through family and employer connections (Portes and Bach, 1985). Today many authors emphasize the role of information and 'cultural capital' (knowledge of other countries, capabilities for organizing travel, finding work and adapting to a new environment) in starting and sustaining migratory movements. Informal networks include personal relationships, family and household patterns, friendship and community ties, and mutual help in economic and social matters. Such links provide vital resources for individuals and groups, and may be referred to as 'social capital' (Bourdieu and Wacquant, 1992: 119). Informal networks bind 'migrants and non-migrants together in a complex web of social roles and interpersonal relationships' (Boyd, 1989: 639).

The family and community are crucial in migration networks. Research on Asian migration has shown that migration decisions are usually made not by individuals but by families (Hugo, 1994). In situations of rapid change, a family may decide to send one or more members to work in another region or country, in order to maximize income and survival chances. In many cases, migration decisions are made by the elders (especially the men), and younger people and women are expected to obey patriarchal authority. The family may decide to send young women to the city or overseas, because the labour of the young men is less dispensable on the farm. Young women are also often seen as more reliable in sending remittances. Such motivations correspond with increasing international demand for female labour as factory workers for precision assembly or as domestic servants, contributing to a growing feminization of migration.

Family linkages often provide both the financial and the cultural capital which make migration possible. Typically migratory chains are started by an external factor, such as recruitment or military service, or by an initial movement of young (usually male) pioneers. Once a movement is established, the migrants mainly follow 'beaten paths' (Stahl, 1993), and are helped by relatives and friends already in the area of immigration. Networks based on family or on common origin help provide shelter, work, assistance in coping with bureaucratic procedures and support in personal difficulties. These social networks make the migratory process safer and more manageable for the migrants and their families. Migratory movements, once started, become self-sustaining social processes.

Migration networks also provide the basis for processes of settlement and community formation in the immigration area. Migrant groups develop their own social and economic infrastructure: places of worship, associations, shops, cafés, professionals such as lawyers and doctors, and other services. This is linked to family reunion: as length of stay increases, the original migrants (whether workers or refugees) begin to bring in their spouses and children, or found new families. People start to see their life perspectives in the new country. This process is especially linked to the situation of migrants' children: once they go to school in the new country, learn the language, form peer group relationships and develop bicultural or transcultural identities, it becomes more and more difficult for the parents to return to their homelands.

The intermediate 'meso-structures' have been attracting increasing attention from researchers in recent years. Certain individuals, groups or institutions may take on the role of mediating between migrants and political or economic institutions. A 'migration industry' emerges, consisting of recruitment organizations, lawyers, agents, smugglers and other intermediaries (Harris, 1996: 132–6). Such people can be both helpers, and exploiters of migrants. Especially in situations of illegal migration or of oversupply of potential migrants, the exploitative role may predominate: many migrants have been swindled out of their savings and have found themselves marooned without work or resources in a strange country. The emergence of a migration industry with a strong interest in the continuation of migration has often confounded government efforts to control or stop movements.

Macro-, meso- and micro-structures are intertwined in the migratory process, and there are no clear dividing lines between them. No single cause is ever sufficient to explain why people decide to leave their country and settle in another. It is essential to try to understand all aspects of the migratory process, by asking questions such as the following:

1. What economic, social, demographic, environmental or political factors have changed so much that people feel a need to leave their area of origin?
2. What factors provide opportunities for migrants in the destination area?
3. How do social networks and other links develop between the two areas, providing prospective migrants with information, means of travel and the possibility of entry?
4. What legal, political, economic and social structures and practices exist or emerge to regulate migration and settlement?
5. How do migrants turn into settlers, and why does this lead to discrimination, conflict and racism in some cases, but to pluralist or multicultural societies in others?
6. What is the effect of settlement on the social structure, culture and national identity of the receiving societies?
7. How does emigration change the sending area?
8. To what extent do migrations lead to new linkages between sending and receiving societies?

Transnational theory

This last aspect – new linkages between societies based on migration – has attracted much attention in recent years, leading to the emergence of a new body of theory on 'transnationalism' and 'transnational communities'. One aspect of globalization is rapid improvement in technologies of transport and communication, making it increasingly easy for migrants to maintain close links with their areas of origin. These developments also facilitate the growth of circulatory or repeated mobility, in which people migrate regularly between a number of places where they have economic, social or cultural linkages. Debates on transnationalism were stimulated by the work of Basch *et al.* (1994), which argued that 'deterritorialized nation-states' were emerging, with potentially serious consequences for national identity and international politics. Portes defines transnational activities as

> those that take place on a recurrent basis across national borders and that require a regular and significant commitment of time by participants. Such activities may be conducted by relatively powerful actors, such as representatives of national governments and multinational corporations, or may be initiated by more modest individuals, such as immigrants and their home country kin and relations. These activities are not limited to economic enterprises, but include political, cultural and religious initiatives as well. (Portes, 1999: 464)

The notion of a transnational community puts the emphasis on human agency. In the context of globalization, transnationalism can extend previous face-to-face communities based on kinship, neighbourhoods or workplaces into far-flung virtual communities, which communicate at a distance. Portes and his collaborators emphasize the significance of transnational business communities (whether of large-scale enterprises or of small ethnic entrepreneurs), but also note the importance of political and cultural communities. They distinguish between *transnationalism from above* – activities 'conducted by powerful institutional actors, such as multinational corporations and states' – and *transnationalism from below* – activities 'that are the result of grass-roots initiatives by immigrants and their home country counterparts' (Portes *et al.*, 1999: 221). Transnational communities can develop countervailing power to contest the power of corporations, governments and intergovernmental organizations. Indeed, informal linkages in the form of migration networks often undermine official migration policies which ignore the interests of migrants.

The term *transmigrant* may be used to identify people whose existence is shaped through participation in transnational communities based on migration (Glick-Schiller 1999: 203). Inflationary use of the term should be avoided: the majority of migrants still do not fit the pattern. Temporary labour migrants who sojourn abroad for a few years, send back remittances, communicate with their family at home and visit them occasionally are not transmigrants. Nor are permanent migrants who leave forever, and simply retain loose contact with their homeland. The key defining feature is that transnational activities are a central part of a person's life. Where this applies to a group of people, one can speak of a transnational community.

Transnational communities are not new, although the term is. The diaspora concept goes back to ancient times, and was used for peoples displaced or dispersed by force (e.g. the Jews; African slaves in the New World). It was also applied to certain trading groups such as Greeks in Western Asia and Africa, or the Arab traders who brought Islam to South-East Asia, as well as to labour migrants (Indians in the British Empire; Italians since the 1860s) (Cohen, 1997; Van Hear, 1998). The term diaspora often has strong emotional connotations, while the notion of a transnational community is more neutral. The new factor is the rapid proliferation of transnational communities under conditions of globalization (Vertovec, 1999: 447). Transnationalism is likely to go on growing, and transnational communities will become an increasingly important way to organize activities, relationships and identity for the growing number of people with affiliations in two or more countries.

From migration to settlement

Although each migratory movement has its specific historical patterns, it is possible to generalize on the social dynamics of the migratory process. It is necessary, however, to differentiate between economically-motivated migration and forced migration. Most economic migrations start with young, economically-active people. They are often 'target-earners', who want to save enough in a higher-wage economy to improve conditions at home, by buying land, building a house, setting up a business, or paying for education or dowries. After a period in the receiving country, some of these 'primary migrants' return home, but others prolong their stay, or return and then remigrate. This may be because of relative success when migrants find living and working conditions in the new country better than in the homeland. But it may also be because of relative failure when migrants find it impossible to save enough to achieve their aims, necessitating a longer sojourn. As time goes on, many erstwhile temporary migrants send for spouses, or find partners in the new country. With the birth of children, settlement takes on a more permanent character.

It is this powerful internal dynamic of the migratory process that often confounds expectations of the participants and undermines the objectives of policy-makers in both sending and receiving countries. In many migrations, there is no initial intention of family reunion and permanent settlement. However, when governments try to stop flows – for instance, because of a decline in the demand for labour – they may find that the movement has become self-sustaining. What started off as a temporary labour flow is transformed into family reunion, undocumented migration or even asylum-seeker flows. This is a result of the maturing of the migratory movement and of the migrants themselves as they pass through the life cycle. It may also be because dependency on migrant workers in certain sectors has become a structural feature of the economy.

The failure of policy-makers and analysts to see international migration as a dynamic social process is at the root of many political and social problems. The source of this failure has often been a one-sided focus on economic models of migration, which mistakenly claim that migration is an individual response to market factors. This has led to the belief that migration can be turned on and off like a tap, by changing policy settings which influence the costs and benefits of mobility for migrants. Migration may continue due to social factors, even when the economic factors which initiated the movement have been completely transformed.

Such developments are well illustrated by the Western European experience of 'guestworker' type movements from the Mediterranean basin from 1945 to 1973. Other situations in which social factors have led to unexpected outcomes include migrations from former colonies to the UK, France and the Netherlands, and migration from Europe, Latin America and Asia to the USA, Australia and Canada (see Chapter 4). One lesson of the last half-century is that it is extremely difficult for countries with democratic rights and strong legal systems to prevent migration turning into settlement. The situation is somewhat different in labour-recruiting countries which lack effective human rights guarantees, such as the Gulf states or some East and South-East Asian countries. The social dynamics of the migratory process do exist, but restrictions by the receiving governments may hinder family reunion and permanent settlement (Chapters 6 and 7).

The dynamics are different in the case of refugees and asylum seekers. They leave their countries because persecution, human rights abuse and generalized violence makes life there unsustainable. Most forced migrants remain in the neighbouring countries of first asylum – which are usually poor and often politically unstable themselves. Onward migration to countries which offer better economic and social opportunities is only possible for a small minority. However, there is evidence of selectivity: it is mainly those with financial resources, human capital (especially education) and social networks in destination countries who are able to migrate onwards (Zolberg and Benda, 2001). This onward migration is motivated both by the imperative of leaving a country of origin where life has become perilous, and by the hope of building a better life elsewhere. Attempts by policy-makers to make clear distinctions between economic and forced migrants are hampered by these 'mixed motivations'.

This has led to the notion of the 'migration-asylum nexus', which points to the complex links between the varying reasons for migration. Labour migrants, permanent settlers and refugees move under different conditions and legal regimes. Yet all these population movements are symptomatic of modernization and globalization. Colonialism, industrialization and integration into the world economy destroy traditional forms of production and social relations, and lead to the reshaping of nations and states. Underdevelopment, impoverishment, poor governance, endemic conflict and human rights abuse are closely linked. These conditions lead both to economically-motivated migration and to politically-motivated flight.

The formation of ethnic minorities

The long-term effects of immigration on society emerge in the later stages of the migratory process when migrants settle permanently and form distinct groups. Outcomes can be very different, depending on the actions of the receiving state and society. At one extreme, openness to settlement, granting of citizenship and gradual acceptance of cultural diversity may allow the formation of *ethnic communities*, which are seen as part of a multicultural society. At the other extreme, denial of the reality of settlement, refusal of citizenship and rights to settlers, and rejection of cultural diversity may lead to formation of *ethnic minorities*, whose presence is widely regarded as undesirable and divisive. Most countries of immigration have tended to lie somewhere between these two extremes.

Critics of immigration portray ethnic minorities as a threat to economic well-being, public order and national identity. Yet these ethnic minorities may in fact be the creation of the very people who fear them. Ethnic minorities may be defined as groups which

(a) have been assigned a subordinate position in society by dominant groups on the basis of socially-constructed markers of phenotype (that is, physical appearance or 'race'), origins or culture;
(b) have some degree of collective consciousness (or feeling of being a community) based on a belief in shared language, traditions, religion, history and experiences.

An ethnic minority is therefore a product of both 'other-definition' and of 'self-definition'. *Other-definition* means ascription of undesirable characteristics and assignment to inferior social positions by dominant groups. *Self-definition* refers to the consciousness of group members of belonging together on the basis of shared cultural and social characteristics. The relative strength of these processes varies. Some minorities are mainly constructed through processes of exclusion (which may be referred to as *racism*) by the majority. Others are mainly constituted on the basis of cultural and historical consciousness (or *ethnic identity*) among their members. The concept of the ethnic minority always implies some degree of marginalization or exclusion, leading to situations of actual or potential conflict. Ethnicity is rarely a theme of political significance when it is simply a matter of different group cultural practices.

Ethnicity

In popular usage, ethnicity is usually seen as an attribute of minority groups, but most social scientists argue that everybody has ethnicity, defined as a sense of group belonging, based on ideas of common origins, history, culture, experience and values (see Fishman, 1985: 4; Smith, 1986: 27). These ideas change only slowly, which gives ethnicity durability over generations and even centuries. But that does not mean that ethnic consciousness and culture within a group are homogeneous and static. Cohen and Bains argue that ethnicity, unlike race 'refers to a real process of historical individuation – namely the linguistic and cultural practices through which a sense of collective identity or "roots" is produced and transmitted from generation to generation, *and is changed in the process*' (Cohen and Bains, 1988: 24–5, emphasis in original).

The origins of ethnicity may be explained in various ways. Geertz, for example, sees ethnicity as a 'primordial attachment', which results 'from being born into a particular religious community, speaking a particular language, or even a dialect of a language and following particular social practices. These congruities of blood, speech, custom and so on, are seen to have an ineffable, and at times, overpowering coerciveness in and of themselves' (Geertz, 1963, quoted from Rex, 1986: 26–7). In this approach, ethnicity is not a matter of choice; it is pre-social, almost instinctual, something one is born into.

By contrast, many anthropologists use a concept of 'situational' ethnicity. Members of a specific group decide to 'invoke' ethnicity, as a criterion for self-identification, in a situation where such identification is necessary or useful. This explains the variability of ethnic boundaries and changes in salience at different times. The markers chosen for the boundaries are also variable, generally emphasizing cultural characteristics, such as language, shared history, customs and religion, but sometimes including physical characteristics (Wallman, 1986: 229). In this view there is no essential difference between the drawing of boundaries on the basis of cultural difference or of phenotypical difference (popularly referred to as 'race'). The visible markers of a phenotype (skin colour, features, hair colour, and so on) correspond to what is popularly understood as 'race'. We avoid using the term 'race' as far as possible, since there is increasing agreement among biologists and social scientists that there are no measurable characteristics among human populations that allow classification into 'races'. Genetic variance within any one population is greater than alleged differences between different populations. 'Race' is thus a social construction produced by the process we refer to as racism.

Similarly, some sociologists see ethnic identification or mobilization as rational behaviour, designed to maximize the power of a group in a situation of market competition. Such theories have their roots in Max Weber's concept of 'social closure', whereby a status group establishes rules and practices to exclude others, in order to gain a competitive advantage (Weber, 1968: 342). For Weber (as for Marx), organization according to 'affective criteria' (such as religion, ethnic identification or communal consciousness) was in the long run likely to be superseded by organization according to economic interests (class) or bureaucratic rationality. Nonetheless, the instrumental use of these affiliations could be rational if it led to successful mobilization.

Other sociologists reject the concept of ethnicity altogether, seeing it as 'myth' or 'nostalgia', which cannot survive against the rational forces of economic and social integration in large-scale industrial societies (Steinberg, 1981). Yet it is hard to ignore the growing significance of ethnic mobilization, so that many attempts have been made to show the links between ethnicity and power. Studies of the 'ethnic revival' by the US sociologists Glazer and Moynihan (1975) and Bell (1975) emphasize the instrumental role of ethnic identification: phenotypical and cultural characteristics are used to strengthen group solidarity, in order to struggle more effectively for market advantages, or for increased allocation of resources by the state. Bell sees ethnic mobilization as a substitute for the declining power of class identification in advanced industrial societies; the decision to organize on ethnic lines seems to be an almost arbitrary 'strategic choice'. This does not imply that markers, such as skin colour, language, religion, shared history and customs, are not real, but rather that the decision to use them to define an ethnic group is not predetermined.

Whether ethnicity is 'primordial', 'situational' or 'instrumental' need not concern us further here. The point is that ethnicity leads to identification with a specific group, but its visible markers – phenotype, language, culture, customs, religion, behaviour – may also be used as criteria for exclusion by other groups. Ethnicity only takes on social and political meaning when it is linked to processes of boundary drawing between dominant groups and minorities. Becoming an ethnic minority is not an automatic result of immigration, but rather the consequence of specific mechanisms of marginalization, which affect different groups in different ways.

Racism

Racism towards certain groups is to be found in virtually all immigration countries. Racism may be defined as the process whereby social groups categorize other groups as different or inferior, on the basis of phenotypical or cultural markers. This process involves the use of economic, social or political power, and generally has the purpose of legitimating exploitation or exclusion of the group so defined.

Racism means making (and acting upon) predictions about people's character, abilities or behaviour on the basis of socially constructed markers of difference. The power of the dominant group is sustained by developing structures (such as laws, policies and administrative practices) that exclude or discriminate against the dominated group. This aspect of racism is generally known as institutional or structural racism. Racist attitudes and discriminatory behaviour on the part of members of the dominant group are referred to as informal racism. Many social scientists now use the term 'racialization' to refer to public discourses which imply that a range of social or political problems are a 'natural' consequence of certain ascribed physical or cultural characteristics of minority groups. Racialization can be used to apply to the social construction of a specific group as a problem, or in the wider sense of the 'racialization of politics' or the 'racialization of urban space'.

In some countries, notably Germany and France, there is reluctance to speak of racism. Euphemisms such as 'hostility to foreigners', 'ethnocentrism' or 'xenophobia' are used. But the debate over the label seems sterile: it is more important to understand the phenomenon and its causes. Racism operates in different ways according to the specific history of a society and the interests of the dominant group. In many cases, supposed biological differences are not the only markers: culture, religion, language or other factors are taken as indicative of phenotypical differences. For instance, anti-Muslim racism in Europe is based on cultural symbols which, however, are linked to phenotypical markers (such as Arab or African features).

The historical explanation for racism in Western Europe and in post-colonial settler societies (like Australia) lies in traditions, ideologies and cultural practices, which have developed through ethnic conflicts associated with nation building and colonial expansion (compare Miles, 1989). The reasons for the recent increase in racism lie in fundamental economic and social changes which question the optimistic view of progress embodied in Western thought. Since the early 1970s, economic restructuring and increasing international cultural interchange have been experienced by many sections of the population as a direct threat to their livelihood, social conditions and identity. Since these changes have coincided with the arrival of new ethnic minorities, the tendency has been to perceive the newcomers as the cause of the threatening changes: an interpretation eagerly encouraged by the extreme right, but also by many mainstream politicians.

Moreover, the very changes which threaten disadvantaged sections of the population have also weakened the labour movement and working-class cultures, which might otherwise have provided some measure of protection. The decline of working-class parties and trade unions and the erosion of local communicative networks have created the social space for racism to become more virulent (Wieviorka, 1995; Vasta and Castles, 1996). (We lay no claim to originality with regard to this definition and discussion of racism. It is oriented towards current sociological debates, which have generated a large body of literature. See, for example, CCCS (1982); Rex and Mason (1986); Cohen and Bains (1988); Miles (1989); Wieviorka (1991, 1992); Solomos (1993); Goldberg and Solomos (2002). There is no unanimity among social scientists about the correct definition and explanations of racism, but we have no space for a more detailed discussion of these matters here.)

Ethnicity, class, gender and life cycle

Racial and ethnic divisions are only one aspect of social differentiation. Others include social class, gender and position in the life cycle. None of these distinctions is reducible to any other, yet they constantly cross-cut and interact, affecting life chances, lifestyles, culture and social consciousness. Immigrant groups and ethnic minorities are just as heterogeneous as the rest of the population. The migrant is a gendered subject, embedded in a wide range of social relationships.

In the early stages of post-1945 international labour mobility, the vital nexus appeared to be that between migration and class. Migration was analysed in terms of the interests of various sectors of labour and capital (Castles and Kosack, 1985) or of the incorporation of different types of workers into segmented labour markets (Piore, 1979). International migration continues to be an important factor helping to shape labour market patterns and class relations (see Chapter 8). However, there has been a growing awareness of the crucial links between class, ethnicity and gender.

Even in the early stages, the role of women in maintaining families and reproducing workers in the country of origin was crucial to the economic benefits of labour migration. Moreover, a large proportion of migrant workers were female. As Phizacklea (1983: 5) pointed out, it was particularly easy to ascribe inferiority to women migrant workers, just because their primary roles in patriarchal societies were defined as wife and mother, dependent on a male breadwinner. They could therefore be paid lower wages and controlled more easily than men. Since the 1970s, restructuring and unemployment have made full employment more the exception than the rule for some minorities. Very high rates of unemployment among ethnic minority youth have meant that 'they are not the unemployed, but the never employed' (Sivanandan, 1982: 49). Members of ethnic minorities have experienced racism from some white workers and therefore find it hard to define their political consciousness in class terms.

Migrant women's work experience often remains distinct from that of men. They tend to be overrepresented in the least desirable occupations, such as repetitive factory work and lower-skilled positions in the personal and community services sectors. However, there has been some mobility into white-collar jobs in recent years, partly as a result of the decline of manufacturing. Professional employment is often linked to traditional caring roles. Minority women have experienced casualization of employment and increasing unemployment (which often does not appear in the statistics due to their status as 'dependants').

Complex patterns of division of labour on ethnic and gender lines have developed (Waldinger *et al.*, 1990). In a study of the fashion industry in European countries, Phizacklea (1990: 72–93) argued that this industry was able to survive, despite the new global division of labour, through the development of 'subcontracting webs': large retail companies were able to put pressure for lower prices on small firms controlled by male ethnic entrepreneurs, whose market position was constrained by racial discrimination. These in turn were able to use both patriarchal power relations and the vulnerable legal position of women immigrants to enforce extremely low wages and poor working conditions in sweatshops and outwork. Collins *et al.* (1995: 180–1) present a similar picture of the links between racialization and gender in ethnic small business in Australia.

Racism, sexism and class domination are three specific forms of 'social normalization and exclusion' which are intrinsic to capitalism and modernity, and which have developed in close relationship to each other (Balibar, 1991: 49). Racism and sexism both involve predicting social behaviour on the basis of allegedly fixed biological or cultural characteristics. According to Essed, racism and sexism 'narrowly intertwine and combine under certain conditions into one, hybrid phenomenon. Therefore it is useful to speak of *gendered racism* to refer to the racist oppression of Black women as structured by racist and ethnicist perceptions of gender roles' (Essed, 1991: 31, emphasis in original).

Anthias and Yuval-Davis (1989) analyse links between gender relations and the construction of the nation and the ethnic community. Women are not only the biological reproducers of an ethnic group, but also the 'cultural carriers' who have the key role in passing on the language and cultural symbols to the young (see also Vasta, 1990, 1992). In nationalist discourses women serve as the symbolic embodiment of national unity and distinctiveness. They nurture and support the (male) warrior-citizens. In defeat and suffering, the nation is portrayed as a woman in danger. Such symbolism legitimates the political inferiority of women: they embody the nation, while the men represent it politically and militarily (Lutz *et al.*, 1995).

The role of gender in ethnic closure is evident in immigration rules which still often treat men as the principal immigrants while women and children are mere 'dependants'. Britain has used gender-specific measures to limit the growth of the black population. In the 1970s, women from the Indian subcontinent coming to join husbands or fiancés were subjected to 'virginity tests' at Heathrow Airport. The authorities also sought to prevent Afro-Caribbean and Asian women from bringing in husbands, on the grounds that the 'natural place of residence' of the family was the abode of the husband (Klug, 1989: 27–9). In many countries, women who enter as dependants do not have an entitlement to residence in their own right and may face deportation if they get divorced.

The stages of the life cycle – childhood, youth, maturity, middle age, old age – are also important determinants of economic and social positions, culture and consciousness. There is often a gulf between the experiences of the migrant generation and those of their children, who have grown up and gone to school in the new country. Ethnic minority youth become aware of the contradiction between the prevailing ideologies of equal opportunity and the reality of discrimination and racism in their daily lives. This can lead to the emergence of counter-cultures and political radicalization. In turn, ethnic minority youth are perceived as a 'social time bomb' or a threat to public order, which has to be contained through social control institutions such as the police, schools and welfare bureaucracies (see Chapter 10).

Culture, identity and community

In the context of globalization, culture, identity and community often serve as a focus of resistance to centralizing and homogenizing forces (Castells, 1997). These have become central themes in debates on the new ethnic minorities. First, as already outlined, cultural difference serves as a marker for ethnic boundaries. Second, ethnic cultures play a central role in community formation: when ethnic groups cluster together, they establish their own neighbourhoods, marked by distinctive use of private and public spaces. Third, ethnic neighbourhoods are perceived by some members of the majority group as confirmation of their fears of a 'foreign takeover'. Ethnic communities are seen as a threat to the dominant culture and national identity. Fourth, dominant groups may see migrant cultures as primordial, static and regressive. Linguistic and cultural maintenance is taken as proof of inability to come to terms with an advanced industrial society. Those who do not assimilate 'have only themselves to blame' for their marginalized position.

For ethnic minorities, culture plays a key role as a source of identity and as a focus for resistance to exclusion and discrimination. Reference to the culture of origin helps people maintain self-esteem in a situation where their capabilities and experience are undermined. But a static, primordial culture cannot fulfil this task, for it does not provide orientation in a hostile environment. The dynamic nature of culture lies in its capacity to link a group's history and traditions with the actual situation in the migratory process. Migrant or minority cultures are constantly recreated on the basis of the needs and experience of the group and its interaction with the actual social environment (Schierup and Ålund, 1987; Vasta *et al.*, 1992). An apparent regression, for instance to religious fundamentalism, may be precisely the result of a form of modernization which has been experienced as discriminatory, exploitative and destructive of identity.

It is therefore necessary to understand the development of ethnic cultures, the stabilization of personal and group identities, and the formation of ethnic communities as facets of a single process. This process is not self-contained: it depends on constant interaction with the state and the various institutions and groups in the country of immigration, as well as with the society of the country of origin. Immigrants and their descendants do not have a static, closed and homogeneous ethnic identity, but instead dynamic *multiple identities*, influenced by a variety of cultural, social and other factors.

The concept of *national culture and identity* has become highly questionable. Increasing global economic and cultural integration is leading to a simultaneous homogenization and fragmentation of culture. As multinational companies take over and repackage the artefacts of local cultures it becomes possible to consume all types of cultural products everywhere, but at the same time these lose their meaning as symbols of group identity. National or ethnic cultures shed their distinctiveness and become just another celebration of the cultural dominance of the international industrial apparatus. Hence the constant search for new subcultures, styles and sources of identity, particularly on the part of youth.

Gilroy sees the focus of this recreation of culture in the social movements of local communities, as well as in youth sub-cultures. He argues that legacies of anti-colonial struggles have been reshaped in Britain in the reproduction of classes and 'races' which become youth culture:

> The institutions they create: temples, churches, clubs, cafés and blues dances confound any Eurocentric idea of where the line dividing politics and culture should fall. The distinction between public and private spheres cuts across the life of their households and communities in a similar manner. Traditional solidarity mediates and adapts the institutions of the British political system against which it is defined. (Gilroy, 1987: 37)

Culture is becoming increasingly politicized in all countries of immigration. As ideas of racial superiority lose their ideological strength, exclusionary practices against minorities increasingly focus on issues of cultural difference. At the same time, the politics of minority resistance crystallize more and more around cultural symbols. Yet these symbols are only partially based on imported forms of ethnicity. Their main power as definers of community and identity comes from the incorporation of new experiences of ethnic minority groups in the immigration country.

State and nation

Large-scale migrations and growing diversity may have important effects on political institutions and national identity. In the contemporary world, nation-states (of which there are some 200) are the predominant form of political organization. They derive their legitimacy from the claim of representing the aspirations of their people (or citizens). This implies two further claims: that there is an underlying cultural consensus which allows agreement on the values or interests of the people, and that there is a democratic process for the will of the citizens to be expressed. Such claims are often empty slogans, for most countries are marked by heterogeneity, based on ethnicity, class and other cleavages. Only a minority of countries consistently use democratic mechanisms to resolve value and interest conflicts. Nonetheless, the democratic nation-state has become a global norm.

Immigration of culturally diverse people presents nation-states with a dilemma: incorporation of the newcomers as citizens may undermine myths of cultural homogeneity; but failure to incorporate them may lead to divided societies, marked by severe inequality and conflict. This problem arises from the character of the nation-state, as it developed in Western Europe and North America in the context of modernization, industrialization and colonialism. Pre-modern states based their authority on the absolute power of a monarch over a specific territory. Within this area, all people were subjects of the monarch (rather than citizens). There was no concept of a national culture which transcended the gulf between aristocratic rulers and peasants. The modern nation-state, by contrast, implies a close link between cultural belonging and political identity (Castles and Davidson, 2000).

A *state*, according to Seton-Watson (1977: 1), 'is a legal and political organization, with the power to require obedience and loyalty from its citizens'. The state regulates political, economic and social relations in a bounded territory. Most modern nation-states are formally defined by a constitution and laws, according to which all power derives from the people (or nation). It is therefore vital to define who belongs to the people. Membership is marked by the status of citizenship, which lays down rights and duties. Non-citizens are excluded from at least some of these. Citizenship is the essential link between state and nation, and obtaining citizenship is of central importance for newcomers to a country.

Seton-Watson describes a *nation* as 'a community of people, whose members are bound together by a sense of solidarity, a common culture, a national consciousness' (Seton-Watson, 1977: 1). Such essentially subjective phenomena are difficult to measure. Moreover, it is not clear how a nation differs from an ethnic group, which is defined in a very similar way (see above). Anderson provides an answer with his definition of the nation: 'it is an imagined political community – and imagined as both inherently limited and sovereign' (Anderson, 1983: 15). This concept points to the political character of the nation and its links with a specific territory: an ethnic group that attains sovereignty over a bounded territory becomes a nation and establishes a nation-state. As Smith (1991: 14) puts it: 'A nation can ... be defined as a named human population sharing an historic territory, common myths and historical memories, a mass, public culture, a common economy and common legal rights and duties for all members.'

Anderson (1983) regards the nation-state as a modern phenomenon, whose birthdate is that of the US Constitution of 1787. Gellner (1983) argues that nations could not exist in pre-modern societies, owing to the cultural gap between elites and peasants, while modern industrial societies require cultural homogeneity to function, and therefore generate the ideologies needed to create nations. However, both Seton-Watson (1977) and Smith (1986) argue that the nation is of much greater antiquity, going back to the ancient civilizations of East Asia, the Middle East and Europe. All these authors seem to agree that the nation is essentially a belief system, based on collective cultural ties and sentiments. These convey a sense of identity and belonging, which may be referred to as national consciousness.

Specific to the modern nation-state is the linking of national consciousness with the principle of democracy: every person classified as a member of the national community has an equal right to participate in the formulation of the political will. This linking of nationality and citizenship is deeply contradictory. In liberal theory, all citizens are meant to be free and equal persons who are treated as homogeneous within the political sphere. This requires a separation between a person's political rights and obligations, and their membership of specific groups, based on ethnicity, religion, social class or regional location. The political sphere is one of universalism, which means abstraction from cultural particularity and difference. Difference is to be restricted to the 'non-public identity' (Rawls, 1985: 232–41).

This conflicts with the reality of nation-state formation, however, in which being a citizen depends on membership in a certain national community, usually based on the dominant ethnic group of the territory concerned. Thus a citizen is always also a member of a nation, a national. Nationalist ideologies demand that ethnic group, nation and state should be facets of the same community and have the same boundaries – every ethnic group should constitute itself as a nation and should have its own state, with all the appropriate trappings: flag, army, Olympic team and postage stamps. In fact, such congruence has rarely been achieved: nationalism has always been an ideology trying to achieve such a condition, rather than an actual state of affairs.

The construction of nation-states has involved the spatial extension of state power, and the territorial incorporation of hitherto distinct ethnic groups. These may or may not coalesce into a single nation over time. Attempts to consolidate the nation-state can mean exclusion, assimilation or even genocide for minority groups. It is possible to keep relatively small groups in situations of permanent subjugation and exclusion from the 'imagined community'. This has applied, for instance, to Jews and gypsies in various European countries, to indigenous peoples in settler colonies and to the descendants of slaves and contract workers in some areas of European colonization. Political domination and cultural exclusion is much more difficult if the subjugated nation retains a territorial base, like the Scots, Welsh and Irish in the UK, or the Basques in Spain.

The experience of 'historical minorities' has helped to mould structures and attitudes, which affect the conditions for new immigrant groups. The pervasive fear of 'ghettos' or 'ethnic enclaves' indicates that minorities seem most threatening when they concentrate in distinct areas. For nationalists, an ethnic group is a potential nation which does not (yet) control any territory, or have its own state. Most modern states have made conscious efforts to achieve cultural and political integration of minorities.

Mechanisms include citizenship itself, centralized political institutions, the propagation of national languages, universal education systems and creation of national institutions like the army, or an established church (Schnapper, 1991, 1994). The problem is similar in character everywhere, whether the minorities are 'old' or 'new': how can a nation be defined, if not in terms of a shared (and single) ethnic identity? How are core values and behavioural norms to be laid down, if there is a plurality of cultures and traditions?

Coping with diversity has become even more difficult in the era of globalization. In the nation-states of the nineteenth and early twentieth centuries, politics, the economy, social relations and culture were all organized within the same boundaries. Even movements for change, such as the labour movement or left-wing parties, based their strategies on the nation-state. Globalization has destabilized this model. The dynamics of economic life now transcend borders, and have become increasingly uncontrollable for national governments. De-industrialization of the older industrial nations has led to profound social changes. The nation-state is still the basic unit for defence, public order and welfare, but its room for autonomous action is severely reduced. No government can pursue policies which ignore the imperatives of global markets. The nexus between power and national boundaries is declining.

Citizenship

The states of immigration countries have had to devise a range of policies and institutions to respond to the problems of increased ethnic diversity (see Aleinikoff and Klusmeyer, 2000, 2001). These relate to certain central issues: defining who is a citizen, how newcomers can become citizens and what citizenship means. In principle the nation-state only permits a single membership, but immigrants and their descendants have a relationship to more than one state. They may be citizens of two states, or they may be a citizen of one state but live in another. These situations may lead to 'divided loyalties' and undermine the cultural homogeneity which is the nationalist ideal. Thus large-scale settlement inevitably leads to a debate on citizenship.

Citizenship designates the equality of rights of all citizens within a political community, as well as a corresponding set of institutions guaranteeing these rights (Bauböck, 1991: 28). However, formal equality rarely leads to equality in practice. For instance, citizenship has always meant something different for men than for women, because the concept of the citizen has been premised on the male family-father, who represents his woman and children (Anthias and Yuval-Davis, 1989). The citizen has generally been defined in terms of the cultures, values and interests of the majority ethnic group. Finally, the citizen has usually been explicitly or implicitly conceived in class terms, so that gaining real participatory rights for members of the working class has been one of the central historical tasks of the labour movement. The history of citizenship has therefore been one of conflicts over the real content of the category in terms of civil, political and social rights (Marshall, 1964).

The first concern for immigrants, however, is not the exact content of citizenship, but how they can obtain it, in order to achieve a legal status formally equal to that of other residents. Access has varied considerably in different countries, depending on the prevailing concept of the nation. We can distinguish the following ideal types of citizenship:

1. The *imperial model*: definition of belonging to the nation in terms of being a subject of the same power or ruler. This notion pre-dates the French and American revolutions. It allowed the integration of the various peoples of multi-ethnic empires (the British, the Austro-Hungarian, the Ottoman). This model remained formally in operation in the UK until the Nationality Act of 1981, which created a modern type of citizenship for the first time. It also had some validity for the former Soviet Union. The concept almost always has an ideological character, in that it helps to veil the actual dominance of a particular ethnic group or nationality over the other subject peoples.
2. The *folk or ethnic model*: definition of belonging to the nation in terms of ethnicity (common descent, language and culture), which means exclusion of minorities from citizenship and from the community of the nation. (Germany came close to this model until the introduction of new citizenship rules in 2000.)
3. The *republican model*: definition of the nation as a political community, based on a constitution, laws and citizenship, with the possibility of admitting newcomers to the community, providing they adhere to the political rules and are willing to adopt the national culture. This assimilationist approach dates back to the French and American revolutions. France is the most obvious current example.
4. The *multicultural model*: definition of the nation as a political community, based on a constitution, laws and citizenship, with the possibility of admitting newcomers, who may maintain cultural difference and form ethnic communities providing they adhere to the political rules. This pluralist or multicultural approach became dominant in the 1970s and 1980s in Australia, Canada and Sweden, and was also influential in other Western countries. However, there was a move away from multiculturalism in many places in the 1990s.

All these ideal types have one factor in common: they are premised on citizens who belong to just one nation-state. Migrant settlement is seen as a process of transferring primary loyalty from the state of origin to the new state of residence. This process, which may be long-drawn-out and even span generations, is symbolically marked by naturalization and acquisition of citizenship of the new state. Transnational theory (see above) argues that this no longer applies for growing groups of migrants who form transnational communities and maintain strong cross-border affiliations – possibly over generations. This is seen as a challenge to traditional models of national identity. Thus an additional ideal type of citizenship seems to be emerging:

5. The *transnational model*: social and cultural identities which transcend national boundaries, leading to multiple and differentiated forms of belonging. Transnationalism could have important consequences for democratic institutions and political belonging in future. This corresponds with the fact that, through globalization, a great deal of political and economic power is shifting to transnational corporations and international agencies which are not currently open to democratic control (Castles and Davidson, 2000). The survival of democracy may depend on finding ways of including people with multiple identities in a range of political communities. It also means ensuring citizen participation in new locations of power, whether supra- or sub-national, public or private.

The applicability of these models to specific countries will be discussed in more detail in Chapter 10. In fact, the models are neither universally accepted nor static even within a single country (Bauböck and Rundell, 1998: 1273). Moreover, the distinction between citizens and non-citizens is becoming less clear-cut. Immigrants who have been legally resident in a country for many years can often obtain a special status, tantamount to 'quasi-citizenship'. This may confer such rights as secure residence status; rights to work, seek employment and run a business; entitlements to social security benefits and health services; access to education and training; and limited political rights, such as the rights of association and of assembly. In some countries, long-term foreign residents have voting rights in local elections. Such arrangements create a new legal status, which is more than that of a foreigner, but less than that of a citizen. Hammar (1990: 15–23) has suggested the term *denizen* for people 'who are foreign citizens with a legal and permanent resident status'. This applies to millions of long-term foreign residents in Western Europe, many of whom were actually born in their countries of residence.

A further element in the emergence of quasi-citizenship is the development of international human rights standards, as laid down by bodies like the UN, the International Labour Organization (ILO) and the World Trade Organization (WTO). A whole range of civil and social rights are legally guaranteed for citizens and non-citizens alike in the states which adopt these international norms (Soysal, 1994). However, the legal protection provided by international conventions can be deficient when states do not incorporate the norms into their national law, despite ratifying the conventions.

The EU provides the furthest-going example for transnational citizenship. The 1991 Maastricht Treaty established the legal notion of Citizenship of the European Union, which embraced the following individual rights:

- freedom of movement and residence in the territory of member states;
- the right to vote and to stand for office in local elections and European Parliament elections in the state of residence;
- the right to diplomatic protection by diplomats of any EU state in a third country;
- the right to petition the European Parliament and the possibility to appeal to an ombudsman (Martiniello, 1994: 31).

However, EU citizens living in another member state do not have the right to vote in elections for the national parliament of that state. People dependent on social security do not have a right to settle in another member country; and access to public employment is still generally restricted to nationals (Martiniello, 1994: 41). For the time being, it seems more appropriate to treat EU citizenship as a case of quasi-citizenship. The limited character is made even clearer by the fact that an 'EU passport' is legally still a passport of one of the member countries. So far, EU citizenship has done nothing for the majority of immigrants, who come from outside the EU. However, the process of European integration is continuing: the 1997 Treaty of Amsterdam (Article 63) established community competence in the areas of migration and asylum, and principles for a common policy were laid down by the European Council meeting at Tampere in 1999. The new policy – planned to come into force in 2004 – may mean common entry criteria for immigrants and refugees, and freedom of movement within the EU for legally-resident third country nationals.

The long-term question is whether democratic states can successfully operate with a population differentiated into full citizens, quasi-citizens and foreigners. The central principle of the democratic state is that all members of civil society should be incorporated into the political community. That means granting full citizenship to all permanent residents. Migrations are likely to continue and there will be increasing numbers of people with affiliations to more than one society. Dual or multiple citizenship will become increasingly common. In fact, nearly all immigration countries have changed their citizenship rules over the last 40–50 years – sometimes several times. More and more countries accept dual citizenship (at least to some extent). A major focus of reform is the introduction of measures to integrate the second generation into the political community through birthright citizenship or easier naturalization (see Aleinikoff and Klusmeyer, 2000; Castles and Davidson, 2000: Chapter 4). The consequence is that the meaning of citizenship is likely to change, and that the exclusive link to one nation-state will become more tenuous. This could lead to some form of 'transnational citizenship', as Bauböck (1994) suggests. But that in turn raises the question of how states will regulate immigration if citizenship becomes more universal.

Conclusion

This chapter has been concerned with some of the theoretical explanations of migration and ethnic minority formation. One central argument is that migration and settlement are closely related to other economic, political and cultural linkages being formed between different countries in an accelerating process of globalization. International migration – in all its different forms – must be seen as an integral part of contemporary world developments. It is likely to grow in volume in the years ahead, because of the strong pressures for continuing global integration.

A second argument is that the migratory process has certain internal dynamics based on the social networks which are at its core. These internal dynamics can lead to developments not initially intended either by the migrants themselves or by the states concerned. The most common outcome of a migratory movement, whatever its initial character, is settlement of a significant proportion of the migrants, and formation of ethnic communities or minorities in the new country. Thus the emergence of societies which are more ethnically and culturally diverse must be seen as an inevitable result of initial decisions to recruit foreign workers, or to permit immigration.

A third argument is that increasing numbers of international migrants do not simply move from one society to another, but maintain recurring and significant links in two or more places. They form transnational communities which live across borders. This trend is facilitated by globalization, both through improvements in transport and communications technology, and through diffusion of global cultural values. Transnational communities currently embrace only a minority of migrants, but may in the long run have enormous consequences for social identity and political institutions.

The fourth argument concerns the nature of ethnic minorities and the process by which they are formed. Most minorities are formed by a combination of other-definition and self-definition. Other-definition refers to various forms of exclusion and discrimination (or racism). Self-definition has a dual character. It includes assertion and recreation of ethnic identity, centred upon pre-migration cultural symbols and practices. It also includes political mobilization against exclusion and discrimination, using cultural symbols and practices in an instrumental way. When settlement and ethnic minority formation take place at times of economic and social crisis, they can become highly politicized. Issues of culture, identity and community can take on great significance, not only for immigrants, but also for the receiving society as a whole.

The fifth argument focuses on the significance of immigration for the nation-state. It seems likely that increasing ethnic diversity will contribute to changes in central political institutions, such as citizenship, and may affect the very nature of the nation-state.

These theoretical conclusions help to explain the growing political salience of issues connected with migration and ethnic minorities. The migratory movements of the last 50 years have led to irreversible changes in many countries. Continuing migrations will cause new transformations, both in the societies already affected and in further countries now entering the international migration arena. The more descriptive accounts which follow will provide a basis for further discussion of these ideas. Chapters 3–7 are mainly concerned with the early stages of the migratory process, showing how initial movements give rise to migratory chains and long-term settlement. Chapters 8–11 are concerned mainly with the later stages of the migratory process. They discuss the ways in which settlement and minority formation affect the economies, societies and political systems of immigration countries.

Guide to further reading

Amongst the many recent works on globalization, the following are useful as introductions: Castells (1996, 1997, 1998), Held *et al.* (1999), Bauman (1998) and Cohen and Kennedy (2000). Two recent works provide overviews of international migration theory: Massey *et al.* (1998) presents a systematic discussion and critique (based on two earlier articles: Massey *et al.* 1993 and 1994), while Brettell and Hollifield (2000) contains chapters addressing the contributions of some of the main social scientific disciplines to the study of migration. Boyle *et al.* (1998) is a good introductory text written by geographers. An earlier, but still valuable, compendium on migration theories is to be found in a special issue of *International Migration Review* (1989, 23:3). Kritz *et al.* (1992) is an excellent collection on migration systems theory. Phizacklea (1983), Morokvasic (1984) and Lutz *et al.* (1995) have edited useful collections on the relationship between migration and gender. Sassen (1988) gives an original perspective on the political economy of migration, while Borjas (1990) presents the neo-classical view.

Goldberg and Solomos (2002) is a comprehensive collection of essays on various aspects of racial and ethnic studies. Rex and Mason (1986) provides detailed expositions of theoretical approaches to race and ethnic relations. Mosse (1985), Cohen and Bains (1988), Miles (1989), Balibar and Wallerstein (1991), Essed (1991) and Wieviorka (1995) are good on racism. Anderson (1983), Gellner (1983) and Ignatieff (1994) provide stimulating analyses of nationalism, while Smith (1986, 1991) discusses the relationship between ethnicity and nation. Analyses of the relationship between migration and citizenship are to be found in Bauböck (1991, 1994), Bauböck and Rundell (1998), Aleinikoff and Klusmeyer (2000, 2001) and Castles and Davidson (2000). Gutmann (1994), Schnapper (1994), Soysal (1994) and Kymlicka (1995) present various perspectives on the same theme. DeWind (1997) is a collection of articles on the changing character of immigrant incorporation in the USA. Good introductions to the emerging field of transnational communities include Basch *et al.* (1994), Cohen (1997), Portes *et al.* (1999), Vertovec (1999) and Faist (2000). Van Hear (1998) discusses transnational theory from the perspective of refugee movements. Zolberg and Benda (2001) is very useful for understanding the links between economic migration and refugee movements.

Chapter 3

International Migration before 1945

The post-1945 migrations may be new in scale and scope, but population movements in response to demographic growth, climatic change and the development of production and trade have always been part of human history. Warfare, conquest, formation of nations and the emergence of states and empires have all led to migrations, both voluntary and forced. The enslavement and deportation of conquered people was a frequent early form of labour migration. From the end of the Middle Ages, the development of European states and their colonization of the rest of the world gave a new impetus to international migrations of many different kinds.

In Western Europe, 'migration was a long-standing and important facet of social life and the political economy' from about 1650 onwards, playing a vital role in modernization and industrialization (Moch, 1995: 126; see also Moch, 1992). The centrality of migration is not adequately reflected in prevailing views on the past: as Gérard Noiriel (1988: 15–67) has pointed out, the history of immigration has been a 'blind spot' of historical research in France. This applies equally to other European countries. Denial of the role of immigrants in nation building has been crucial to the creation of myths of national homogeneity. This was obviously impossible in classical countries of immigration such as the USA. It is only in very recent times that French, German and British historians have started serious investigation of the significance of immigration. Box 3.1 provides an illustration of the significance of migration in early processes of nation building.

Individual liberty is portrayed as one of the great moral achievements of capitalism, in contrast with earlier societies where liberty was restricted by traditional bondage and servitude. Neo-classical theorists portray the capitalist economy as being based on free markets, including the labour market, where employers and workers encounter each other as free legal subjects, with equal rights to make contracts. International migration is portrayed as a market in which workers make the free choice to move to the area where they will receive the highest income (compare Borjas, 1990: 9–18). But this harmonious picture often fails to match reality. As Cohen (1987) has shown, capitalism has made use of both free and 'unfree' workers in every phase of its development. Labour migrants have frequently been unfree workers, either because they are taken by force to the place where their labour is needed, or because they are denied rights enjoyed by other workers, and cannot therefore compete under equal conditions. Even where migration is voluntary and unregulated, institutional and informal discrimination may limit the real freedom and equality of the workers concerned.

Since economic power is usually linked to political power, mobilization of labour often has an element of coercion, sometimes involving violence, military force and bureaucratic control. Examples are the slave economy of the Americas; indentured colonial labour in Asia, Africa and the Americas; mineworkers in southern Africa in the nineteenth and twentieth centuries; foreign workers in Germany and France before the Second World War; forced labourers in the Nazi war economy; 'guestworkers' in post-1945 Europe, and 'illegals' denied the protection of law in many countries today.

One important theme is not dealt with here because it requires more intensive treatment than is possible in the present work: the devastating effects of international migration on the indigenous peoples of colonized countries. European conquest of Africa, Asia, America and Oceania led either to the domination and exploitation of native peoples or to genocide, both physical and cultural. Nation building – particularly in the Americas and Oceania – was based on the importation of new populations. Thus immigration contributed to the exclusion and marginalization of aboriginal peoples. One starting point for the construction of new national identities was the idealization of the destruction of indigenous societies: images such as 'how the West was won' or the struggle of Australian pioneers against the Aborigines became powerful myths. The roots of racist stereotypes – today directed against new immigrant groups – often lie in historical treatment of colonized peoples. Nowadays there is increasing realization that appropriate models for inter-group relations have to address the needs of indigenous populations, as well as those of immigrant groups.

Colonialism

European colonialism gave rise to various types of migration. One was the large outward movement from Europe, first to Africa and Asia, then to the Americas, and later to Oceania. Europeans migrated, either permanently or temporarily, as sailors, soldiers, farmers, traders, priests and administrators. Some of them had already migrated within Europe: Lucassen (1995) has shown that around half the soldiers and sailors of the Dutch East India Company in the seventeenth and eighteenth centuries were not Dutch but 'transmigrants', mainly from poor areas of Germany. The

Box 3.1 Migration and nation in French history

Ancient Gaul encompassed much of the area of modern-day France. At the collapse of the western Roman Empire in the fifth century AD, Gaul was inhabited by a patchwork of culturally and politically diverse peoples, including Roman citizens and soldiers, slaves, settled Germanic tribes and more recent arrivals. There were multiple centres of political power. Celts from the west of Britain moved across the English Channel to what is now Brittany, to escape the invading Saxons. These Celts fought with the embryonic Frankish state, from which the medieval French kingdom would emerge.

Norse raiders wreaked havoc upon the Frankish territory and, from 900 AD, they settled in the area now called Normandy. The expansion of the Frankish state and its steady incorporation of adjacent lands and peoples was a long process, and French identity and consciousness emerged slowly. Life for most inhabitants of medieval France was encapsulated by the village and its environs, but there was awareness of the exterior world. To the inhabitants of the Frankish state, the people of Brittany, Normandy or Languedoc were foreigners.

But there were also newcomers: traders and artists from Italy, mercenaries, itinerant clergy, scholars and musicians, Muslim slaves from North Africa, the Eastern Mediterranean and Spain, as well as Jews and gypsies. Jews lived interspersed with the rest of the population and most appear to have spoken the local language. During the Crusades, Jews became scapegoats and victims of violence and persecution. Enforced residential segregation – ghettos – became commonplace. In 1306, the French king, Philip the Fair, ordered the expulsion of the Jews, who by that time numbered about 100 000, allowing him to seize Jewish possessions. But in 1315 economic considerations led King Louis X to reopen the doors of the French kingdom to Jews. It was only with the French Revolution of 1789 that Jews gained legal equality with the

→

mortality of these migrant workers through shipwreck, warfare and tropical illnesses was very high, but service in the colonies was often the only chance to escape from poverty. Such overseas migrations helped to bring about major changes in the economic structures and the cultures of both the European sending countries and the colonies.

An important antecedent of modern labour migration is the system of chattel slavery, which formed the basis of commodity production in the plantations and mines of the New World from the late seventeenth century to the mid-nineteenth century. The production of sugar, tobacco, coffee, cotton and gold by slave labour was crucial to the economic and political power of Britain and France – the dominant states of the eighteenth century – and played a major role for Spain, Portugal and the Netherlands as well. By 1770 there were nearly 2.5 million slaves in the Americas,

→

Christian population as citizens. However, some people continued to regard Jews as foreigners to the French nation. Even today, the propaganda of the Front National (FN) has marked anti-Semitic overtones.

The gypsies, also called the Rom or the Tzigane, are the descendants of a people who emigrated from the area of present-day India. Travelling in groups of 50 to 100, they spread throughout the kingdom, hawking their wares. There were soon manifestations of hostility towards them. French cities such as Angers banned them in 1498, followed soon after by King François I's edict prohibiting them from entering his kingdom. The gypsies returned and became part of French society, but they were never fully accepted by some people. Like the Jews, they were singled out for extermination by the Nazis during the Second World War. The roots of twentieth-century genocide were deeply etched in the history of immigration to European countries. Jews and gypsies have been perhaps the most enduring targets of European racism.

The fifteenth century was a turning point at which early modern states emerged. This is the dawn of the Age of Discovery in which Europeans circumnavigated the globe, beginning a long process which eventually brought the world under European domination. By the eighteenth century, the 'divine right of kings' was being questioned. The ideas that gave rise to the 1789 French Revolution included the principle of popular sovereignty, the concept of the nation-state and the idea that every human being belongs to a state. These ideas are particularly significant for our theme: international migration would be meaningless in a world not organized into nation-states. One of the key attributes of sovereignty is the idea, now universally accepted, that states have the authority to regulate movement into and out of the territory of the state. Illegal immigration has become such a politically volatile issue today partly because it is seen as violating one of the main prerogatives of sovereign states.

Source: Lequin (1988).

producing a third of the total value of European commerce (Blackburn, 1988: 5). The slave system was organized in the notorious 'triangular trade': ships laden with manufactured goods, such as guns or household implements, sailed from ports such as Bristol and Liverpool, Bordeaux and Le Havre, to the coasts of West Africa. There Africans were either forcibly abducted or were purchased from local chiefs or traders in return for the goods. Then the ships sailed to the Caribbean or the coasts of North or South America, where the slaves were sold for cash. This was used to purchase the products of the plantations, which were then brought back for sale in Europe.

An estimated 15 million slaves were taken to the Americas before 1850 (Appleyard, 1991: 11). For the women, hard labour in the mines, plantations and households was frequently accompanied by sexual

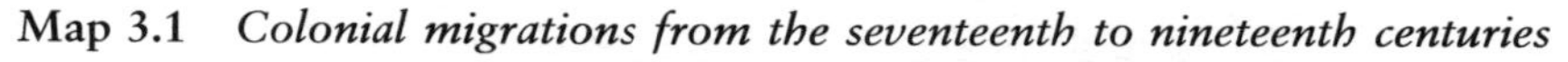

Map 3.1 *Colonial migrations from the seventeenth to nineteenth centuries*

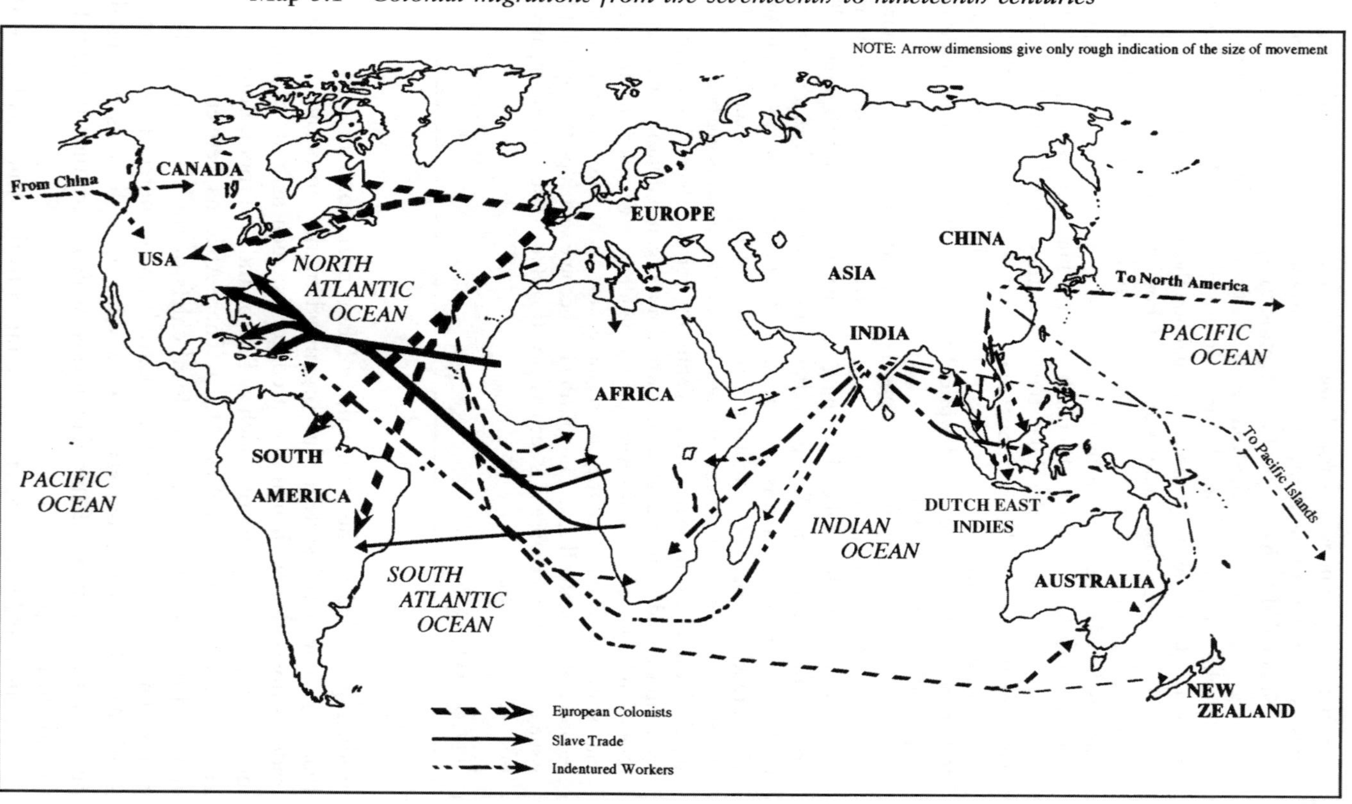

exploitation. The children of slaves remained the chattels of the owners. Slavery was not abolished until 1834 in British colonies, 1863 in Dutch colonies and 1865 in the southern states of the USA (Cohen, 1991: 9). Despite slave rebellions and the abolition of the Atlantic traffic by the great powers between 1807 and 1815, slavery continued to grow in economic significance. The number of slaves in the Americas doubled from 3 million in 1800 to 6 million in 1860, with corresponding growth in the area of plantation agriculture in the south-western USA, Cuba and Brazil (Blackburn, 1988: 544).

Slavery had existed in many pre-capitalist societies, but the colonial system was new in character. Its motive force was the emergence of global empires, which began to construct a world market, dominated by merchant capital. Slaves were transported great distances by specialized traders, and bought and sold as commodities. Slaves were economic property and were subjected to harsh forms of control to maximize their output. The great majority were exploited in plantations which produced for export, as part of an internationally integrated agricultural and manufacturing system (Fox-Genovese and Genovese, 1983; Blackburn, 1988).

In the latter half of the nineteenth century, slaves were replaced by indentured workers as the main source of plantation labour. Indenture (or the 'coolie system') involved recruitment of large groups of workers, sometimes by force, and their transportation to another area for work. British colonial authorities recruited workers from the Indian sub-continent for the sugar plantations of Trinidad, British Guiana and other Caribbean countries. Others were employed in plantations, mines and railway construction in Malaya, East Africa and Fiji. The British also recruited Chinese 'coolies' for Malaya and other colonies. Dutch colonial authorities used Chinese labour on construction projects in the Dutch East Indies. Up to 1 million indentured workers were recruited in Japan, mainly for work in Hawaii, the USA, Brazil and Peru (Shimpo, 1995).

According to Potts (1990: 63–103) indentured workers were used in 40 countries by all the major colonial powers. She estimates that the system involved from 12 to 37 million workers between 1834 and 1941, when indentureship was finally abolished in the Dutch colonies. Indentured workers were bound by strict labour contracts for a period of several years. Wages and conditions were generally very poor, workers were subject to rigid discipline and breaches of contract were severely punished. Indentured workers were often cheaper for their employers than slaves (Cohen, 1991: 9–11). On the other hand, work overseas offered an opportunity to escape poverty and repressive situations, such as the Indian caste system. Many workers remained as free settlers in East Africa, the Caribbean, Fiji and elsewhere, where they could obtain land or set up businesses (Cohen, 1995: 46).

Indenture epitomized the principle of divide and rule, and a number of post-colonial conflicts (for example, hostility against Indians in Africa and Fiji, and against Chinese in South-East Asia) have their roots in such divisions. The Caribbean experience shows the effect of changing colonial labour practices on dominated peoples: the original inhabitants, the Caribs and Arawaks, were wiped out completely by European diseases and violence. With the development of the sugar industry in the eighteenth century, Africans were brought in as slaves. After emancipation in the nineteenth century, these generally became small-scale subsistence farmers, and were replaced with indentured workers from India. Upon completion of their indentures, many Indians settled in the Caribbean, bringing in dependants. Some remained labourers on large estates, while others became established as a trading class, mediating between the white ruling class and the black majority.

Industrialization and migration to North America and Australia before 1914

The wealth accumulated in Western Europe through colonial exploitation provided much of the capital which was to unleash the industrial revolutions of the eighteenth and nineteenth centuries. In Britain, profits from the colonies were invested in new forms of manufacture, as well as encouraging commercial farming and speeding up the enclosure of arable land for pasture. The displaced tenant farmers swelled the impoverished urban masses available as labour for the new factories. This emerging class of wage labourers was soon joined by destitute artisans, such as hand-loom weavers, who had lost their livelihood through competition from the new manufactures. Herein lay the basis of the new class which was crucial for the British industrial economy: the 'free proletariat' which was free of traditional bonds, but also of ownership of the means of production.

However, from the outset, unfree labour played an important part. Throughout Europe, draconian poor laws were introduced to control the displaced farmers and artisans, the 'hordes of beggars' who threatened public order. Workhouses and poorhouses were often the first form of manufacture, where the disciplinary instruments of the future factory system were developed and tested. In Britain, 'parish apprentices', orphan children under the care of local authorities, were hired out to factories as cheap unskilled labour. This was a form of forced labour, with severe punishments for insubordination or refusal to work.

The peak of the industrial revolution was the main period of British migration to America: between 1800 and 1860, 66 per cent of migrants to the USA were from Britain, and a further 22 per cent were from Germany. From 1800 to 1930, 40 million Europeans migrated permanently overseas, mainly to North and South America and Australia (Decloîtres, 1967: 22). From 1850 to 1914 most migrants came from Ireland, Italy, Spain and Eastern Europe, areas in which industrialization came later. America offered the dream of becoming an independent farmer or trader in new lands of opportunity. Often this dream was disappointed: the migrants became wage-labourers building roads and railways across the vast expanses of the New World, 'cowboys', gauchos or stockmen on large ranches, or factory workers in the emerging industries of the north-eastern USA. However, many settlers did eventually realize their dream, becoming farmers, white-collar workers or business people, while others were at least able to see their children achieve education and upward social mobility.

The USA is generally seen as the most important of all immigration countries. An estimated 54 million people entered between 1820 and 1987 (Borjas, 1990: 3). The peak period was 1861 to 1920, during which 30 million people came. Until the 1880s, migration was unregulated: anyone who could afford the ocean passage could come to seek a new life in America. However, American employers did organize campaigns to attract potential workers, and a multitude of agencies and shipping companies helped organize movements. Many of the migrants were young single men, hoping to save enough to return home and start a family. But there were also single women, couples and families. Racist campaigns led to exclusionary laws to keep out Chinese and other Asians from the 1880s. For Europeans and Latin Americans, entry remained free until 1920 (Borjas, 1990: 27). The census of that year showed that there were 13.9 million foreign-born people in the USA, making up 13.2 per cent of the total population (Briggs, 1984: 77).

Slavery had been a major source of capital accumulation in the early USA, but the industrial take-off after the Civil War (1861–5) was fuelled by mass immigration from Europe. At the same time the racist 'Jim Crow' system was used to keep the now nominally free African-Americans in the plantations of the southern states, since cheap cotton and other agricultural products were central to industrialization. The largest immigrant groups from 1860 to 1920 were Irish, Italians and Jews from Eastern Europe, but there were people from just about every other European country, as well as from Mexico. Patterns of settlement were closely linked to the emerging industrial economy. Labour recruitment by canal and railway companies led to settlements of Irish and Italians along the construction routes. Some groups of Irish, Italians and Jews settled in the east coast ports of arrival, where work was available in construction, transport and factories. The same was true of the Chinese on the west coast. Some Central and Eastern European peoples became concentrated in the midwest, where the development of heavy industry at the turn of the century provided work opportunities (Portes and Rumbaut, 1996: 29–32). The American working class thus developed through processes of chain migration which led to patterns of ethnic segmentation.

Canada received many loyalists of British origin after the American Revolution. From the late eighteenth century there was immigration from Britain, France, Germany and other Northern European countries. Many African-Americans came across the long frontier from the USA to escape slavery: by 1860, there were 40 000 black people in Canada. In the nineteenth century, immigration was stimulated by the gold rushes, while rural immigrants were encouraged to settle the vast prairie areas. Between 1871 and 1931, Canada's population increased from 3.6 million to 10.3 million. Immigration from China, Japan and India also began in the late nineteenth century. Chinese came to the west coast, particularly to British Columbia, where they helped build the Canadian Pacific Railway. From 1886 a series of measures was introduced to stop Asian immigration (Kubat, 1987: 229–35). Canada received a large influx from Southern and Eastern Europe over the 1895 to 1914 period. But in 1931, four preferred classes of immigrants were designated: British subjects with adequate financial means from the UK, Ireland and four other domains of the crown; US citizens; dependants of permanent residents of Canada; and agriculturists. Canada discouraged migration from Southern and Eastern Europe, while Asian immigration was prohibited from 1923 to 1947.

For Australia, immigration has been a crucial factor in economic development and nation building ever since British colonization started in 1788. The Australian colonies were integrated into the British Empire as suppliers of raw materials such as wool, wheat and gold. The imperial state took an active role in providing workers for expansion through convict transportation (another form of unfree labour) and the encouragement of free settlement. Initially there were large male surpluses, especially in the frontier areas, which were often societies of 'men without women'. But many female convicts were transported, and there were special schemes to bring out single women as domestic servants and as wives for settlers.

When the surplus population of Britain became inadequate for labour needs from the mid-nineteenth century, Britain supported Australian employers in their demand for cheap labour from elsewhere in the Empire: China, India and the South Pacific Islands. The economic interests of Britain came into conflict with the demands of the nascent Australian labour movement. The call for decent wages came to be formulated in racist (and sexist) terms, as a demand for wages 'fit for white men'. Hostility towards Chinese and other Asian workers became violent. The exclusionary boundaries of the emerging Australian nation were drawn on racial lines, and one of the first Acts of the new Federal Parliament in 1901 was the introduction of the White Australia Policy (see de Lepervanche, 1975).

Labour migration within Europe

In Europe, overseas migration and intra-European migration took place side-by-side. Of the 15 million Italians who emigrated between 1876 and 1920, nearly half (6.8 million) went to other European countries (mainly France, Switzerland and Germany: see Cinanni, 1968: 29). As Western Europeans went overseas in the (often vain) attempt to escape proletarianization, workers from peripheral areas, such as Poland, Ireland and Italy, were drawn in as replacement labour for large-scale agriculture and industry.

As the earliest industrial country, Britain was the first to experience large-scale labour immigration. The new factory towns quickly absorbed labour surpluses from the countryside. Atrocious working and living conditions led to poor health, high infant mortality and short life expectancy. Low wage levels forced both women and children to work, with disastrous results for the family. Natural increase was inadequate to meet labour needs, so Britain's closest colony, Ireland, became a labour

Map 3.2 *Labour migrations connected with industrialization, 1850–1920*

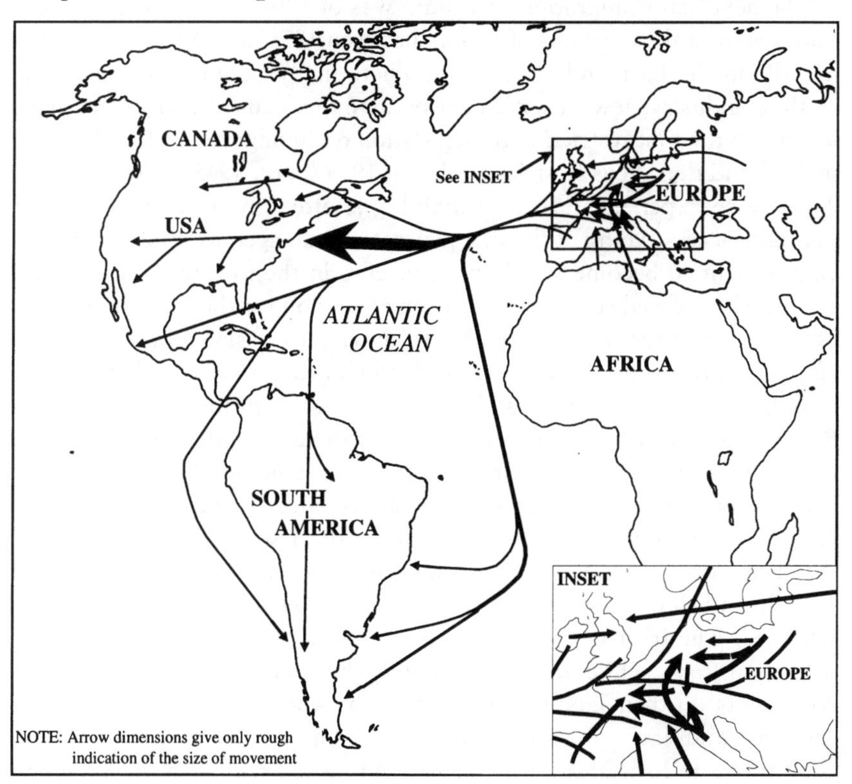

source. The devastation of Irish peasant agriculture through absentee landlords and enclosures, combined with the ruin of domestic industry through British competition, had led to widespread poverty. The famines of 1822 and 1846–7 triggered massive migrations to Britain, the USA and Australia.

By 1851 there were over 700 000 Irish in Britain, making up 3 per cent of the population of England and Wales and 7 per cent of the population of Scotland (Jackson, 1963). They were concentrated in the industrial cities, especially in the textile factories and the building trades. Irish 'navvies' (a slang term derived from 'navigators') dug Britain's canals and built its railways. Engels (1962) described the appalling situation of Irish workers, arguing that Irish immigration was a threat to the wages and living conditions of English workers (see also Castles and Kosack, 1973: 16–17). Hostility and discrimination against the Irish was marked right into the twentieth century. This was true of Australia too, where Irish immigration accompanied British settlement from the outset. In both countries it was the active role played by Irish workers in the labour movement which was finally to overcome this split in the working class just in time for its replacement by new divisions after 1945, when black workers came to Britain and Southern Europeans to Australia.

The next major migration to Britain was of 120 000 Jews, who came as refugees from the pogroms of Russia between 1875 and 1914. Most settled initially in the East End of London, where many became workers in the clothing industry. Jewish settlement became the focus of racist campaigns, leading to the first restrictionary legislation on immigration: the Aliens Act of 1905 and the Aliens Restriction Act of 1914 (Foot, 1965; Garrard, 1971). The Jewish experience of social mobility is often given as an example of migrant success. Many of the first generation managed to shift out of wage employment to become small entrepreneurs in the rag trade or the retail sector. They placed strong emphasis on education for their children. Many of the second generation were able to move into business or white-collar employment, paving the way for professional careers by the third generation. Interestingly one of Britain's newer immigrant groups – Bengalis from Bangladesh – now live in the same areas of the East End, often working in the same sweatshops, and worshipping in the same buildings (synagogues converted to mosques). However, they are isolated by racism and violence, and show little sign at present of repeating the Jewish trajectory. It seems that British racism today is more rigid than a century ago.

Irish and Jewish migrant workers cannot be categorized as 'unfree workers'. The Irish were British subjects, with the same formal rights as other workers, while the Jews rapidly became British subjects. The constraints on their labour market freedom were not legal but economic (poverty and lack of resources made them accept inferior jobs and conditions) and social (discrimination and racism restricted their freedom of movement). It is in Germany and France that one finds the first large-scale use of the status of 'foreigner' to restrict workers' rights.

The heavy industries of the Ruhr, which emerged in the mid-nineteenth century, attracted agricultural workers away from the large estates of Eastern Prussia. Conditions in the mines were hard, but still preferable to semi-feudal oppression under the Junkers (large landowners). The workers who moved west were of Polish ethnic background, but had Prussian (and later German) citizenship, since Poland was at that time divided up between Prussia, the Austro-Hungarian Empire and Russia. By 1913, it was estimated that 164 000 of the 410 000 Ruhr miners were of Polish background (Stirn, 1964: 27). The Junkers compensated for the resulting labour shortages by recruiting 'foreign Poles' and Ukrainians as agricultural workers. Often workers were recruited in pairs – a man as cutter and a woman as binder – leading to so-called 'harvest marriages'. However, there was fear that settlement of Poles might weaken German control of the eastern provinces. In 1885, the Prussian government deported some 40 000 Poles and closed the frontier. The landowners protested at the loss of up to two-thirds of their labour force (Dohse, 1981: 29–32), arguing that it threatened their economic survival.

By 1890, a compromise between political and economic interests emerged in the shape of a system of rigid control. 'Foreign Poles' were recruited as temporary seasonal workers only, not allowed to bring dependants and forced to leave German territory for several months each year. At first they were restricted to agricultural work, but later were permitted to take industrial jobs in Silesia and Thuringia (but not in western areas such as the Ruhr). Their work contracts provided pay and conditions inferior to those of German workers. Special police sections were established to deal with 'violation of contracts' (that is, workers leaving for better-paid jobs) through forcible return of workers to their employers, imprisonment or deportation. Thus police measures against foreigners were deliberately used as a method to keep wages low and to create a split labour market (Dohse, 1981: 33–83).

Foreign labour played a major role in German industrialization with Italian, Belgian and Dutch workers alongside the Poles. In 1907, there were 950 000 foreign workers in the German Reich, of whom nearly 300 000 were in agriculture, 500 000 in industry and 86 000 in trade and transport (Dohse, 1981: 50). The authorities did their best to prevent family reunion and permanent settlement. Both in fact took place, but the exact extent is unclear. The system developed to control and exploit foreign labour was a precursor both of forced labour in the Nazi war economy and of the 'guestworker system' in the German Federal Republic from about 1955.

The number of foreigners in France increased rapidly from 381 000 in 1851 (1.1 per cent of total population) to 1 million (2.7 per cent) in 1881, and then more slowly to 1.2 million (3 per cent) in 1911 (Weil, 1991b:

Appendix, Table 4). The majority came from neighbouring countries: Italy, Belgium, Germany and Switzerland, and later from Spain and Portugal. Movements were relatively spontaneous, though some recruitment was carried out by farmers' associations and mines (Cross, 1983: Chapter 2). The foreign workers were mainly men who carried out unskilled manual work in agriculture, mines and steelworks (the heavy, unpleasant jobs that French workers were unwilling to take).

The peculiarity of the French case lies in the reasons for the shortage of labour during industrialization. Birth rates fell sharply after 1860. Peasants, shopkeepers and artisans followed 'Malthusian' birth control practices, which led to small families earlier than anywhere else (Cross, 1983: 5–7). According to Noiriel (1988: 297–312) this *grève des ventres* (belly strike) was motivated by resistance to proletarianization. Keeping the family small meant that property could be passed on intact from generation to generation, and that there would be sufficient resources to permit a decent education for the children. Unlike Britain and Germany, France therefore saw relatively little overseas emigration during industrialization. The only important exception was the movement of settlers to Algeria, which France invaded in 1830. Rural-urban migration was also fairly limited. The 'peasant worker' developed: the small farmer who supplemented subsistence agriculture through sporadic work in local industries. Where people did leave the countryside it was often to move straight into the new government jobs that proliferated in the late nineteenth century: straight from the primary to the tertiary sector.

In these circumstances, the shift from small to large-scale enterprises, made necessary by international competition from about the 1880s, could only be made through the employment of foreign workers. Thus labour immigration played a vital role in the emergence of modern industry and the constitution of the working class in France. Immigration was also seen as important for military reasons. The nationality law of 1889 was designed to turn immigrants and their sons into conscripts for the impending conflict with Germany (Schnapper, 1994: 66). From the mid-nineteenth century to the present, the labour market has been regularly fed by foreign immigration, making up, on average, 10–15 per cent of the working class. Noiriel estimates that without immigration the French population today would be only 35 million instead of over 50 million (Noiriel, 1988: 308–18).

The inter-war period

At the onset of the First World War, many migrants returned home to participate in military service or munitions production. However, labour shortages soon developed in the combatant countries. The German authorities prevented 'foreign Polish' workers from leaving the country, and recruited labour by force in occupied areas of Russia and Belgium (Dohse, 1981: 77–81). The French government set up recruitment systems for workers from the North African and Indo-Chinese colonies, and from China (about 225 000 in all). They were housed in barracks, paid minimal wages and supervised by former colonial overseers. Workers were also recruited in Portugal, Spain, Italy and Greece for French factories and agriculture (Cross, 1983: 34–42). Britain, too, recruited colonial workers during the conflict, although in smaller numbers. All the warring countries also made use of the forced labour of prisoners of war.

The period from 1918 to 1945 was one of reduced international labour migrations. This was partly because of economic stagnation and crisis, and partly because of increased hostility towards immigrants in many countries. Migration to Australia, for example, fell to low levels as early as 1891, and did not grow substantially until after 1945. Southern Europeans who came to Australia in the 1920s were treated with suspicion. Immigrant ships were refused permission to land and there were 'anti-Dago' riots in the 1930s. Queensland passed special laws, prohibiting foreigners from owning land, and restricting them to certain industries (de Lepervanche, 1975).

In the USA, 'nativist' groups claimed that Southern and Eastern Europeans were 'unassimilable' and that they presented threats to public order and American values. Congress enacted a series of laws in the 1920s designed to limit drastically entries from any area except north-west Europe (Borjas, 1990: 28–9). This national-origins quota system stopped large-scale immigration to the USA until the 1960s. But the new mass production industries of the Fordist era had a substitute labour force at hand: black workers from the South. The period from about 1914 to the 1950s was that of the 'Great Migration' in which African-Americans fled segregation and exploitation in the southern states for better wages and – they hoped – equal rights in the north-east, midwest and west. Often they simply encountered new forms of segregation in the ghettoes of New York or Chicago, and new forms of discrimination, such as exclusion from the unions of the American Federation of Labor.

Meanwhile, Americanization campaigns were launched to ensure that immigrants learned English and became loyal US citizens. During the Great Depression, Mexican immigrants were repatriated by local governments and civic organizations, with some cooperation from the Mexican and US governments (Kiser and Kiser, 1979: 33–66). Many of the nearly 500 000 Mexicans who returned home were constrained to leave, while others left because there was no work. In these circumstances, little was done to help Jews fleeing the rise of Hitler. There was no concept of refugee in US law, and it was difficult to build support for admission of Jewish refugees when millions of US citizens were unemployed. Anti-Semitism was also a factor, and there was never much of a prospect for large numbers of European Jews to find safe haven before the Second World War.

France was the only Western European country to experience substantial immigration in the inter-war years. The 'demographic deficit' had been exacerbated by war losses: 1.4 million men had been killed and 1.5 million permanently handicapped (Prost, 1966: 538). There was no return to the pre-war free movement policy; instead the government and employers refined the foreign labour systems established during the war. Recruitment agreements were concluded with Poland, Italy and Czechoslovakia. Much of the recruitment was organized by the Société générale d'immigration (SGI), a private body set up by farm and mining interests. Foreign workers were controlled through a system of identity cards and work contracts, and were channelled into jobs in farming, construction and heavy industry. However, most foreign workers probably arrived spontaneously outside the recruiting system. The non-communist trade union movement cooperated with immigration, in return for measures designed to protect French workers from displacement and wage cutting (Cross, 1983: 51–63; Weil, 1991b: 24–7).

Just under 2 million foreign workers entered France from 1920 to 1930, about 567 000 of them recruited by the SGI (Cross, 1983: 60). Some 75 per cent of French population growth between 1921 and 1931 is estimated to have been the result of immigration (Decloîtres, 1967: 23). In view of the large female surplus in France, mainly men were recruited, and a fair degree of intermarriage took place. By 1931, there were 2.7 million foreigners in France (6.6 per cent of the total population). The largest group were Italians (808 000), followed by Poles (508 000), Spaniards (352 000) and Belgians (254 000) (Weil, 1991b: Appendix, Table 4). North African migration to France was also developing. Large colonies of Italians and Poles sprang up in the mining and heavy industrial towns of the north and east of France: in some towns, foreigners made up a third or more of the total population. There were Spanish and Italian agricultural settlements in the south-west.

In the depression of the 1930s, hostility towards foreigners increased, leading to a policy of discrimination in favour of French workers. In 1932 maximum quotas for foreign workers in firms were fixed. They were followed by laws permitting dismissal of foreign workers in sectors where there was unemployment. Many migrants were sacked and deported, and the foreign population dropped by half a million by 1936 (Weil, 1991b: 27–30). Cross concludes that in the 1920s foreign workers 'provided a cheap and flexible workforce necessary for capital accumulation and economic growth; at the same time, aliens allowed the French worker a degree of economic mobility'. In the 1930s, on the other hand, immigration 'attenuated and provided a scapegoat for the economic crisis' (Cross, 1983: 218).

In Germany, the crisis-ridden Weimar Republic had little need of foreign workers: by 1932 their number was down to about 100 000, compared with nearly a million in 1907 (Dohse, 1981: 112). Nonetheless, a new system of regulation of foreign labour developed. Its principles were: strict state control of labour recruitment, employment preference for nationals, sanctions against employers of illegal migrants and unrestricted police power to deport unwanted foreigners (Dohse, 1981: 114-17). This system was partly attributable to the influence of the strong labour movement, which wanted measures to protect German workers, but it confirmed the weak legal position of migrant workers. Box 3.2 describes the use of forced foreign labour during the Second World War.

Box 3.2 Forced foreign labour in the Nazi war economy

The Nazi regime recruited enormous numbers of foreign workers – mainly by force – to replace the 11 million German workers conscripted for military service. The occupation of Poland, Germany's traditional labour reserve, was partly motivated by the need for labour. Labour recruitment offices were set up within weeks of the invasion, and the police and army rounded up thousands of young men and women (Dohse, 1981: 121). Forcible recruitment took place in all the countries invaded by Germany, while some voluntary labour was obtained from Italy, Croatia, Spain and other 'friendly or neutral countries'. By the end of the war, there were 7.5 million foreign workers in the Reich, of whom 1.8 million were prisoners of war. It is estimated that a quarter of industrial production was carried out by foreign workers in 1944 (Pfahlmann, 1968: 232). The Nazi war machine would have collapsed far earlier without foreign labour.

The basic principle for treating foreign workers declared by Sauckel, the Plenipotentiary for Labour, was that: 'All the men must be fed, sheltered and treated in such a way as to exploit them to the highest possible extent at the lowest conceivable degree of expenditure' (Homze, 1967: 113). This meant housing workers in barracks under military control, the lowest possible wages (or none at all), appalling social and health conditions, and complete deprivation of civil rights. Poles and Russians were compelled, like the Jews, to wear special badges showing their origin. Many foreign workers died through harsh treatment and cruel punishments. These were systematic; in a speech to employers, Sauckel emphasized the need for strict discipline: 'I don't care about them [the foreign workers] one bit. If they commit the most minor offence at work, report them to the police at once, hang them, shoot them. I don't care. If they are dangerous, they must be liquidated' (Dohse, 1981: 127).

The Nazis took exploitation of rightless migrants to an extreme which can only be compared with slavery, yet its legal core – the sharp division between the status of national and foreigner – was to be found in both earlier and later foreign labour systems.

Conclusion

Contemporary migratory movements and policies are often profoundly influenced by historical precedents. This chapter has described the key role of labour migration in colonialism and industrialization. Labour migration has always been a major factor in the construction of a capitalist world market. In the USA, Canada, Australia, the UK, Germany and France (as well as in other countries not discussed here) migrant workers have played a role which varies in character according to economic, social and political conditions. But in every case the contribution of migration to industrialization and population building was important and sometimes even decisive.

To what extent does the theoretical model of the migratory process suggested in Chapter 2 apply to the historical examples given? Involuntary movements of slaves and indentured workers do not easily fit the model, for the intentions of the participants played little part. Nonetheless some aspects apply: labour recruitment as the initial impetus, predominance of young males in the early stages, family formation, long-term settlement and emergence of ethnic minorities. Worker migrations to England, Germany and France in the nineteenth and twentieth centuries fit the model well. Their original intention was temporary, but they led to family reunion and settlement. As for migrations to America and Australasia in the nineteenth and early twentieth centuries, it is generally believed that most migrants went with the intention of permanent settlement. But many young men and women went in order to work for a few years and then return home. Some did return, but in the long run the majority remained in the New World, often forming new ethnic communities. Here, too, the model seems to fit.

Clearly the study of migrant labour is not the only way of looking at the history of migration. Movements caused by political or religious persecution have always been important, playing a major part in the development of countries as diverse as the USA and Germany. It is often impossible to draw strict lines between the various types of migration. Migrant labour systems have always led to some degree of settlement, just as settler and refugee movement have always been bound up with the political economy of capitalist development.

The period from about 1850 to 1914 was an era of mass migration in Europe and North America. Industrialization was a cause of both emigration and immigration (sometimes in the same country, as the British case shows). After 1914, war, xenophobia and economic stagnation caused a considerable decline in migration, and the large-scale movements of the preceding period seemed to have been the results of a unique and unrepeatable constellation. When rapid and sustained economic growth got under way after the Second World War, the new age of migration was to take the world by surprise.

Guide to further reading

Cohen (1987) provides a valuable overview of migrant labour in the international division of labour, while Potts (1990) presents a history of migration which leads from slavery and indentured labour up to modern guestworker systems. Blackburn (1988) and Fox-Genovese and Genovese (1983) analyse slavery and its role in capitalist development. Archdeacon (1983) examines immigration in US history, showing how successive waves of entrants have 'become American'. For German readers, Dohse (1981) gives an interesting historical analysis of the role of the state in controlling migrant labour in Germany. Cross (1983) gives a detailed account of the role of migrant workers in French industrialization, de Lepervanche (1975) shows how ethnic divisions played a central role in the formation of the Australian working class, while Homze (1967) describes the extreme exploitation of migrant labour practised by the Nazi war machine. Moch (1992) is good on earlier European migration experiences, while many contributions in Cohen (1995) are on the history of migration.

Chapter 4

Migration to Highly-developed Countries since 1945

Since the end of the Second World War, international migrations have grown in volume and changed in character. There have been two main phases. In the first, from 1945 to the early 1970s, the chief economic strategy of large-scale capital was concentration of investment and expansion of production in the existing highly-developed countries. As a result, large numbers of migrant workers were drawn from less-developed countries into the fast-expanding industrial areas of Western Europe, North America and Australia. The end of this phase was marked by the 'oil crisis' of 1973–4. The ensuing recession gave impetus to a restructuring of the world economy, involving capital investment in new industrial areas, altered patterns of world trade, and introduction of new technologies. The result was a second phase of international migration, starting in the mid-1970s and gaining momentum in the 1980s and 1990s. This phase involved complex new patterns of migration, affecting both old and new receiving countries. This chapter will discuss post-1945 migratory movements to highly-developed countries, including Europe, North America and Australia. Labour migration to Japan, which did not become significant until the mid-1980s, will be discussed in Chapter 7, in the context of Asian regional migration.

Migration in the long boom

A detailed review of the literature is not possible here. For Europe the account is based mainly on our own works: Castles and Kosack (1973); M. J. Miller (1981); Castles *et al.* (1984); Castles (1986); Castles (1989). For the USA, we recommend Briggs (1984); Portes and Rumbaut (1996). For Australia, see Collins (1991). For useful overviews, see Kritz *et al.* (1983); Cohen (1987); *International Migration Review* (1989). Precise references will only be given where absolutely necessary.

Between 1945 and the early 1970s, three main types of migration led to the formation of new, ethnically distinct populations in advanced industrial countries:

- migration of workers from the European periphery to Western Europe, often through 'guestworker systems';

- migration of 'colonial workers' to the former colonial powers;
- permanent migration to North America and Australia, at first from Europe and later from Asia and Latin America

The precise timing of these movements varied: they started later in Germany and ended earlier in the UK, while migration to the USA grew rapidly after the immigration reforms of 1965 and, unlike migrations to Western Europe and Australia, did not decline at all in the mid-1970s. These three types, which all led to family reunion and other kinds of chain migration, will be examined here. There were also other types of migration which will not be dealt with here, since they did not contribute decisively to the formation of ethnic minorities:

- mass movements of European refugees at the end of the Second World War (post-1945 refugee movements were most significant in the case of Germany, see Chapter 9);
- return migrations of former colonists to their countries of origin as colonies gained their independence.

Foreign workers and 'guestworker' systems

All the highly industrialized countries of Western Europe used temporary labour recruitment at some stage between 1945 and 1973, although this sometimes played a smaller role than spontaneous entries of foreign workers. The rapidly expanding economies were able to utilize the labour reserves of the less-developed European periphery: the Mediterranean countries, Ireland and Finland. In some cases the economic backwardness was the result of former colonization (Ireland, Finland, North Africa). In the case of Southern Europe, underdevelopment resulted from antiquated political and social structures, reinforced by wartime devastation.

Immediately after the Second World War, the British government brought in 90 000 mainly male workers from refugee camps and from Italy through the European Voluntary Worker (EVW) scheme. EVWs were tied to designated jobs, had no right to family reunion, and could be deported for indiscipline. The scheme was fairly small and only operated until 1951, because it was easier to make use of colonial workers (see below). A further 100 000 Europeans entered Britain on work permits between 1946 and 1951, and some European migration continued subsequently, though it was not a major flow (Kay and Miles, 1992).

Belgium also started recruiting foreign workers immediately after the war. They were mainly Italian men, and were employed in the coal mines and the iron and steel industry. The system operated until 1963, after which foreign work-seekers were allowed to come of their own accord. Many brought in dependants and settled permanently, changing the ethnic composition of Belgium's industrial areas.

Map 4.1 *Global migrations, 1945–73*

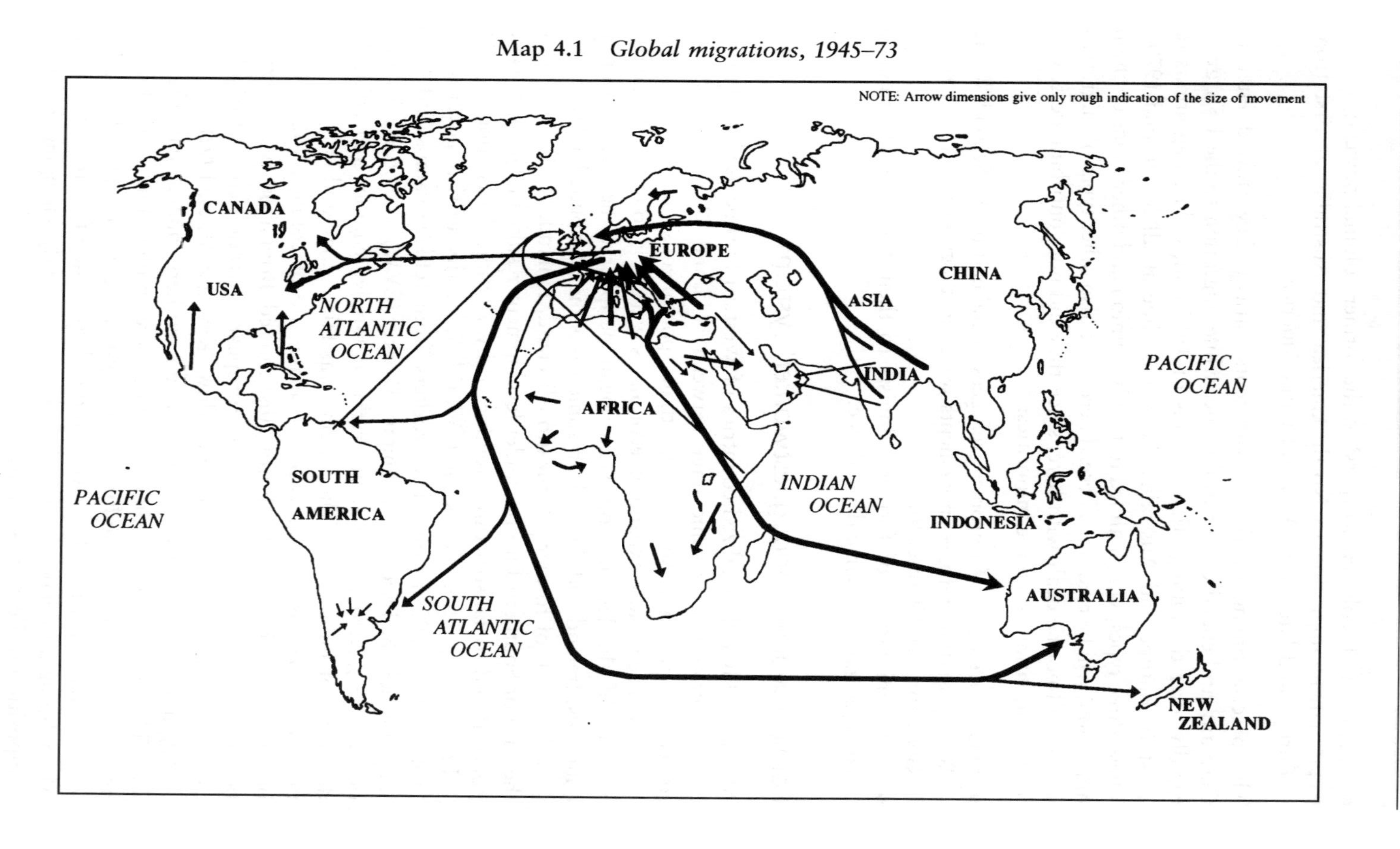

France established an Office National d'Immigration (ONI) in 1945 to organize recruitment of workers from Southern Europe. Migration was seen as a solution to post-war labour shortages and to what the French termed their 'demographic insufficiency'. In view of continuing low birth rates and war losses, massive family settlement was envisaged. ONI also coordinated the employment of up to 150 000 seasonal agricultural workers per year, mainly from Spain. By 1970, 2 million foreign workers and 690 000 dependants resided in France. Many found it easier to come as 'tourists', get a job and then regularize their situation. This applied particularly to Portuguese and Spanish workers, escaping their respective dictatorships, who generally lacked passports. By 1968, ONI statistics revealed that 82 per cent of the aliens admitted by the ONI came as 'clandestines'. In any case, ONI had no jurisdiction over French citizens from overseas departments and territories, or from certain former colonies (see below).

Switzerland pursued a policy of large-scale labour import from 1945 to 1974. Foreign workers were recruited abroad by employers, while admission and residence were controlled by the government. Job changing, permanent settlement and family reunion were forbidden to seasonal workers until the mid-1960s. Considerable use was also made of cross-frontier commuters. Swiss statistics include both these groups as part of the labour force but not of the population: 'guestworkers' par excellence. Swiss industry became highly dependent on foreign workers, who made up nearly a third of the labour force by the early 1970s. The need to attract and retain workers, coupled with diplomatic pressure from Italy, led to relaxations on family reunion and permanent stay, so that Switzerland, too, experienced settlement and the formation of migrant communities.

The examples could be continued: the Netherlands brought in 'guestworkers' in the 1960s and early 1970s, Luxembourg's industries were highly dependent on foreign labour, and Sweden employed workers from Finland and from Southern European countries. Another case worth mentioning is that of Italy, in which migration from the underdeveloped south was crucial to the economic take-off of the northern industrial triangle between Milan, Turin and Genoa in the 1960s: this was internal migration, but very similar in its economic and social character to foreign worker movements in other European countries. The key case for understanding the 'guestworker system' was the Federal Republic of Germany (FRG), which set up a highly organized state recruitment apparatus (see Box 4.1).

In the FRG we see in the most developed form all the principles – but also the contradictions – of temporary foreign labour recruitment systems. These include the belief in temporary sojourn, the restriction of labour market and civil rights, the recruitment of single workers (men at first, but with increasing numbers of women as time went on), the inability to

Box 4.1 The German 'guestworker' system

The German Government started recruiting foreign workers in the mid 1950s. The Federal Labour Office (*Bundesanstalt für Arbeit*, or BfA) set up recruitment offices in the Mediterranean countries. Employers requiring foreign labour paid a fee to the BfA, which selected workers, testing occupational skills, providing medical examinations and screening police records. The workers were brought in groups to Germany, where employers had to provide initial accommodation. Recruitment, working conditions and social security were regulated by bilateral agreements between the FRG and the sending countries: first Italy, then Spain, Greece, Turkey, Morocco, Portugal, Tunisia and Yugoslavia.

The number of foreign workers in the FRG rose from 95 000 in 1956 to 1.3 million in 1966 and 2.6 million in 1973. This massive migration was the result of rapid industrial expansion and the shift to new methods of mass production, which required large numbers of low-skilled workers. Foreign women workers played a major part, especially in the later years: their labour was in high demand in textiles and clothing, electrical goods and other manufacturing sectors.

German policies conceived migrant workers as temporary labour units, which could be recruited, utilized and sent away again as employers required. To enter and remain in the FRG, a migrant needed a residence permit and a labour permit. These were granted for restricted periods, and were often valid only for specific jobs and areas. Entry of dependants was discouraged. A worker could be deprived of his or her permit for a variety of reasons, leading to deportation.

However, it was impossible to prevent family reunion and settlement. Often officially recruited migrants were able to get employers to request their wives or husbands as workers. Competition with other labour-importing countries for labour led to relaxation of restrictions on entry of dependants in the 1960s. Families became established and children were born. Foreign labour was beginning to lose its mobility and social costs (for housing, education and health care) could no longer be avoided. When the Federal government stopped labour recruitment in November 1973, the motivation was not only the looming 'oil crisis', but also the belated realization that permanent immigration was taking place.

prevent family reunion completely, the gradual move towards longer stay, the inexorable pressures for settlement and community formation. The FRG took the system furthest, but its central element – the legal distinction between the status of citizen and of foreigner as a criterion for determining political and social rights – was to be found throughout Europe (see Hammar, 1985a).

Multinational and bilateral agreements were also used to facilitate labour migration. Free movement of workers within the EC, which came into force in 1968, was relevant mainly for Italian workers going to Germany, while the Nordic Labour Market affected Finns going to

Table 4.1 *Minority population in the main Western European countries of immigration, 1950–75 (thousands)*

Country	*1950*	*1960*	*1970*	*1975*	*Per cent of total population 1975*
Belgium	354	444	716	835	8.5
France	2128	2663	3339	4196	7.9
Germany (FRG)	548	686	2977	4090	6.6
Netherlands	77	101	236	370	2.6
Sweden	124	191	411	410	5.0
Switzerland	279	585	983	1012	16.0
UK	1573	2205	3968	4153	7.8

Notes: Figures for all countries except the UK are for foreign residents. They exclude naturalized persons and immigrants from the Dutch and French colonies. UK data are Census figures for 1951, 1961 and 1971 and estimates for 1975. The 1951 and 1961 data are for overseas-born persons, and exclude children born to immigrants in the UK. The 1971 and 1975 figures include children born in the UK, with both parents born abroad.

Source: Castles *et al.*, 1984: 87–8 (where detailed sources are given).

Sweden. The EC arrangements were the first step towards creating a 'European labour market', which was to become a reality in 1993. However, in the 1960s and early 1970s labour movement within the Community was actually declining, owing to gradual equalization of wages and living standards within the EC, while migration from outside the Community was increasing. Table 4.1 shows the development of minority populations arising from migration in selected Western European countries up to 1975.

Colonial workers

Migration from former colonies was important for Britain, France and the Netherlands. Britain had a net inflow of about 350 000 from Ireland, its traditional labour reserve, between 1946 and 1959. Irish workers provided manual labour for industry and construction, and many brought in their families and settled permanently. Irish residents in Britain enjoyed all civil rights, including the right to vote. Immigration of workers from the New Commonwealth (former British colonies in the Caribbean, the Indian sub-continent and Africa) started after 1945 and grew during the 1950s. Some workers came as a result of recruitment by London Transport, but most migrated spontaneously in response to labour demand. By 1951, there were 218 000 people of New Commonwealth origin (including Pakistan; which subsequently left the Commonwealth), a figure which increased to 541 000 in 1961. Entry of workers from the New Commonwealth almost stopped after 1962, partly owing to the introduction of severe restrictions through the Commonwealth Immigrants Act of 1962, and partly as the result of the early onset of economic stagnation in Britain.

However, most of the Commonwealth immigrants had come to stay, and family reunion continued, until it in turn was restricted by the 1971 Immigration Act. The population of New Commonwealth origin increased to 1.2 million in 1971 and 1.5 million in 1981. Most Afro-Caribbean and Asian immigrants and their children in Britain enjoyed formal citizenship (although this no longer applies to those admitted since the 1981 Nationality Act). Their minority status was not defined by being foreign, but by widespread institutional and informal discrimination. Most black and Asian workers found unskilled manual jobs in industry and the services, and a high degree of residential segregation emerged in the inner cities. Educational and social disadvantage became a further obstacle to mobility out of initial low-status positions. By the 1970s, the emergence of ethnic minorities was inescapable.

France experienced large-scale spontaneous immigration from its former colonies, as well as from Southern Europe. By 1970 there were over 600 000 Algerians, 140 000 Moroccans and 90 000 Tunisians. Many black workers were also coming in from the former West African colonies of Senegal, Mali and Mauritania. Some of these migrants came before independence when they were still French citizens. Others came later through preferential migration arrangements, or illegally. Migration from Algeria was regulated by bilateral agreements which accorded Algerian migrants a unique status. Moroccans and Tunisians, by contrast, were admitted through ONI. Many people also came from the overseas departments and territories such as Guadeloupe, Martinique and Réunion. They were French citizens, so there were no migration statistics, though estimates put their number at 250 000 to 300 000 in 1972. All these migrations were initially male-dominated, but with increasing proportions of women as the movement matured. Non-European immigrants in France were relegated to the bottom of the labour market, often working in highly exploitative conditions. Housing was frequently segregated, and very poor in quality; indeed, shanty towns (known as *bidonvilles*) appeared in France in the 1960s. Extreme-right groups began to subject non-European immigrants to a campaign of racial violence: 32 North Africans were murdered in 1973.

The Netherlands had two main inflows from former colonies. Between 1945 and the early 1960s up to 300 000 'repatriates' from the former Dutch East Indies (now Indonesia) entered the Netherlands. Although most had been born overseas and many were of mixed Dutch and Indonesian parentage, they were Dutch citizens. The official policy of assimilation appears to have worked well in this case, and there is little evidence of racism or discrimination against this group. The exception is the roughly 32 000 Moluccans, who wanted to return to their homeland if it could achieve independence from Indonesia. They remained segregated in camps, and rejected integration into Dutch society. In the late 1970s, their disaffection led to several violent incidents. After 1965, increasing numbers of black workers came to the Netherlands from the Caribbean territory of Surinam. A peak was reached in the two years leading up to independence in 1975, at which time Surinamese (except those already living in the Netherlands) lost their Dutch citizenship. By the late 1970s there were estimated to be 160 000 Surinamese in the Netherlands.

Permanent migration to North America and Australia

Large-scale migration to the USA developed later than in Western Europe, owing to the restrictive legislation enacted in the 1920s. Intakes averaged 250 000 persons annually in the 1951–60 period, and 330 000 annually during 1961–70: a far cry from the average of 880 000 immigrants per year from 1901 to 1910. The 1970 Census showed that the number of overseas-born people had declined to 9.6 million, only 4.7 per cent of the population (Briggs, 1984: 7). The 1965 amendments to the Immigration and Nationality Act were seen as part of the civil rights legislation of the period, designed to remove the discriminatory national-origins quota system. They were not expected or intended to lead to large-scale non-European immigration (Borjas, 1990: 29–33). In fact the amendments created a system of worldwide immigration, in which the most important criterion for admission was kinship with US citizens or residents. The result was a dramatic upsurge in migration from Asia and Latin America.

US employers, particularly in agriculture, also recruited temporary migrant workers, mainly men, in Mexico and the Caribbean. Organized labour was highly critical, arguing that domestic workers would be displaced and wages held down. Government policies varied: at times, systems of temporary labour recruitment, such as the Mexican *Bracero* Program of the 1940s, were introduced. In other periods recruitment was formally prohibited, but tacitly tolerated, leading to the presence of a large number of illegal workers. Significantly, the 1952 amendments to US immigration law included the so-called 'Texas Proviso', which was interpreted as barring punishment of employers who hired unauthorized foreign labour.

Canada followed policies of mass immigration after 1945. At first only Europeans were admitted. Most entrants were British, but Eastern and Southern Europeans soon played an increasing role. The largest immigrant streams in the 1950s and 1960s were of Germans, Italians and Dutch. The introduction of a non-discriminatory 'points system' for screening potential migrants after the 1966 White Paper opened the door for non-European migrants. The main source countries in the 1970s were Jamaica,

India, Portugal, the Philippines, Greece, Italy and Trinidad (Breton *et al.*, 1990: 14–16). Throughout the period, family entry was encouraged, and immigrants were seen as settlers and future citizens.

Australia initiated a mass immigration programme after 1945, because policy-makers believed that the population of 7.5 million needed to be increased for both economic and strategic reasons. (For more details and sources, see Collins (1991); Castles *et al.* (1992c).) The policy, summed up in the popular slogan 'populate or perish', was one of permanent, family immigration. The initial target was 70 000 migrants per year and a ratio of ten British migrants to every 'foreigner'. However, it proved impossible to attract enough British migrants. The Department of Immigration began recruiting refugees from the Baltic and Slavic countries, who were perceived as both 'racially acceptable' and anti-communist. Gradually the concept of 'acceptable European races' widened to include Northern Europeans and then Southern Europeans. By the 1950s, the largest sources of migrants were Italy, Greece and Malta. Non-Europeans were not admitted at all, as the White Australia Policy was still in force. Despite the policy of family migration, there was a male surplus among entrants, leading to schemes to encourage single women to come from Britain and elsewhere. It was not until 1975 that women were allowed to migrate as heads of families.

Immigration was widely regarded as the motor of post-war growth: from 1947 to 1973 it provided 50 per cent of labour force growth, giving Australia the highest rate of increase of any Organization for Economic Cooperation and Development (OECD) country. By the late 1960s, it was becoming hard to attract Southern European migrants, and many were returning to their homelands in response to economic developments there. The response was further liberalization of family reunions, recruitment in Yugoslavia and Latin America, and some relaxations of the White Australia Policy. By the 1970s, Australian manufacturing industry relied heavily on migrant labour and factory jobs were popularly known as 'migrant work'.

Comparative perspectives

One common feature in the migratory movements of the 1945–73 period is the predominance of economic motivations. Foreign worker migrations to Western Europe were caused primarily by economic considerations on the part of migrants, employers and governments. The same is true of temporary worker recruitment for US agriculture. Economic motives played a major part in Australia's post-war migration programme, although population building was also a consideration. The colonial workers who migrated to Britain, France and the Netherlands generally had economic reasons, although for the governments political considerations (such as the desire to maintain links with former colonies) also played a part. Permanent migration to the USA was probably the movement in which economic factors were least dominant. Yet the migrants themselves often had economic motivations, and their labour played a major role in US economic growth. Of course there were also refugee migrations, in which economic motivations were secondary. The overwhelmingly economic motivation for migration was to become less clear-cut in the post-1973 period.

How important was labour migration for the economies of the receiving countries? Some economists have argued that it was crucial to expansion. Migrants replaced local workers, who were able to obtain more highly-skilled jobs during the boom. Without the flexibility provided by immigration, bottlenecks in production and inflationary tendencies would have developed. However, other economists have argued that immigration reduced the incentive for rationalization, keeping low-productivity firms viable and holding back the shift to more capital-intensive forms of production. Such observers also claim that social capital expenditure on housing and social services for immigrants reduced the capital available for productive investment. Overall there is little doubt that the high net immigration countries, like the FRG, Switzerland, France and Australia, had the highest economic growth rates in the 1945–73 period. Countries with relatively low net immigration (like the UK and the USA at this time) had much lower growth rates. (See Castles and Kosack (1973: Chapter 9) and Castles *et al.* (1984: Chapter 2) for more discussion of these issues.) Thus the argument that immigration was economically beneficial in this period is convincing.

Another general feature of the 1945–73 period was growing diversity of areas of origin, and increasing cultural difference between migrants and receiving populations. At the beginning of the period, most migrants to all main receiving countries came from various parts of Europe. As time went on, increasing proportions came from Asia, Africa and Latin America. This trend was to become even more marked in the following period.

A comparison of the situation of colonial workers with that of guest-workers is instructive. The differences are obvious: colonial workers were citizens of the former colonial power, or had some preferential entitlement to enter and live there. They usually came spontaneously, often following lines of communication built up in the colonial period. Once they came in, they generally had civil and political rights; most (though by no means all) intended to stay permanently. On the other hand, guestworkers and other foreign workers were non-citizens. Their rights were severely restricted. Most came because they were recruited; some came spontaneously and were able to regularize their situation; others came illegally and worked without documentation. Generally they were seen as temporary workers who were expected to leave after a few years.

There are also similarities, however, especially in the economic and social situations of the two categories. Both became overwhelmingly concentrated in low-skilled manual work, mainly in industry and construction. Both tended to suffer substandard housing, poor social conditions and educational disadvantage. Over time, there was a convergence of legal situations, with family reunion and social rights of foreign workers improving, while the colonial migrants lost many of their privileges. Finally, both groups were affected by similar processes of marginalization, leading to a degree of separation from the rest of the population and an ethnic minority position.

Migrations in the period of global economic restructuring

The curbing of organized recruitment of manual workers by industrialized countries in the early 1970s was a reaction to a fundamental restructuring of the world economy. The subsequent period has been marked by:

(a) changes in global investment patterns, with increased capital export from developed countries and establishment of manufacturing industries in some previously underdeveloped areas;
(b) the micro-electronic revolution, which has reduced the need for manual workers in manufacturing;
(c) erosion of traditional skilled manual occupations in highly-developed countries;
(d) expansion in the services sector, with demand for both highly-skilled and low-skilled workers;
(e) growing informal sectors in the economies of developed countries;
(f) casualization of employment, growth in part-time work, increasingly insecure conditions of employment;
(g) increased differentiation of labour forces on the basis of gender, age and ethnicity, through mechanisms which push many women, young people and members of minorities into casual or informal-sector work, and which force workers with outmoded skills to retire early.

These transformations have had dramatic effects in Africa, Asia and Latin America. In some places, rapid industrialization and social change have taken place, leading to the emergence of 'newly industrializing countries' (NICs). In countries belonging to the Organization of Petroleum Exporting Countries (OPEC), reinvestment of oil profits after 1973 led to industrialization and social change. But in large areas of Africa, Latin America and Asia, post-colonial development strategies have failed. Many countries are marked by rapid population growth, overuse and destruction of natural resources, uncontrolled urbanization, political instability, falling living standards, poverty and even famine. Thus the idea of the 'Third

World' as an area with common economic problems and development perspectives has lost its meaning, and has largely been replaced by the idea of a 'South–North divide'. Economic crisis and social change in the South is generating new pressures for migration to the North. (The use of the terms South–North and East–West is Eurocentric, but will be retained here because it has become part of general usage. It is important to understand that these concepts refer not primarily to geographical locations but to economic, social and cultural divisions.)

These developments have led to considerable shifts in migratory patterns and to new forms of migration. The main trends include:

(a) broad decline of government-organized labour migration to Western Europe followed by emergence of a second generation of temporary foreign worker policies in the 1990s;
(b) family reunion of former foreign workers and colonial workers, and formation of new ethnic minorities;
(c) transition of many Southern and Central European countries from countries of emigration to countries of immigration;
(d) continuation of migration to the 'classical immigration countries' of North America and Oceania, but with considerable shifts in the areas of origin and the forms of migration;
(e) new migratory movements (both internal and international) connected with economic and social change in the NICs;
(f) recruitment of foreign labour, mainly from less-developed countries, by oil-rich countries;
(g) development of mass movements of refugees and asylum seekers, generally moving from South to North, but also (especially after the collapse of the Soviet Bloc) from East to West;
(h) increasing international mobility of highly-qualified personnel, in both temporary and permanent flows;
(i) proliferation of illegal migration and legalization policies.

These movements will be examined in more detail below and in Chapter 5. The main population flows of the post-1973 period are shown in Map 1.1.

Migrants and minorities in Western Europe

The post-1973 period was one of consolidation and demographic normalization of immigrant populations in Western Europe. Recruitment of both foreign workers and colonial workers largely ceased. After 1990, however, a new generation of temporary foreign worker programmes developed. But their scale was small compared to those of the 1945–73 period. For colonial migrants in Britain, France and the Netherlands, trends to family reunion and permanent settlement continued. At the same time, the settlement process and the emergence of second and third generations, born in Western Europe, led to internal differentiation and the development of community structures and consciousness. By the 1980s, colonial migrants and their descendants had become clearly visible social groups.

Permanent settlement generally had not been envisaged for the foreign workers. When the German government stopped recruitment in 1973 and other governments followed suit, they hoped that the now unwanted 'guests' would go away. Many Western European states proclaimed themselves 'zero immigration countries'. In fact some foreign workers did go home, but many stayed. Governments initially tried to prevent family reunion, but with little success. In the end, it was grudgingly accepted as a human right. In several countries, the law courts played a major role in preventing policies deemed to violate the protection of the family contained in national constitutions. Foreign populations changed in structure. In the FRG, for instance, the number of foreign men declined slightly between 1974 and 1981, but the number of foreign women increased by 12 per cent, while the number of children aged up to 15 grew by 52 per cent (Castles, Booth and Wallace, 1984: 102). Instead of declining, as policy-makers had expected, the total foreign population of the FRG remained fairly constant at about 4 million in the late 1970s, only to increase again to 4.5 million in the early 1980s and to over 5 million prior to German reunification in 1990. In 1999, 7.3 million foreigners resided in reunited Germany (OECD, 2001: 172). Table 4.2 gives information on the growth of foreign populations in some European immigration countries.

By 1995 the total foreign population of European OECD countries was 19.4 million, of whom only 6.7 million were EU citizens. There were 2 million North Africans, 2.6 million Turks and 1.4 million people from former Yugoslavia (OECD, 1997: 30). The foreigners who left after 1973 were mainly those from the more developed countries, where there was some prospect of work for returnees. Those who stayed were from less-developed areas, in particular Turkey and North Africa. It was above all the non-European groups who experienced socio-economic exclusion through discrimination and racism, like the former colonial worker groups.

Intra-EC movement did continue after 1973. It was increasingly an individual migration, mainly of skilled workers or highly-qualified personnel. By the late 1980s it was becoming customary to treat the EC (from 1993, the EU) as a single labour market, and to see intra-EU mobility as analogous to internal migration within a national economy.

In the second half of the 1980s, there was a resurgence of migration to Western Europe. The main driving force was economic and political problems in the countries of origin. The new migrants came as workers (both legal and illegal) but increasingly also as asylum seekers. Many were

Table 4.2 *Foreign resident population in selected OECD countries (thousands)*

Country	*1980*	*1985*	*1990*	*1995*	*1999*	*Percentage of total population 1999*
Austria	283	272	413	724	748	10.0[a]
Belgium	–	845	905	910	900	8.8
Denmark	102	117	161	223	259	4.9
France	3714[b]	–	3597	–	3263	5.6[c]
Germany	4453	4379	5242	7174	7344	8.9
Ireland	–	79	80	94	126[d]	3.3[d]
Italy	299	423	781	991	1520[e]	2.6[e]
Luxembourg	94	98	–	138	159	36.6
Netherlands	521	553	692	757	651	4.1
Norway	83	102	143	161	179	4.0
Portugal	–	–	108	168	191	2.0
Spain	–	242	279	500	801	2.0
Sweden	422	389	484	532	487	5.5
Switzerland	893	940	1100	1331	1400	19.2
UK	–	1731	1875	2060	2208	3.8

[a] Figure for 1998
[b] Figure for 1982
[c] For metropolitan France only
[d] Figure for April 2000
[e] Figure for December 2000

Notes: These figures are for foreign population. They therefore exclude naturalized immigrants (particularly important for France, the UK and Sweden). They also exclude immigrants from colonies or former colonies with the citizenship of the immigration country (particularly important for France, the Netherlands and the UK). The figures for the UK in this table are not comparable with the birthplace figures given in Table 4.1. The figures for Germany refer to the area of the old Federal Republic up to 1990, and to the whole of united Germany thereafter. Some of the calculations are our own.

Sources: OECD (1992: 131, 1997: 29, 2000, 2001); Lebon (2000: 7); Strozza and Venturini (2002: 265).

from Africa, Asia and Latin America, but in the late 1980s the crises in the Soviet Union and Eastern Europe led to new East–West movements.

By the mid-1990s, immigration flows to many Western Europe countries had stabilized and in some cases declined from the peak levels of 1991–2. However, new debate began about the future need for immigrants. Low fertility was leading to demographic decline and an ageing population (Münz, 1996). Germany began to recruit new types of 'guestworkers' from Eastern Europe, often under conditions even more onerous than the old 'guestworker system' (Rudolph, 1996). Yet Eastern Europe offers no long-term demographic reserves: the total fertility rate is low and life expectancy is actually declining in some areas due to environmental factors. (The total fertility rate is the average number of births per woman during her lifetime, assuming constant fertility. The rate needed to maintain a constant population is just over two children per woman.) The highly fertile and young underemployed populations of North Africa and Turkey appear in an ambivalent light to many Europeans. On the one hand they are seen as a source of workers for factories and building sites, and of carers for the aged; on the other hand there are fears of being 'swamped' by new influxes.

The debate gained new impetus in 2000 through a United Nations Population Division Report on *Replacement Migration* (UNPD, 2000), which gave detailed calculations on the decline in fertility in developed countries, and how this would affect population size and ageing. The Report showed that, on current trends, populations might decline dramatically. For instance, the Italian population could fall from 57 million in 2000 to just 41 million by 2050. This led to the question whether increased immigration could compensate for demographic change, by maintaining the size of the general population, the working-age population, or the ratio of workers to dependants. The UNPD showed that extremely high levels of immigration would be needed to achieve such objectives, and implied that these would be neither desirable nor realistic in the context of other social and political goals. This conclusion was shared by most commentators and policy-makers. However, the strong public impact of the Report did open up new debates on the need for both skilled and unskilled labour migration to developed countries, and helped undermine the myth of 'zero immigration policies'.

An upward trend in the overall pattern of international migration to OECD-area states became discernible by 1997 and was confirmed in 1998 and 1999 (OECD, 2001: 13). A number of factors contributed. In the 15 EU states, 342 000 asylum applications were made in 1999 as compared to 227 000 in 1996. There were also growing movements of temporary and highly-skilled foreign workers. A number of governments perceived actual or looming shortages of highly-skilled workers in the information and communications sectors. Concurrently, population ageing affected labour supply and demand in sectors like health, education and home services (OECD, 2001: 14). Germany's adoption of a new immigration law in 2002 authorizing increased immigration, particularly of highly-skilled foreign workers, and the US decision to authorize expanded availability of H-1B visas for highly-skilled professionals from abroad were emblematic.

Southern Europe

Italy, Spain, Portugal and Greece comprise a distinctive sub-group of EU states with regard to international migration. Until 1973, they were viewed as lands of emigration. Then, at somewhat different junctures and to various degrees, they underwent migration transitions (see Chapter 7 for explanation of this term) becoming significant lands of both emigration and immigration. In the post-Cold War period, their roles as lands of emigration have diminished, whereas their roles as lands of immigration have become more pronounced. They have come to share many of the concerns and characteristics of their EU partner states to the north, yet remain demarcated by the key role played by the underground economy in shaping inflows, the preponderance of illegal migration in overall migration and by weak governmental capacity to regulate international migration (Reyneri, 2001).

In Italy, numbers of foreigners with residence permits doubled between 1981 and 1991, from 300 000 to 600 000, and then doubled again over the next decade to 1.4 million. Inclusive of aliens under age 18 who live with their parents and therefore do not hold residence permits, the total legally-resident alien population reached an estimated 1.5 million or 2.6 per cent of Italy's resident population by 2001 (Strozza and Venturini, 2002: 265). There are estimated to be an additional 300 000 undocumented foreign residents (Calavita, 2003). Most legally-resident aliens arrived illegally or violated visa conditions and were subsequently legalized. The four major legalizations authorized since 1986 have had variable requirements. The 1998 procedure, dubbed Law 40, was open to employed aliens and to family members of legally-resident aliens. As many as 300 000 persons were expected to gain residency compared with 118 700 in 1987–88, 217 700 in 1990 and 147 900 in 1996 (OECD, 2000: 56). As in the case of recent Spanish legalizations, many aliens who legalize in Italy subsequently lose their legal residency due to loss of employment, non-renewal of permits or administrative problems and therefore frequently legalize again. A fifth amnesty was under way in 2002, but linked to new laws to cut immigration introduced by the Centre–Right coalition government. It was designed to facilitate deportation of those who do not qualify.

The upsurge in immigration has coincided with persistently high levels of unemployment at the national level, a dramatic decrease in fertility and acute crises in neighbouring areas like Bosnia, Kosovo and Albania. Nevertheless, the prevalent pattern appears employer demand-driven from the underground economy, which is assumed to be much more pervasive in Italy and other Southern European countries than in Northern Europe. Most immigrants move to areas of Italy where employment is available, not to areas with high unemployment (Reyneri, 2001).

The composition of the legally-resident foreign population has evolved considerably in recent years. East Europeans now outnumber North Africans, although Moroccans remain the single largest community, comprising 11.6 per cent of all aliens with residence permits. Albanians comprise 10 per cent of the resident permit holders. Various nationality groups vary sharply in terms of gender with immigrants from Africa and the Mediterranean littoral being predominantly male while immigrants from East Asia and from Latin America are mainly women (Strozza and Venturini, 2002: 271).

Spain, which still has 2.5 million expatriated citizens, similarly underwent a remarkable transformation, with the foreign population growing from 279 000 in 1990 (0.7 per cent of the total population) to 801 000 in 1999 (2 per cent) (OECD, 2001: 282). While the foreign population still represented a small fraction of the total population, its significance was disproportionate. In 1991, Spanish political parties agreed to 'depoliticize' migration issues. But the pact unravelled in 1999 when the governing Popular Party broke ranks and ran on a platform favouring a more restrictive policy. From 1991 to 1999, successive legalizations and quota admissions resulted in 0.5 million requests for legalization. Some 273 634 persons were legalized, including 105 861 Moroccans. Under a new law adopted in 2000, an additional 243 392 persons applied for legalization. As in Italy in the 1990s, immigration to Spain became a dominant socio-economic and political issue.

A major concern has been the increase in *pateras*, makeshift craft used by migrants from Africa to cross over to Spain, often with fatal results. Provisional, incomplete statistics for 2000 revealed a huge increase in the number of persons apprehended for *patera* usage, from 3569 in 1999 to 14 893 in 2000 (Lopez Garcia, 2001: 129). Bilateral immigration issues frayed Moroccan-Spanish relations. When the leaders of the EU met to discuss immigration policy matters in Spain in 2002, North African migrants staged protests to voice complaints that they were being displaced by workers from Eastern European countries like Romania. Violent attacks against North African farm workers in the El Ejido area of Andalusia in 2000 followed the murder of a Spanish woman and subsequent arrest of a young mentally-unbalanced North African (Lluch, 2002).

The key trends affecting migration to Portugal in the post-Cold War period included growth in unskilled workers of foreign origin, especially in construction and domestic services and expansion of the informal economy (Baganha and Reyneri, 2001). The foreign resident population rose from 108 000 (1.1 per cent of the total population), to 191 000 in 1999 (1.9 per cent (OECD, 2001: 282). In addition, Portugal experienced entries from former colonies (especially Mozambique, Angola, the Cape Verde Islands and East Timor) following the end of colonial wars due to the 1974 revolution. These African and Asian immigrants were entitled to Portuguese citizenship and usually speak Portuguese. Many of them are well integrated, and form a privileged group compared to later non-EU European immigrants (such as Ukrainians or Romanians). This is an interesting inversion of more customary racial hierarchies in Europe. Legalizations for undocumented foreign residents were authorized in

1992, 1996 and 2001. Even though some 4 million out of the 10 million citizens of Portugal continued to reside abroad, immigrants outnumbered emigrants after 1993. East Europeans arrived in large numbers and many of them were trafficked. Concern grew over exploitation of migrants (Dieux, 2002).

Until 1990, international migration to Greece mainly involved repatriation of ethnic Greeks from abroad and arrivals of refugees in transit. In the post-Cold War period, immigration has soared and foreigners constituted 8 per cent of the total population of nearly 11 million and 13 per cent of the workforce by 2001 (Fakiolas, 2002: 281). Within a decade, despite high unemployment, Greece became one of the EU states most affected by international migration that was mainly illegal. Deeply entrenched corruption in Greek public administration contributed significantly to the influx (Fakiolas, 2002: 283). Greece welcomed the repatriation of Pontian Greeks, mainly from areas of the former Soviet Union, and some 150 000 entered between 1989 and 1999. Special measures and procedures also applied to ethnic Greeks from Albania. A number of ethnicity certificates issued legally by Greek consulates abroad were determined to be fraudulent (Fakiolas, 2002: 285–6).

Albanian Muslims, meanwhile, encountered harsher treatment. Some 2.4 million aliens have been deported since 1990 and 80 per cent of them were Albanians (Fakiolas, 2002: 290). The unusually high incidence of expulsions and deportations reflected the weak capacity of the Greek state to regulate international migration through measures like enforcement of laws against illegal employment of aliens, and the absence of legalization policy until 1998. Greek trade unions finally succeeded in achieving a legalization policy but it was poorly administered (Papantoniou-Frangouli and Leventi, 2000). Some 37 000 aliens applied but many others did not know about the procedure or declined to participate, mainly out of fear of expulsion. This held particularly true of Albanians. Many found it difficult to find jobs covered by social security or employers willing to declare them to the social security administration and thereby be compelled to pay payroll taxes (Papantoniou-Frangouli and Leventi, 2002: 955). The legalization nevertheless increased the non-Greek, non-EU immigrant population by a factor of 14 (Fakiolas, 2002: 292).

The first decade of the post-Cold War period thus witnessed a dramatic transformation of Southern Europe due mainly to illegal migration. Legalization procedures enabled millions of migrants to achieve legal status but did not alter underlying processes fostering illegal migration. Hence the need for recurrent legalization policies. Many of the migrants belonged to elite youth circles in the lands of emigration in Eastern Europe and North Africa. They possessed the resources, both personal and material, to attempt typically gruelling and dangerous migration to societies with which they were largely unfamiliar (Reyneri, 2001: 12).

Central and Eastern Europe

The bulk of the states comprising this area are seeking to join the EU. Poland, Hungary, the Czech Republic, Estonia and Slovenia began negotiating their accession in 1998, while Bulgaria, Lithuania, Latvia, the Slovak Republic and Romania followed suit in 2000. The two waves of EU candidates differ in many respects, with the first group enjoying a markedly higher level of economic development. The vast region that extends from the Oder-Neisse boundary between Poland and Germany to the Eurasian steppes of the Russian Federation and from the Baltic States south-eastward to the Mediterranean and Black seas is extremely heterogeneous. States like the Ukraine, Yugoslavia and Belarus clearly cannot be termed highly developed but rather comprise the eastern hinterland to the emerging European Union of some 25 member states expected by 2007.

The negotiations over EU accession have resulted in dramatic changes in immigration policies in the candidate states as they are required to accept all the *acquis communautaire* (community legislation) of the EU including the Schengen Agreement of 14 June 1985 which sought to abolish passport controls at internal boundaries between signatory states and to create a common external border. Hence, the candidate states for EU enlargement have sought to harmonize their immigration-related policies with an emergent EU immigration policy which is slated to be fashioned by 2004 but whose contours remain difficult to discern. Most importantly, the EU candidate states have had to impose visa requirements upon citizens from neighbouring states that are not candidates for EU membership.

Formerly, travel between states like the Ukraine and Poland was not contingent on visa-issuance – a legacy of communist rule. To meet pre-accession requirements, the applicant states have to pour precious budgetary resources into the development of a governmental capacity to fulfil external border control functions. Indeed, the driving force behind eastward enlargement of the European Union has been the desire to create a migration buffer zone in Central and Eastern Europe which would displace eastward many of the problems and issues associated with border control (Nygård and Stacher, 2001). In return for the grant of visa-free entry to the Schengen-area EU states, candidate states have had to sign readmission agreements to take back third country nationals who cross their territory and illegally enter states like Germany and Austria.

With the collapse of the Iron Curtain in 1989–90, many feared mass, uncontrollable migration from the former Warsaw Bloc area. Movements of the magnitude feared did not materialize, although 1.2 million persons did emigrate in the first three years. Most of those emigrating were ethnic minorities like Germans and Jews who benefited from preferential treatment in German and Israeli immigration policies. Between 1990 and

1997, there was total net immigration of 2.4 million people from Central and Eastern Europe into the EU which amounted to roughly half of all immigration to the EU during the same period. Some 1.8 million came from countries that formerly comprised the Soviet Union (Hönekopp, 1999: 6–7). But after 1993, emigration from Central and Eastern Europe to the EU declined. Significant outflows of various categories of temporary foreign workers and of tourists who take up temporary employment in the EU continued. Total numbers of officially admitted temporary foreign workers from Central and Eastern Europe in Germany fluctuated between 200 000 and 300 000 per year. Germany and Austria received the bulk of such workers who were admitted under bilateral agreements (Hönekopp, 1999: 22).

After 1993, the more economically advanced EU candidate states like Poland, Hungary and the Czech Republic became immigration lands overnight. They were generally poorly prepared to regulate international migration, lacking laws and administrative agencies to deal with various categories of migrants. Official statistics concerning international migration did not reflect unregistered migration of 'tourists' who found employment in the underground or informal economy. Poland was thought to have received an estimated 800 000 Ukrainians who took up employment in 1995 (Okólski, 2001: 115). Ukrainians mainly worked in agriculture and construction but were also engaged in trading activities.

Disparities in levels of economic development, wages and opportunity played a major role in intra-regional migrations. Unemployment in states like Belarus and the Ukraine runs much higher than officially recorded. Perhaps half of all Ukrainians seeking employment are unemployed (Bedzir, 2001). And many employed persons are unable to live on the income derived from their jobs in Belarus or Romania. Hence they seek to supplement their incomes through temporary employment abroad, mainly as tourists who take up short-term, unregistered or off-the-books employment (Wallace and Stola, 2001: 8).

Most of the states in the region recorded huge increases in border crossings in the 1990s. Poland and the Czech Republic each register 0.25 billion per year (Stola, 2001: 84). However, it is the prevalence of transit migration and tourism for employment that demarcated the region in the first decade of the post-Cold War period (Stola, 2001: 80). Transit migration involved third country nationals moving through Central and Eastern Europe to points west. There were three major streams. The first involved citizens from countries of the former Warsaw Pact who, until recently, could enter legally without a visa and then attempted to migrate illegally to the EU. Many Gypsies (or Roma) from countries like Romania participated. A second stream comprised refugees from conflicts in the Western Balkans, most notably in Bosnia and Croatia in 1991 to 1993 and Kosovo in 1999. Those fleeing affected neighbouring states unevenly, with

Hungary and the Czech Republic receiving many more of them than Poland. Some of the Kosovars admitted to refugee reception centres in Poland ended up in Germany (Stola, 2001: 89). The third stream consisted of people from distant lands in Africa and Asia. The USSR had served as a barrier to migrants from afar. When it disintegrated its successor states became an easy-to-cross bridge between poles of economic inequality (Stola, 2001: 89). Smugglers and traffickers of aliens proliferated in this environment and remain deeply entrenched, despite counter-measures (IOM, 2000a).

Within the area of the former Soviet Union there also were significant movements of populations between successor states. By 1996 4.2 million persons had repatriated, mainly ethnic Russians going to the Russian Federation. Additionally, there were nearly 1 million refugees from various conflicts and some 700 000 ecological refugees, mainly from areas affected by the Chernobyl disaster (Wallace and Stola, 2001: 15).

The category of false tourist or tourism for employment included many petty traders. Overall, the post-Cold War period has resulted in an enormously complex and incompletely understood migration pattern. Some of that complexity was evident in statistics on foreigners in nearby states. In Hungary, most foreigners came from Romania; in Poland and the Czech Republic, most came from the Ukraine; and in the Slovak Republic, most came from the Czech Republic (OECD, 2001: 69, 74). Most migratory movements were thought to be short term or 'pendular' in nature and that would not be unusual in the early stages of a migration process. The real question concerns what will happen with migrant populations after enlargement occurs. Most likely, the dynamics defining the region will be greatly affected. Much will depend on the learning curve of public administrations in these countries, how quickly they can absorb and implement policies consistent with those practised in states like Germany.

Studies of human trafficking in Central and Eastern Europe suggested that there was a considerable distance to travel. A major concern involved embedded corruption and the inability of governments to hire and train sufficient personnel because salaries were so low (IOM, 2000a: 174–6). The Yugoslav Republic, Albania, Bosnia-Herzegovina and the former Yugoslav Republic of Macedonia would not be part of the EU enlargement process and their proximity and instability greatly complicated the regional migration picture. Insufficient development of governmental capacity to regulate international migration clouded accession prospects for many of the candidate countries despite their participation in inter-governmental fora like the Budapest Process, and the provision of EU assistance to build governmental capacity to regulate migration (Geddes, 2000: 9 and IOM, 2000: 117). Austria made known that it would oppose enlargement until

Hungary and the Czech Republic made their borders more secure (Hárs *et al.*, 2001: 269).

Numbers of asylum seekers in the Czech Republic and Hungary in 1998, 4086 and 7386 respectively, were higher than those recorded in EU member states like Finland and Greece, 1272 and 2953 respectively. Most applicants were rejected but most were not returned home. In some instances, this resulted from their homeland being insecure or because the migrants lacked official documents. In other cases, the cost of returning the failed applicants was prohibitive. Central and Eastern European states sometimes had no diplomatic or consular presence in distant states to facilitate repatriation. The nationalities posing the greatest problems in terms of readmission were the Federal Republic of Yugoslavia, Afghanistan, China, Vietnam, Sri Lanka, Iraq and India (Lazcko, 2001).

Despite the enormity of the transformations wrought by international migration in the post-Cold War period, the domestic politics of Central and Eastern European states were relatively unaffected. There was little politicization of migration issues and political elites did not appear to pay much attention to them, even though the all-important bid for EU accession by candidate states hinged upon reform of immigration policies (Drbohlav, 2001: 216; Hárs *et al.*, 2001: 264; Stola, 2001: 190–91). This situation began to change in Hungary in late 1997. In March 1998, the Secretary of State of the Ministry of the Interior declared: 'Schengen has become the ultimate element of the accession. There is no chance for derogation' (Hárs *et al.*, 2001: 268). Subsequently, the coalition of conservative parties campaigned on the need for strong border controls and revised visa policies. They won the elections and soon began implementing the reforms.

North America and Australia

Migration to the USA grew steadily after 1970. Total immigration, which refers to aliens granted legal permanent resident status, rose from 4.5 million from 1971–1980, to 7.3 million from 1981–1990, and to 9.1 million from 1991–2000. In fiscal year 2000, 850 000 immigrants were admitted. The number of foreign-born residents and children of immigrants in the USA also rose dramatically over the 30-year period, increasing to 56 million from 34 million in 1970 (Scott, 2002). Persons born in Mexico comprised more than a quarter of the foreign-born population in 2000. Most immigrants intended to reside in just six states (California, New York, Florida, Texas, Illinois and New Jersey) but the 1990s witnessed a spatial diffusion of migrants from the six gateway states to rural America. Some 840 000 persons naturalized in fiscal year 1999 but there was a backlog of more than 1 million naturalization applications pending (INS, 2002a: 12).

The 1965 amendments to the Immigration and Nationality Act had unexpected results (see Borjas, 1990: 26–39). US residents of Latin American and Asian origin were able to use family reunion provisions to initiate processes of chain migration, which brought about a major shift in ethnic composition. In the 1951–60 period, Europeans made up 53 per cent of new immigrants, compared with 40 per cent from the Americas and only 8 per cent from Asia. In 1999, Europeans were only 15 per cent of all immigrants, while 46 per cent came from the Americas (excluding Canada) and 30 per cent from Asia. Mexico has been the largest single source country for the USA for many years. In 1999, 23 per cent of all immigrants were Mexican (INS, 2002a: 22). Other important source countries in 1999 (20 000–32 000 immigrants each) were China, the Philippines, Vietnam and India.

The 1990 Immigration Act was designed to increase the number of immigrants admitted on the basis of skills, but also maintained levels of family reunion and refugee resettlement, leading to an overall increase in intakes. The Act also established new admissions programmes for the groups considered to have been adversely affected by the 1965 amendments (OECD, 1995: 130). A world-wide lottery now distributes 55 000 entry visas annually at random, contingent upon certain basic qualifications. Nationals of the 11 countries which have sent the largest numbers of legal immigrants to the USA are excluded. Applicants simply have to write their names and those of their spouse and children on a piece of paper and mail it together with photographs to the processing centre in the USA.

Much of the Mexican immigration started as temporary (and frequently illegal) labour movements across the southern border. Farmworkers from Mexico (and to a lesser extent from other Central American and Caribbean countries) have long played a crucial role for US agribusiness, which has opposed any measures for effective control, such as sanctions against employers of illegal workers. The number of undocumented residents was estimated at almost 9 million in 2000, compared with 5 million in 1996. The Immigration Reform and Control Act of 1986 (IRCA) introduced a limited amnesty which led to about 3 million applications (over 70 per cent from Mexicans) (Borjas, 1990: 61–74). Most applicants (about 2.7 million) were granted resident alien status, giving the right to bring in dependants, and thus establishing new migration chains. IRCA also imposed sanctions against employers of illegal workers. Of the nearly 850 000 legal immigrants in fiscal year 2000, over half were former undocumented migrants, refugees or asylum seekers who were living in the USA and who adjusted their status through the INS (INS, 2002b: 2).

Illegal immigration soared due to the ease of falsifying identification documents which undercut enforcement of employer sanctions. By 1996 public outrage about illegal immigration led to the adoption of a new law which significantly increased INS personnel, reinforced physical barriers to illegal entry (such as high fences, video surveillance and more border patrols) and enabled building of additional detention facilities. Measures were also introduced to deny welfare benefits to illegal residents. The INS has received the largest budget increase of any Federal Government agency in recent years. Supporters of stronger enforcement measures believe that illegal residence and employment would be even more extensive in their absence, while critics regard greater enforcement as misguided and in some instances as counter-productive and dangerous. A growing number of illegal entrants have perished because heightened border controls led them to attempt more dangerous forms of entry in remote areas. More than 71 000 criminal aliens were removed in fiscal year 2000 and over 1.8 million aliens were apprehended (INS, 2002b: 2). Clearly immigration remains one of the major forces which shape US society. The volume of entries, the concentration of settlement in certain areas and the changing ethnic composition are all factors likely to bring about considerable social and cultural change.

Canada remains one of the very few countries in the world with an active and expansive immigration policy, which aims at permanent settlement. The overall number of entries grew from 89 000 in 1983 to nearly 190 000 in 1999. In 1990, the Canadian government announced a five-year immigration plan, designed to maintain the principles of family reunion and support for refugees, while at the same time increasing entries of skilled workers. Admissions rose from 214 000 in 1990 to 255 000 in 1993, but then fell to 212 000 in 1995, due to poor economic conditions. The 5 million resident aliens made up 17 per cent of the Canadian population at the 1996 Census (OECD, 2001: 147). However, concentrations were much higher in cities like Toronto and Vancouver. Entries from Asia, Africa and the Middle East have grown rapidly, while European migration has declined. In 1999, 51 per cent of immigrants came from Asia and the Pacific whereas 21 per cent came from Europe (OECD, 2001: 143). Over 50 per cent of immigrants in 1999 were skilled workers or business people and less than a third entered on the basis of family reunion.

Australians, by contrast, have become more sceptical about immigration, and entry policies have become more restrictive since 1996. The emphasis is on skills, while family reunion has been limited. Australia has also dramatically changed its position on humanitarian entries, moving away from its traditional generous admission policy, to one based on exclusion and mandatory detention of asylum seekers. These changes are described in detail in Chapter 9, and will therefore not be presented here. More information on ethnic communities (i.e. the later stages of the migratory process) in many of the immigration countries are to be found in Chapter 10.

Conclusion

This overview of international migration to developed countries since 1945 may lay no claim to completeness. Rather we have tried to show some of the main trends, and to link them to various phases in the global political economy. The upsurge in migratory movements in the post-1945 period, and particularly since the mid-1980s, indicates that international migration has become a crucial part of global transformations. It is linked to the internationalization of production, distribution and investment and, equally important, to the globalization of culture. The end of the Cold War and the collapse of the Soviet bloc added new dimensions to global restructuring. One was the redirection of some investment of the advanced capitalist countries away from the South towards Eastern Europe. Another dimension was the growth of East–West migration, with previously isolated countries entering global migratory flows.

Many of the large-scale migrations have been primarily economic in their motivations. Labour recruitment and spontaneous labour migration were particularly significant in the 1945–73 period. In the following years, other types of migration, such as family reunion and refugee and asylum-seeker movements, took on greater importance. Even migrations in which non-economic motivations have been predominant have had significant effects on the labour markets and economies of both sending and receiving areas. But it is equally true that no migration can ever be adequately understood solely on the basis of economic criteria. Economic causes of migration have their roots in processes of social, cultural and political change. And the effect on both sending and receiving societies is always more than just economic: immigration changes demographic and social structures, affects political institutions and helps to reshape cultures.

In the early 1990s, Western Europe was gripped by fears of uncontrolled influxes from the East and South. By 1995 this scenario had receded, due both to changes in sending countries, and the tightening of entry rules and border controls. In the second edition of this book (published in 1998) we noted a slowdown in migration to developed countries, but argued that this might be a passing phase, like that of the late 1970s, which could pave the way for even greater future movements. This has indeed proved the case, with significant increases in entries since about 1997, as well as diversification of migratory types. The main growth has been in asylum, illegal migration, and skilled migration.

Much depends on political decisions and government actions, including allocation of resources to enforce entry laws, and international cooperation on the management of migration. States and regional bodies can and do affect migratory outcomes, and there is reason to believe that control measures have some effect in reducing 'unwanted' migrations. However, control policies may also have unforeseen and negative consequences –

such as the growth of illegal movements and people smuggling. In the long run, it is the measures to address the 'root causes' which really count, and there is a long way to go before these are likely to be effective. These are the themes of the next chapter.

Guide to further reading

Castles and Kosack (1973 and 1985) is a comparative study of immigrant workers in France, Germany, Switzerland and the UK during the phase of mass labour recruitment from 1945 to 1973. Castles *et al.* (1984) continue the story for the period following the ending of recruitment in 1973-4. Portes and Rumbaut (1996) give a detailed analysis of recent immigrant settlement in the USA, while Collins (1991) gives a valuable account of post-war immigration to Australia. Hammar (1985a) provides a comparative study of the position of immigrants in Western European countries. The *International Migration Review* Special Issues 23:3 (1989) and 26:2 (1992) also provide comparative material. The OECD reports (mentioned in the Guide to further reading for Chapter 1) give useful data on migratory movements and immigrant populations. Harris (1996) gives a popular overview of global migration, while Stalker (1994) presents a global picture and provides very useful tables and charts. Zolberg *et al.* (1989) examine global refugee movements, while UNHCR (1995) provides more recent comprehensive information. Regulation issues are discussed in Teitelbaum and Weiner (1995).

Important edited books on migration in Europe include Rocha-Trindade (1993), Fassman and Münz (1994), OECD (2001) and Messina (2002). Concerning Central Europe, see Wallace and Stola (2001), and for Southern Europe see Baganha (1997), Luso-American Development Foundation (1999), King *et al.* (2000) and King (2001). Horowitz and Noiriel (1992) and Togman (2002) engage in comparisons of France and the USA.

Chapter 5

The State and International Migration: The Quest for Control

Around 1970, a new phase of international migration to highly-developed countries began to take shape. One defining feature was proliferating illegal migration. It was at this juncture that key industrial democracies like France, Germany and the USA embarked on what can be termed a quest for control, a sustained effort to prevent illegal migration and abuse or circumvention of immigration regulations.

This chapter examines key components of government strategies the better to regulate international migration, including enforcement of employer sanctions, legalization or amnesty programmes, temporary foreign worker admissions programmes, asylum policies, measures against human smuggling and trafficking, regional integration measures and strategies to prevent illegal migration by the fostering of sustainable development through overseas investment or assistance. Cooperation on migration has become a core element of international politics. The emphasis has shifted from national migration control policies based on (often short-term) economic and political interests, to broader international strategies for migration management. The chapter focuses on the USA and Europe, but many of the issues raised are relevant to other migration areas, which will be discussed in Chapters 6 and 7.

The overall assessment that emerges is mixed. What governments do matters a great deal. But significant levels of illegal migration persist throughout the world, and the estimated population of illegally-resident aliens in the USA expanded significantly in the 1990s. The US experience over the last three decades in particular has raised doubts about the willingness and capacity of democratic governments to regulate international migration. However, the events of 11 September 2001 threw into sharp relief shortcomings in immigration control in the USA. The USA, and many other states around the world, began to reform policies and procedures to ensure better security and this may result in more credible and coherent policies in the future. (OECD, 2001: 14).

As explained in Chapters 1 and 2, the current phase of intense globalization has had manifold repercussions upon the sovereign state. Some analysts interpret the persistence of illegal migration in spite of governmental efforts to curb it as evidence of the decline or obsolescence of the national state (Hobsbawn, 1994). Some students of transnationalism view its manifestations as an erosion of state sovereignty as if in a zero-sum relationship between the two. But such a view ignores considerable evidence of states deliberately fostering transnationalism better to achieve goals (Smith, 2001). Other analysts stress the adaptiveness of national states to changing circumstances and interpret deepening of regional integration as in the European context as a strategy for saving the state (Milward, 1992). Still other theorists have written of new forms of state, whether a conglomerate state, in which national governments still matter a great deal in conjunction with other levels of governance, or a global state that is centred on the transatlantic area and dominated by the USA (Shaw, 2000; Habermas, 2001). Some have also argued that the post-Cold War world is dominated by an empire centered in Washington, DC. Perhaps unsurprisingly, support for legalization drives by undocumented aliens or asylum seekers denied asylum features at the forefront of resistance to empire (Hardt and Negri, 2000: 400). The autonomy of international population movements and the difficulties and contradictions encountered by empire in regulating them is viewed as the wellspring for new forms of society and socio-political consciousness for the labouring masses, the multitudes.

Despite the significance, both theoretical and practical, ascribed to unregulated migration and governmental efforts to curb it, study of governmental strategies to prevent illegal migration has been surprisingly sparse. Understanding of governmental capacity to regulate international migration is at best described as fragmentary and limited. A major limitation derives from the very nature of illegal migration, its opaqueness to measurement and scrutiny. Nevertheless, much more is known about illegal movements and matters like legalizations and human trafficking than was the case at the outset of the era of the quest for control, circa 1970.

Employer sanctions

Laws punishing employers for unauthorized hiring of aliens constituted the centerpiece of many immigration policy reforms in the 1970s in the USA and Western Europe. Such 'employer sanctions' are often coupled with 'legalization programmes', which give work and residence permits to former undocumented workers who fulfil certain conditions. These carrot-and-stick measures, it is argued, remove the motivation for undocumented work, since employers run serious risks of punishment and workers are better off with legal status, and no longer undercut local wage levels. However, in practice, these programmes may come up against powerful interests: employers may have the political clout to prevent effective enforcement, while migrant workers may reject legalization for fear that it will make it harder for them to find jobs.

Most Western European states adopted employer sanctions in the 1970s. The UK constituted a major exception – British officials claimed that illegal migration did not pose a significant problem. But this perception changed by the 1990s. Enforcement strategies in key states like Germany and France initially differed importantly, with German authorities approaching illegal alien employment as an aspect of a broader problem of illegal employment linked to the growth of the underground economy, whereas the French approach was less comprehensive and focused on punishment of illegal alien employment. By the late 1980s, the French approach changed and largely converged with the German approach.

The major exception to the overall pattern until 1986 was the USA. Concern over illegal migration had grown in the 1960s and 1970s in the wake of termination of the *Bracero* Program with Mexico in 1964. The creation of *maquiladores* (assembly plants just south of the US-Mexico border) was supposed to provide alternative employment opportunities for former *braceros* (migrant labourers) but many simply continued to come to the USA for employment. Indeed, there had been a great deal of illegal migration during the 1942 to 1964 period in which some 5 million Mexicans were authorized to perform temporary services of labour in the USA. In 1954, 1 million Mexicans were deported during Operation Wetback. One scholar noted that 'the *bracero* program, instead of diverting the flow of wetbacks into legal channels, as Mexican officials had hoped, actually stimulated unlawful emigration' (Select Commission on Immigration and Refugee Policy, 1981: 470).

President Carter appointed a Select Commission on Immigration and Refugee Policy (SCIRP). In 1981 SCIRP recommended employer sanctions, implementation of a counterfeit-resistant employment identification document and a legalization policy. It did not recommend expansion of temporary foreign worker admissions. The IRCA of 6 November 1986 made hiring of unauthorized aliens a punishable offence, but did not introduce counterfeit-resistant employment documents. Instead, an assortment of documents could be used to satisfy the requirements of the new I-9 form that had to be filled out at each hiring; many of these documents could easily be forged or obtained fraudulently. Enforcement of employer sanctions would begin after an education period that would familiarize employers with their new obligations (Montwieler, 1987).

The concept was hotly debated in the USA, unlike in Western Europe where employer sanctions were supported by mainstream political parties. Some Hispanic advocacy groups alleged that imposition of employer sanctions would increase employment discrimination suffered by minorities. Others viewed the I-9 requirement as another burdensome task imposed by government upon US employers. Some feared that the enactment of sanctions would disrupt entire industries, such as labour-intensive agriculture, and result in crops rotting in fields or in much higher food prices. Many of the arguments made against employer sanctions were groundless but they nevertheless had an effect. It became very clear that there was not a political consensus in support of their enactment and implementation.

Enforcement of employer sanctions in Western Europe was scarcely issue-free despite the political consensus behind them. Principal barriers to effective enforcement included insufficient personnel, poor coordination between various enforcement agencies, inadequate judicial follow-up to enforcement and the slow adaptation of employers and illegal employees to enforcement measures. Western European governments constantly revised the relevant legal texts and fine-tuned enforcement strategies.

The French government, for instance, established an inter-agency mission to coordinate enforcement of new laws against manpower trafficking in 1976. It issued an annual report summarizing French government counter-action. By 1989, the mission was renamed the Interministerial Delegation for Combating Illegal Work (Délégation interministérielle à lutte contre le travail illegal), reflecting a new approach very similar to Germany's, which focused more broadly on the prevention of all forms of illegal employment (Miller, 1994). Innovations in the 1990s included an obligation on employers to notify the government of planned new hirings of labour prior to the onset of employment. Most transgressions of French laws against illegal employment in the 1990s did not involve aliens. Only 6 per cent of illegal hires in 1997 were aliens without documents, compared with 17 per cent in 1992 (Marie, 2000: 119). But aliens figured disproportionately among employers cited for illegal employment of other aliens and among those cited for illegal work. Many aliens authorized for legal employment in France nevertheless chose or were forced to work illegally.

A number of analysts have been dismissive of the efforts by European governments to enforce employer sanctions. They thereby ignore the development of a credible state capacity to deter and punish illegal employment of aliens since 1970 in several Western and Northern European states. The overall record of employer sanctions enforcement, however, appears very uneven. It is sometimes suggested that enforcement of employer sanctions in the USA had failed by 1999 because the US government announced it was suspending enforcement. Employer investigations by the INS declined from 7537 cases completed and 17 552 arrests in 1997 to 3898 cases completed and 2849 arrests in 1999 (INS, 2002a: 214). However, the US never truly had a credible employer sanctions regime because of the ease of circumvention of the 1986 law. Already by 1994, the Commission for Immigration Reform concluded that the employer sanctions system adopted in 1986 had failed, because many unauthorized foreign workers presented false documents to employers (Martin and Miller, 2000b: 46).

By 2000, the AFL-CIO, the major confederation of trade unions in the USA, announced that it no longer supported enforcement of employer sanctions. The AFL-CIO had been a major proponent of punishing illegal employment of aliens back in the SCIRP era. What had changed? A new leadership came into office in the AFL-CIO which emerged from unions with large numbers of immigrant members, including many illegals. This faction had close ties to the US Conference of Catholic Bishops, which supported a broad legalization of the millions of illegally-resident aliens in the USA. The US Roman Catholic leadership now favoured an approach very close to that advocated by the new president of Mexico, Vincente Fox, by 2000. This evolution bore witness to enduring, rather than diminished, US exceptionalism concerning illegal migration.

Legalization programmes

In the USA, de facto legalization of illegally-employed Mexican workers was routine during the 1942 to 1964 period, when it was commonly referred to as 'drying out wetbacks'. France routinely legalized aliens who took up employment in contravention to ONI procedures after 1947. Between 1945 and 1970, legalization comprised the major mode of legal entry into France (Miller, 1999: 40–1). Thereafter, the French government declared that legalization would be exceptional, but there were recurrent legalizations throughout the 1970s (Miller, 2002).

The election of a Socialist president and leftist majority in the National Assembly in 1981 set the stage for a new French approach to legalization. It was explicitly linked to a planned reinforcement of the government's measures against illegal migration and employment. For the first time, the trade unions and immigrant associations were to be associated with the effort, and there would be a major mobilization of governmental personnel to facilitate processing of applications. The Socialist legalization was supposed to break with past governmental policy, but the overall public policy posture of the government was still clearly opposed to illegal migration.

About 150 000 aliens applied and 130 000 were legalized. Assessments of the legalization varied. It became clear that fraud could become a major problem and that legalization procedures were administratively difficult to implement. Many eligible aliens did not know about the procedure or feared to participate in it. There was reason to believe that the Socialist legalization had a magnet effect, attracting additional illegal immigrants to France. The legalization helped those individuals benefiting from the opportunity but did not alter underlying labour market dynamics fostering illegal migration and employment. The legalization period was repeatedly extended and criteria for eligibility evolved. For French government officials, the Socialist legalization of 1981–83 constituted a notable success, one fit for emulation elsewhere.

The IRCA legalization in the USA differed in key respects from roughly concurrent programmes in Western Europe. The five-year gap between the cut-off date for eligibility for legalization (1 January 1982) and the effective starting date of the programme (4 May 1987) contrasted sharply with the minimal gap between the two dates in European programmes. Different approaches to immediate family members of successful legalization applicants constituted a second key contrast. Two principal and several smaller programmes emerged. The major one was open to all aliens who could prove residency prior to 1 January 1982. The second principal programme was open to aliens who could prove they had worked for 90 days in seasonal agriculture between 1 May 1985 and 1 May 1986. There was also a special procedure open to Haitian and Cuban entrants. Almost 2 million aliens applied for legal status under the major programme and 97 per cent were approved. A Congressional bid to extend the deadline for application was defeated.

There were different rules governing the farmworker programme. The legalization period was longer and beneficiaries underwent quasi-automatic adjustment of status to permanent residency. Moreover, applicants could apply from abroad and commute to work from abroad. Some 1.3 million aliens applied but the rate of approval was much lower than in the other principal programme. Widespread fraud contributed to the lower acceptance rate. In all, about 2.7 million aliens were legalized under IRCA provisions (Kramer, 1999: 43).

After prompting by the US Conference of Catholic Bishops, the US government would later promulgate a 'family fairness doctrine', in effect extending toleration to illegally-resident dependents of legalizing aliens. Initially, district commissioners of the INS were empowered to grant a kind of temporary, protected legal status to such family members out of humanitarian considerations. The Immigration Act of 1990 then included a family fairness provision, which enabled spouses and children of legalized aliens to become permanent resident aliens (Miller, 1989: 143-44).

Despite the differences, transatlantic commonalties emerged. Illegally resident aliens responded slowly and cautiously to legalization. Governments were disappointed and eased rules and regulations to encourage more applications. Perhaps most importantly, legalizations enabled governments to learn more about illegal migration flows and processes. They were generally more complex than had been thought, and principally involved aliens from countries which also sent large numbers of legally-admitted aliens (inclusive of aliens who had been previously legalized).

Beneficiaries of legalizations tended to be young workers employed in sectors with high concentrations of foreign workers (OECD, 2000: 57–9).

The 1990s witnessed recurrent legalizations in France, Spain and Italy and resumption of legalization in the USA under the 245i procedure, which enabled aliens to receive permanent resident alien status, without needing to return home, for a fee of US $1000. Many other states around the world authorized legalizations. Even Switzerland enacted a programme in 2000 for aliens who had entered before 31 December 1992 and were in a distressed state. About 13 000 persons were involved, mainly Sri Lankans (OECD, 2001: 251). Similarly, advocacy of legalization grew in Germany where authorities had long scoffed at legalization as a policy likely to backfire and encourage further illegal migration. Roman Catholic Cardinal Stazinsky of Berlin called for a German legalization policy in 2001, as did the German trade union confederation (Appenzeller *et al.*, 2001). On 22 July 2002, US Congressional minority leader and past Democratic presidential hopeful, Richard Gephardt, announced his intention to introduce legislation to authorize illegal immigrants with an employment history in the USA to obtain legal residency (Hulse, 2002). Gephardt unveiled his plan at the annual meeting of the National Council of La Raza, a Hispanic advocacy organization, and it appeared linked to Democratic hopes of attracting support from Hispanic voters.

Assessment of the significance of legalization policies in the quest for control must be nuanced. On the one hand, legalizations attest to the reality of illegal alien residency and employment. On the other hand, they enable governments to 'bring people out of the shadows' and confer legal residency. They can be interpreted either as evidence of governmental inability to prevent illegal migration or as evidence that sovereign states, for all their recent vicissitudes, can adapt to and cope with international population movements in the current era of globalization. Opponents of legalization typically contend that such policies undermine the rule of law and are inimical to the broad quest for control. There is some evidence that French legalization policies contributed to the emergence of the anti-immigrant National Front. Recent legalizations in Southern Europe have been characterized by post-legalization reversion to irregular status for many migrants (OECD, 2000: 63). Beneficiaries of legalizations generally, however, experience improvements in their overall socio-economic and employment prospects (Laacher, 2002: 66).

Temporary foreign worker admissions programmes

The post-Cold War era witnessed the re-emergence of modest temporary foreign worker policies in a number of settings where governments had curtailed further guestworker recruitment circa 1973. The second generation of European states authorizing temporary foreign worker policies included such former guestworker policy states as Germany, the Netherlands and Sweden, but also Southern European states. There was great irony in the adoption of a temporary foreign worker policy by Italy because the Italian Republic had criticized vociferously the status afforded Italian seasonal workers in Switzerland. Three decades after the renegotiation of the Italo–Swiss bilateral labour agreement in 1964, which finally enabled Italian seasonal workers to achieve residency in Switzerland, Italy was admitting contingents of temporary foreign workers from nearby states. By 1999, there were 20 000 seasonal migrant workers in the country, a tenfold increase since 1992, and this total included some legalized aliens who had been given seasonal contracts (OECD, 2001: 195).

Somehow, policies that were generally viewed as regressive and discriminatory in the 1960s and 1970s could be viewed as innovative and progressive after 1990. In fact, when compared with policies of the guestworker era, many of the post-Cold War temporary worker policies were more stringent. Would second generation temporary foreign worker programmes lead to outcomes similar to those of the guestworker generation? A key difference pertained to volume of admissions: the second generation programmes were very small. Germany in 1999 admitted 40 000 contract workers, 3700 guestworkers and 223 400 seasonal workers. The latter were admitted for a maximum of three months per year (OECD, 2001: 174). This compared to 646 000 new foreign workers entering Germany in 1969 (Castles and Kosack, 1973: 40).

The German resumption of temporary foreign worker admissions came in the context of efforts to support new democratic governments in Central and Eastern Europe and to secure their cooperation in German and EU efforts to curb illegal migration and human trafficking. Foreign policy considerations trumped historical memory. Ever since the cessation of recruitment in 1973, representatives of certain employer groups, especially hotels, restaurants and agriculture, had advocated resumption of foreign worker recruitment, but this time in a manner similar to recruitment of seasonal workers in Switzerland. There was little understanding among Germans that the Swiss government had, in fact, changed its policies towards seasonal workers. By the 1980s, former seasonal workers and their families comprised the single largest component of new resident aliens in Switzerland and, by 1999, only 10 000 seasonal workers were admitted to Switzerland, compared with over 200 000 in 1964 (Miller, 1986: 71; OECD, 2001: 50). Like France, which admitted 7612 seasonal workers in 1999 compared with a past average of 100 000 a year, the Swiss had largely phased out their seasonal foreign worker policy (Tapinos, 1984: 47; Lebon, 2000: 45).

Confusion surrounded Spain's temporary foreign worker policies. Many of the visas supposedly allotted for recruitment of foreign workers from abroad were given instead to illegal aliens in Spain undergoing legalization (Lopez Garcia, 2001: 114–15). The annual contingent for all of Spain varied from 20 000 to 30 000 visas. Authorizations for visas in some localities, such as Almería (the site of several days of violence against Arab workers in 2000) were much lower than the levels requested by employers. This was partly because the procedure required the foreign worker with a pre-contract to return home to receive a valid visa, but also partly because employers often preferred illegally to legally employed foreign workers for whom they were obliged to pay social security taxes (Lluch, 2002: 87–8).

Temporary foreign worker proposals in the USA are similarly advocated as a way of legalizing existing populations of illegally employed foreigners. There was little reason to suppose, however, that the expansion of temporary foreign workers admissions from Mexico, as advocated by US President Bush and Mexican President Fox in 2001, would reduce illegal employment in Delaware or Texas any more than it had in Almería or in the Midi of France back in the 1960s and 1970s. Perhaps the trends in seasonal foreign worker admissions to France and Switzerland made much more sense in the quest for control than the second generation of temporary foreign worker policies born of the post-Cold War era. The situation evoked historical memories, once aptly summarized by W.R. Böhning of the ILO Migrant Workers Branch:

> Temporary worker programs and restrictions are not only morally offensive but politically less and less tenable in Western plural societies. One can save oneself a great deal of domestic, political and administrative commotion and loss of international political capital by adopting from the start a position that is in conformity with the democratic values to which one claims allegiance rather than to have to yield in inauspicious circumstances to domestic and international pressures. (Böhning, 1984: 162)

Refugees and asylum

Since the mid-1980s, there has been a dramatic increase in the numbers of refugees and asylum seekers worldwide. These are 'forced migrants' who move to escape persecution or conflict, rather than 'voluntary migrants' who move for economic or other benefits. Asylum has become a major political issue in Western countries. Sensationalist journalists and right-wing politicians map out dire consequences, such as rocketing crime rates, fundamentalist terrorism, collapsing welfare systems and mass unemployment. They call for strict border control, detention of asylum seekers and deportation of illegals. The public appeal of such polemics is obvious: right-wing electoral successes in countries as disparate as Denmark, Austria, France and Australia can be attributed to fears of mass influxes from the South and East. (This section is partly based on two research papers jointly written by Stephen Castles and Sean Loughna of the Refugee Studies Centre, University of Oxford in 2002. Sean Loughna's contribution is gratefully acknowledged.)

According to the 1951 United Nations Convention Relating to the Status of Refugees, a refugee is a person residing outside his or her country of nationality, who is unable or unwilling to return because of a 'well-founded fear of persecution on account of race, religion, nationality, membership in a particular social group, or political opinion'. About 140 of the world's 190 states have signed the Convention or its 1967 Protocol. Member states undertake to protect refugees and to respect the principle of *non-refoulement* (that is, not to return them to a country where they may be persecuted). Officially recognized refugees are often better off than other forced migrants, as they have a clear legal status and enjoy the protection of a powerful institution: the United Nations High Commissioner for Refugees (UNHCR). Refugees often gain initial refuge in a country neighbouring their place of origin. Where conditions are difficult, the UNHCR may seek to resettle the refugees elsewhere, although traditional resettlement countries (especially in Europe and North America) are increasingly unwilling to accept them.

The global refugee population grew from 2.4 million in 1975 to 10.5 million in 1985 and 14.9 million in 1990. A peak was reached after the end of the Cold War with 18.2 million in 1993. By 2000, the global refugee population had declined to 12.1 million (UNHCR, 1995: 2000a). The broader category of 'people of concern to the UNHCR' (which includes refugees, some internally displaced persons and some returnees) peaked at 27.4 million in 1995, and was down to 21.1 million in 2000. Refugees came mainly from countries hit by war, violence and chaos, such as Afghanistan (2.6 million refugees in 2000), Iraq (572 000), Burundi (524 000), Sierra Leone (487 000), Sudan (468 000), Somalia (452 000), Bosnia (383 000), Angola (351 000), Eritrea (346 000) and Croatia (340 000) (UNHCR, 2000b: 315).

Asylum seekers are people who move across international borders in search of protection, but whose claims for refugee status have not yet been decided. Some observers claim that asylum seekers are not real victims of persecution, but simply economic migrants in disguise. Yet in many conflict situations it is difficult to distinguish between flight because of persecution and departure caused by the destruction of the economic and social infrastructure needed for survival. Asylum seekers live in a drawn-out limbo situation, since determination procedures and appeals may take many years. In some countries, asylum seekers are not allowed to work, and have to exist on meagre welfare handouts. Up to 90 per cent of asylum applications are rejected – yet many asylum seekers cannot be deported because the country of origin will not take them back, or because they have no passports. For some, asylum seekers are a useful source of labour which fuels Western countries' burgeoning informal economies.

Annual asylum applications in Western Europe, Australia, Canada and the USA combined rose from 90 400 in 1983 to 323 050 in 1988, and then surged again with the end of the Cold War to peak at 828 645 in 1992 (UNHCR, 1995: 253). Applications fell sharply to 480 000 in 1995, but began creeping up again to 534 500 in 2000 (OECD, 2001: 280). Nearly the whole of the decline can be explained by falls in asylum applications following changes in refugee law in Germany (438 200 applications in 1992, but only 127 900 in 1995) and Sweden (84 000 in 1992, 9000 in 1995). The UK had relatively few asylum seekers in the early 1990s, with 32 300 in 1992, but numbers increased at the end of the decade to 55 000 in 1998 and 97 900 in 2000 (OECD, 2001: 280). However, only a small proportion of the world's total number of asylum seekers and refugees come to the highly-developed countries; most remain concentrated in the poorest countries.

Another type of forced migrant are the so-called 'environmental refugees': people displaced by environmental change (desertification, deforestation, land degradation, water pollution or inundation), natural disasters (floods, volcanoes, landslides, earthquakes), and man-made disasters (industrial accidents, radioactivity). A 1995 report claimed that there were at least 25 million environmental refugees, that the number could double by 2010, and that as many as 200 million people could eventually be at risk of displacement (Myers and Kent, 1995). Refugee experts reject such apocalyptic visions and argue that their main purpose is to shock Western governments into taking action to protect the environment. For instance, Black argues that there are no environmental refugees as such. While environmental factors do play a part in forced migration, displacements due to environmental factors are always closely linked to social and ethnic conflict, weak states and abuse of human rights. The emphasis on environmental factors is a distraction from central issues of development, inequality and conflict resolution (Black, 1998).

Forced migration has become a major factor in global politics. This is reflected in the changing nature of the international refugee regime since 1945. The international refugee regime consists of a set of legal norms based on humanitarian and human rights law, as well as a number of institutions designed to protect and assist refugees. The core of the regime is the 1951 Convention, and the most important institution is the UNHCR, but many other organizations also play a part: intergovernmental agencies like the International Committee of the Red Cross (ICRC), the World Food Programme (WFP) and the United Nations Children's Fund (UNICEF); as well as hundreds of non-governmental organizations (NGOs) such as OXFAM, CARE International, Médecins sans Frontières (MSF) and the International Rescue Committee (IRC).

The refugee regime was shaped by two major international conflicts: the Second World War, which left over 40 million displaced persons in Europe, and the Cold War. Many of the displaced persons were resettled in Australia, Canada and other countries, where they made an important contribution to post-war economic growth. In the Cold War, offering asylum to those who 'voted with their feet' against communism was a powerful source of propaganda for the West. Since the 'non-departure regime' of the Iron Curtain kept the numbers low, the West could afford to offer a warm welcome to those few who made it. The numbers rose following events like the 1956 Hungarian Revolution and the 1968 Prague Spring, but were still manageable.

Yet, meanwhile, very different refugee situations were developing in the South. The colonial legacy led to weak undemocratic states, underdeveloped economies and widespread poverty in Asia, Africa and Latin America. Northern countries sought to maintain their dominance by influencing new elites, while the Soviet Bloc encouraged revolutionary movements. Many local conflicts became proxy wars in the East–West struggle, with the superpowers providing modern weapons. Such factors gave rise to situations of generalized violence, leading to mass flight (Zolberg *et al.*, 1989). The escalation of struggles against white colonial or settler regimes in Africa from the 1960s, resistance against US-supported military regimes in Latin America in the 1970s and 1980s, and long-drawn-out political and ethnic struggles in the Middle East and Asia – all led to vast flows of refugees.

Northern countries and international agencies responded by claiming that such situations were qualitatively different from the individual persecution for which the 1951 Convention was designed (Chimni, 1998). The solution of permanent resettlement in developed countries was not seen as appropriate – except for Indo-Chinese and Cuban refugees who fitted the Cold War mould. In 1969, the Organization of African Unity (OAU) introduced its own Refugee Convention, which broadened the definition to include people forced to flee their country by war, human rights violations or generalized violence. A similar definition for Latin America was contained in the Cartagena Declaration of 1984.

The UNHCR began to take on new functions as a humanitarian relief organization. It helped run camps and provided food and medical care around the world. It became the 'focal point' for coordinating the activities of the various UN agencies in major emergencies (Loescher, 2001). This expanding role was reflected in UNHCR's budget, which tripled from

US$145 million in 1978 to $510 million in 1980 (UNHCR, 2000a), making it one of the most powerful UN agencies.

By the 1980s, increasing flows of asylum seekers were coming directly to Europe and North America from conflict zones in Latin America, Africa and Asia. Numbers increased sharply with the collapse of the Soviet Bloc. The most dramatic flows were from Albania to Italy in 1991 and again in 1997, and from former Yugoslavia during the wars in Croatia, Bosnia and Kosovo. Many of the 1.3 million asylum applicants arriving in Germany between 1991–5 were members of ethnic minorities (such as Roma) from Romania, Bulgaria and elsewhere in Eastern Europe. At the same time, increasing numbers of asylum seekers were arriving in Europe from the South. The situation was further complicated by ethnic minorities returning to ancestral homelands as well as undocumented workers from Poland, Ukraine and other post-Soviet states.

The early 1990s were thus a period of panic about asylum. Extreme-right mobilization, arson attacks on asylum-seeker hostels and assaults on foreigners were threatening public order. European states reacted with a series of restrictions, which seemed to herald the construction of a 'Fortress Europe' (UNHCR, 2000a; Keeley, 2001):

- Changes in national legislation to restrict access to refugee status.
- Temporary protection regimes instead of permanent refugee status for people fleeing the wars in former Yugoslavia.
- 'Non-arrival policies' designed to prevent people without adequate documentation from entering Western Europe. Citizens of certain states were required to obtain visas before departure. 'Carrier sanctions' were introduced, whereby airline personnel had to check documents before allowing people to embark.
- Diversion policies: by declaring Central European countries such as Poland, Hungary and the Czech Republic to be 'safe third countries', Western European countries could return asylum seekers to these states, if they had used them as transit routes.
- Restrictive interpretations of the 1951 UN Refugee Convention, for instance by excluding persecution through 'non-state actors' (such as the Taliban in Afghanistan).
- European cooperation on asylum and immigration rules, such as the Schengen Convention and the Dublin Convention. The 1997 Treaty of Amsterdam laid down a commitment to introduce common EU policies on immigration and asylum by 2004.

The North American experience was similar. Large numbers of persons from Central America fleeing conflict and persecution in their home countries began arriving in the US in the 1980s. Many of these did so 'illegally' as the USA did not recognize all Central American countries as refugee-producing countries. The open door policy towards Cubans fleeing to the USA, which had been in place since 1959, began to be restricted in the 1980s, and interdiction at sea commenced in the 1990s. Large numbers of Haitians attempting to come to the USA during the 1980s and 1990s were prevented from doing so.

Such restrictive measures – rather than real improvements in human rights – are the main reason why the number of officially recognized refugees worldwide has fallen since 1995. The refugee regime of the rich countries of the North has been fundamentally transformed over the last 20 years. It has shifted from a system designed to welcome Cold War refugees from the East and to resettle them as permanent exiles in new homes, to a 'non-entrée regime', designed to exclude and control asylum seekers from the South.

We will now look briefly at refugees and asylum seekers in various Western countries. Between 1975 and 2000 the USA provided permanent resettlement to over 2 million refugees, including some 1.3 million people from Indo-China. The USA accepted more people for resettlement during this period than the rest of the world put together (UNHCR, 2000b). The total of asylum applications rose from 75 600 in 1990 to a peak of 148 700 in 1995, then declined to 32 700 in 1999 before rising again to 59 400 in 2001. In the early 1990s, with a weak economy and growing numbers of undocumented migrants, there were strong anti-immigrant sentiments in the USA. These sentiments were reflected in the passing of the Illegal Immigration Reform and Immigrant Responsibility Act (IIRIRA) in 1996. The IIRIRA fundamentally changed the way in which the US government processed asylum claims and the rights afforded to asylum seekers. It created a new legal standard for screening asylum seekers arriving at US borders which aimed to determine whether they should be admitted to the asylum procedure. It also authorized 'expedited removals' and the detention of asylum seekers.

For the 1990–2001 period as a whole, six out of the top ten countries of origin of asylum seekers coming to the USA were Latin American or Caribbean countries. However, there were substantial fluctuations. In 1990, the great majority of asylum seekers came from Latin America. In 2001 by contrast, the top country of origin was Mexico, followed by China, while the rest of the top ten included a wide range of areas of origin. The asylum approval rate for Mexicans was only 7 per cent, compared with 64 per cent for Chinese and 57 per cent for asylum seekers overall (USCR, 2001: 275). After the terrorist attacks of 11 September 2001, the USA halted its refugee resettlement programme. Resumption was authorized in November, but the USA only admitted 800 refugees in the last three months of 2001. An Act of October 2001 introduced much stronger detention powers for non-citizens suspected of terrorist activities. Although these measures were not directed specifically against asylum seekers, it was feared that they might lead to an increase in the already substantial use of detention: an average of 3000 asylum seekers were in detention during 2001 (USCR, 2001: 279).

Like the USA, Canada accepted large numbers of people from Indo-China, some 200 000 between 1975 and 1995. During the 1980s, Canada offered resettlement to an average of 21 000 refugees per year. Between 1989 and 1998, resettlement admissions fell from 35 000 to under 9000. However, they rose to 17 000 in 1999 as a result of the humanitarian evacuation programme for refugees from Kosovo (UNHCR, 2000b). The number of asylum seekers coming to Canada declined from 36 700 in 1990 to 20 300 in 1993, increased again to 39 400 in 1999, and then reached its highest level ever of 44 000 in 2001. In contrast to the USA, only one out of the top ten countries of origin for the 1990–2001 period was a Latin American country: Mexico. Most asylum seekers in Canada are from the Indian subcontinent or China, with Sri Lanka topping the list. The general picture is one of considerable and increasing diversity over the whole period. In 2001, Canada decided 22 887 refugee claims, with an approval rate of 58 per cent. The highest approval rates were for Afghanistan (97 per cent), Somalia (92 per cent), Colombia (85 per cent), Sri Lanka (76 per cent), and Democratic Republic of Congo (76 per cent). The lowest success rates were for Hungary (27 per cent) and Mexico (28 per cent). (USCR, 2001: 261).

Australia has a Humanitarian Programme, designed to bring in refugees from overseas, with fairly constant targets of around 12 000 per year since the early 1990s. The number of asylum seekers arriving by boat in Australia without permission averaged only a few hundred per year up to the late 1990s, but went up to 4175 in 1999–2000 and 4141 in 2000–1 (Crock and Saul 2002: 24). Although these numbers are very low compared with other parts of the world, the growth is seen as undermining the tradition of strict government control of entries, which has hitherto been possible because of Australia's remote location. This has led to a marked politicization of refugee issues in Australia since the late 1990s, as will be described in detail in Chapter 9.

In the case of the EU, the top ten countries of origin of asylum seekers for the period 1990–2000 were the Federal Republic of Yugoslavia (FRY), Romania, Turkey, Iraq, Afghanistan, Bosnia-Herzegovina, Sri Lanka, Iran, Somalia and Democratic Republic of the Congo (DRC). The two peaks of asylum seekers from the FRY coincided with the wars in Croatia and Bosnia in 1991–3 and the war in Kosovo in 1998–9, with a total of 836 000 asylum seekers altogether entering from 1990–2000. The next country of origin is Romania, with a total of just under 400 000 concentrated overwhelmingly in the early part of the 1990s, at a time of marked persecution of Roma and other ethnic minorities. Next comes Turkey, with 356,000 asylum seekers quite evenly distributed across the period. Most appear to be Kurds, fleeing violent conflicts involving government forces in areas of supposed support for the Kurdish separatist party, the PKK.

Asylum-seeker flows from all the source countries show considerable fluctuations, linked to the development of internal conflicts and civil wars. Together, the top ten countries of origin accounted for about 2.5 million asylum seekers entering the EU from 1990 to 2000. This is 59 per cent of the total of 4.4 million asylum seekers in the period. Just over one-third of these asylum seekers came from three European countries: FRY, Romania and Bosnia-Herzegovina. The next ten countries of origin are Bulgaria, Pakistan, India, Nigeria, Russia, Vietnam, Algeria, China, Albania and Lebanon. The top 20 countries together make up 77 per cent of all asylum seekers entering the EU in the 11-year period.

In general, the top ten countries of origin of asylum seekers in the EU as a whole are also in the top 15 countries of origin for each individual country. However, there are significant national variations. These appear to be linked to a number of factors. The first is geographical position (or proximity): countries towards the eastern borders of the EU are more likely to receive asylum seekers from Eastern Europe, such as Russians and Bulgarians entering Finland and Austria. Southern EU countries like Greece are more likely to receive asylum seekers from south-eastern Europe (Albania, Romania) or the Middle East (Iraq, Iran, etc.). The second factor is pre-existing links, especially through a former colonial presence. For instance, Belgium is host to many asylum seekers from the DRC, its former colony of the Congo; France has many from Mali and Mauritania.

Regional integration

Has progress in regional integration in the post-Cold War period in Europe and North America helped or hindered the quest for migration control? Transatlantic comparisons need to be grounded in history as the evolution of the two regional integration processes are quite dissimilar. The European project is much older and more far-reaching than NAFTA and there is no inevitable logic that NAFTA will evolve in the same way as the EU has done. They are very different projects.

The EU and its predecessors stretching back to the European Coal and Steel Community (ECSC) of the early 1950s and the EC up to 1993 have always comprised a federalist project with an explicit commitment to eventual supersession of member-state sovereignty through creation of European institutions and governance. The project has always been security driven. Regional integration was above all a strategy to prevent the recurrence of war between member states. The treaty-based strategy adopted was first of all one of creating a European authority over the coal and steel sector, then followed by the creation of a common market, again governed by European institutions. The Single European Act (SEA) of 1986 aimed to achieve a more complete or genuine common market and paved the way for signature of the Treaty on European Union (TEU) in 1991 which resulted in reinforcement and expansion of federalist European institutions within a 15 member-state area. The 1997 Amsterdam Treaty refined and complemented the TEU.

Migration has figured importantly in the history of European integration. The Treaty of Paris of 1951 which created the ECSC barred restrictions on employment based on nationality for citizens of the six member states (Geddes, 2000: 45). The Treaty of Rome of 1957 envisaged the creation of a common market between the six signatory states. Under Article 48, workers from member states were to enjoy freedom of movement if they found employment in another member state. In the 1950s, Italy pushed for regional integration in order to foster employment opportunities for its large unemployed population (Romero, 1993). Other member states of the EC (and also within the predecessor to today's OECD) resisted. By 1968, when Article 48 came into effect, Italy's unemployment problem had eased. Italian citizens nevertheless were the major beneficiaries of Article 48 but relatively little intra-EC labour migration occurred despite expectations (Werner, 1973). It was firmly established that the freedom of labour movement provision applied only to citizens of EC member states, not to third country nationals from outside the EC.

The planned accession of Spain and Portugal to the EC in the mid-1980s sparked an important debate over the likely effects of their entry upon labour mobility. Some feared the rest of the enlarged EC would be flooded with Portuguese and Spanish workers. But at the end of a seven-year transition period, the predicted massive inflow did not occur. Instead, Spain and Portugal had become significant lands of immigration in their own right as EC and private investments poured into both countries. Intra-European capital mobility substituted for intra-European labour mobility (Koslowski, 2000: 117). Negotiations between the EU and Poland over planned accession sparked a similar debate in 2000–1. But, at the end of the proposed transition period, knowledgeable German officials do not anticipate a major influx of Poles. However, this did not stop the conservative candidate for the post of German chancellor in the September 2002 election, Stoiber, from warning of mass influxes from the East due to Germany's new immigration law. The use of the 'race card' as a measure of desperation in elections continues to be a dangerous tradition in European politics.

By 1990, the foreign resident population from other EC states had grown to over 5.5 million in a total population of 370 million, about 1.5 per cent of the EC's total population (Koslowski, 2000: 118). Under the TEU, resident aliens from other EC states were enfranchised to vote in their countries of residence (in local and European, but not national elections) and this was seen as an important aspect of the new European citizenship possessed by all citizens of the 15 member states. More problematic was the status of third country nationals who numbered 11.7 million in 1994, 3.15 per cent of the total EU population (Geddes, 2000: 11). They did not benefit from the freedom of movement extended to EU citizens. Instead, EU member states largely retained their prerogatives over entry, stay and removal of non-EU citizens. The European Commission, the supranational half of the EU's dual executive, favoured granting third country nationals freedom of movement within the European space but this was opposed in the Council of Ministers which represents the interests of member states. Among other objections, several EU member states argued that the European Commission proposal would further devalue the importance of national citizenship, something that European citizenship had not replaced but complemented. The 1997 Amsterdam Treaty gave the Council of Ministers five years to adopt measures to ensure absence of controls for third country nationals crossing internal borders (Geddes, 2000: 121).

The growing support for creation of a more genuine common market led France, Germany, Belgium, Luxembourg and the Netherlands to sign the Schengen Agreement in 1985. They committed themselves to hastening the creation of a border-free Europe, in which EU citizens could circulate freely, along with compensatory external frontier controls. The SEA of 1986 defined the single market as 'an area without internal frontiers in which the free movement of goods, persons, services and capital is ensured within the provisions of this treaty' (Geddes, 2000: 70). Many Europeans, including the governments of several EU member states, baulked at the idea of eliminating internal boundaries, fearing that it would lead to further illegal migration and loss of governmental control over entry and stay of aliens. Indeed, anti-immigrant parties such as the National Front, which were hostile to the EU for nationalist reasons, made opposition to such agreements part of their programmes. However, signatories of the Schengen Agreement retained the prerogative to reimpose frontier controls if warranted by circumstances (several signatories subsequently availed themselves of this option).

On 26 March 1995, the Schengen Agreement finally came into force for those signatory states which had established the necessary procedures: Germany, Belgium, Spain, France, Portugal, Luxembourg and the Netherlands. This meant complete removal of border controls for people moving between these countries. Effectively, the Agreement created a new class of 'Schengen citizens', to be added to the existing categories of EU citizens and non-EU citizens. The UK refused to join Schengen, insisting on continuing its own strict border controls of people coming from the Continent.

Has the elimination of internal borders within the European space resulted in loss of control over international movements of people, other than EU citizens? There is no conclusive evidence either way. The overall effect of the remarkable progress in European regional integration in recent years may be to have made the European quest for control more credible, as EU states participating in the Schengen group have been able to externalize control functions through the creation of a buffer zone in Central and Eastern Europe and of a common border in Southern Europe. The EU remains open to legal migration and porous to illegal migration. But the enlargements, treaties and institutional changes of recent years have been importantly affected by migration control concerns.

So too has NAFTA, although it contains only minor provisions relating specifically to population movements between Canada, Mexico and the USA. The origins of NAFTA had a great deal to do with the sudden progression of regional integration in Europe in the mid-1980s. Rightly or wrongly, many EC trading partners feared that Schengen and the SEA would lead to a Fortress Europe, a zone less accessible to exports from outside the EC. This perception helped hasten the signature of a US-Canadian free trade agreement in 1988. Meanwhile, a provision of IRCA authorized the creation of a Commission for the Study of International Migration and Cooperative Economic Development. This bipartisan group was to study 'push' factors motivating unauthorized migration to the USA from Mexico and other countries of the Western Hemisphere.

The report of the bipartisan Commission in 1990 favoured a more comprehensive approach to the prevention of illegal migration than was evidenced in the 1986 law which supported liberalization of trade with Mexico. President Salinas then approached the administration of President George Bush (Snr) with the idea of enlarging the US-Canadian free trade pact to include Mexico. President Bush referred the Mexican proposal to his National Security Council which gave its support. Thus, there was a security dimension to NAFTA but it differed quite significantly from the centrality of security concerns in European regional integration. NAFTA did not involve any explicit federalist project to foster socio-economic integration giving rise to cooperation and regional governance which would preclude war. The NAFTA treaty signed in 1993 and which came into effect on 1 January 1994 created a free trade area only. Even this much less ambitious project met with considerable political opposition in the USA and also in Mexico (where it helped spark the Zapatista revolt), but not in Canada.

Paradoxically, concerns over international migration figured centrally in NAFTA's genesis but scarcely at all in the text of the treaty. US and Mexican views on illegal migration were still sharply opposed. For Mexico, migration to the USA was driven by the US labour market demand. For the USA, much migration contravened its laws and arose from the paucity of socio-economic opportunities in Mexico. Simply put, the Mexican economy did not generate sufficient growth and jobs to employ its rapidly growing population. So gaping was the chasm between official US and Mexican perception of the illegal migration issue in bilateral relations that it was labelled a 'poison pill' that had to be avoided if negotiations were to proceed.

During the run-up to NAFTA, both the US president (by now, Bill Clinton) and President Salinas hailed the pact as a way to reduce illegal migration. Salinas warned that the USA would either get Mexican tomatoes or Mexican migrants who would pick them in the USA. Such presidential optimism over the migration-reducing impact of NAFTA belied the key research finding of the droves of studies generated by the US Commission studying international migration after IRCA, namely that trade liberalization between the USA and Mexico would diminish illegal migration only over the long term. Philip L. Martin later refined this finding into his theory of a 'migration hump' (Martin, 1993). Illegal migration from Mexico to the USA in fact grew significantly in the wake of NAFTA. Liberalization of the Mexican economy in the 1990s hit the poor and middle class in Mexico very hard. Farmers and their families in the formerly subsidized *ejido* (communal farms) sector were adversely affected and many moved northward, just as Martin had predicted (Martin, 1993: 101).

Overall, NAFTA survived the Mexican economic crash of 1995 (OECD, 1998a: 172–4). A massive US government-led response played a decisive role. NAFTA has resulted in significantly expanded trade between the signatory states and greater socio-economic interdependence. The historic election of Vincente Fox as president of Mexico in 2000 ushered in a new era. As summarized in Box 1.2, the new Mexican president and his newly-elected American counterpart sought a new departure in US-Mexico relations, and more specifically on bilateral migration issues, but their initiative faltered following 11 September 2001.

President Fox and his foreign minister repeatedly referred to the European experience and called for freedom of migration within NAFTA. Apparently the goal is to achieve mobility for workers akin to the rights of European citizens under Article 48 of the Treaty of Rome. Yet, as argued above, the two regional integration projects differ markedly. The North American situation differs also in that the region is so dominated by the US economy and there is a huge economic gap between Mexico and the US, but much less so between the US and Canada (OECD, 1998a: 7). The evocation of a European referent betrayed a grievous misunderstanding of the history of regional integration in Europe. EC member states resisted Italian efforts to export its unemployed northward in the immediate post-war era. The EU decided against the accession of Turkey in large part because of concerns that Turkish membership would result in further mass emigration even after a transition period (Martin, 1991). Morocco's bid, for similar reasons, went nowhere. No EU member state, if placed in the same situation as the USA, would contemplate freedom of movement for workers because socio-economic disparities are too great between the US and Mexico. The notion would be a non-starter.

Regional integration in North America and Europe, thus, has had important implications for governmental control strategies. The two projects and their historical and institutional contexts vary greatly but they comprise a salient dimension of overall strategies to reduce illegal or unwanted migration. It may be that NAFTA will evolve into something more akin to the EU, but the first seven years after the entry into force of the agreement witnessed a significant increase in illegal migration from Mexico to the USA.

The 'migration industry'

One reason why official migration policies often fail to achieve their objectives is the emergence of the so-called 'migration industry'. This term embraces the many people who earn their livelihood by organizing migratory movements as travel agents, labour recruiters, brokers, interpreters, and housing agents. Such people range from lawyers who give advice on immigration law, through to human smugglers who transport migrants illegally across borders (like the 'coyotes' who guide Mexican workers across the Rio Grande, or the Moroccan fishermen who ferry Africans to Spain). Banks become part of the migration industry by setting up special transfer facilities for remittances. Some migration agents are themselves members of a migrant community, helping their compatriots on a voluntary or part-time basis: shopkeepers, priests, teachers and other community leaders often take on such roles. Others are unscrupulous criminals, out to exploit defenceless migrants or asylum seekers by charging them extortionate fees for non-existent jobs. Yet others are police officers or bureaucrats, making money on the side by showing people loopholes in regulations.

The development of the migration industry is an inevitable aspect of the social networks and the transnational linkages which are part of the migratory process (see Chapter 2). Whatever its initial causes, once a migration gets under way, a variety of needs for special services arise. Even when governments initiate labour recruitment, they rarely provide all the necessary infrastructure. In spontaneous or illegal movements, the need for agents and brokers is all the greater. There is a broad range of entrepreneurial opportunities, which are seized upon by both migrants and non-migrants. The role of the agents and brokers is vital: without them, few migrants would have the information or contacts needed for successful migration.

In time, the migration industry may become the main motive force in a migratory movement. If the government concerned then decides to curtail migration, it could run into difficulties. The agents have an interest in the continuation of migration, and go on organizing it, though the form may change (for example, from legal worker recruitment to illegal entry). Harris (1996: 135) characterizes migration agents as 'a vast unseen international network underpinning a global labour market; a horde of termites ... boring through the national fortifications against migration, and changing whole societies'.

Human smuggling and trafficking

A disturbing and increasingly salient element of the migration industry is the rise of organizations devoted to the smuggling and trafficking of migrants. It is important to distinguish between people-trafficking and people-smuggling. Formal definitions are embodied in two international treaties, known as the 'Vienna Protocols' adopted by the UN General Assembly in 2000. According to Ann Gallagher (2002) of the UN High Commission for Human Rights:

> Smuggled migrants are moved illegally for profit; they are partners, however unequal, in a commercial transaction... By contrast, the movement of trafficked persons is based on deception and coercion and is for the purpose of exploitation. The profit in trafficking comes not from the movement but from the sale of a trafficked person's sexual services or labour in the country of destination.

The trafficking of women and children for the sex industry occurs all over the world. Thai and Japanese gangsters collaborate to entice women into prostitution in Japan by claiming that they will get jobs as waitresses or entertainers. Victims of civil war and forced displacement in former Yugoslavia, Georgia or Azerbaijan are sold to brothels in Western Europe. Women in war zones are forced into sex slavery by combatant forces, or sold to international gangs. It is impossible to quantify the number of people affected by trafficking and smuggling, but both are widespread practices. Clients of smuggling gangs include not only economic migrants, but also legitimate refugees, unable to make asylum claims because restrictive border rules prevent them entering countries of potential asylum (Gibney, 2000).

A comparison of trends discernible in Germany and the USA give credence to the widely-shared supposition that human smuggling and trafficking have grown exponentially. German statistics on smugglers and smuggled aliens recorded an increase from 1847 and 1794 respectively in 1990 to 3162 and 12 533 respectively in 1998 (IOM, 2000a: 32; Bundesgrenzschutzamt, 2001). An authoritative US government report revealed increased smuggling: between fiscal years 1997 and 1999, the percentage of aliens smuggled rose from 9 per cent of all border patrol apprehensions to 14 per cent. In fiscal year 1999, the INS arrested 4100 smugglers and over 40 000 smuggled aliens. INS prosecuted 2000 smugglers and 61 per cent were convicted, receiving an average sentence of ten months and an average fine of US$140 (US General Accounting Office, 2000: 2).

Global human trafficking is thought to involve millions of persons. A UN under-secretary general held that 200 million persons worldwide were involved in some manner. He stated 'this is the fastest growing criminal market in the world because of the number of people who are involved, the scale of profits being generated for criminal organizations and because of its multifold nature' (Crossette, 2000; see also Parisot, 1998). The human trafficking industry may generate profits of US$5 to 10 billion per year (Martin and Miller, 2000a: 969). One US study estimated that out of the 700 000 to 2 million women trafficked globally each year, as many as 50 000 were trafficked to the USA (Richard, 1999: 3). The evidence of the disproportionate involvement of women and children in human trafficking to the USA is echoed in numerous reports and surveys from around the world.

The IOM, which made combating human trafficking one of its priorities in the 1990s, gave multiple reasons for the post-Cold War upsurge in human trafficking (IOM, 1999: 4). Migrants driven by war, persecution, violence and poverty search for better opportunities. Sometimes wilfully, they accept the services of the trafficker. But, in other instances, they are deceived into accepting trafficker services by promises of good jobs and salaries. Possibilities for legal immigration have declined. Anti-trafficking legislation is often absent or deficient and enforcement of anti-trafficking laws is insufficient. Many aliens subject to deportation avail themselves of trafficker services

As understanding of the dimensions of the human trafficking phenomenon grew, many government, non-governmental organizations and international organizations responded – a key illustration of the global governance imperative discussed in Chapter 1. Koslowski argued that policy-making evolved in three steps (Kyle and Koslowski, 2001: 342–7). Post-Cold War foreign policy-making was restructured to elevate international migration and transnational crime to 'high politics', that is a question germane to security, while the two were placed together on the agendas of new institutional frameworks. Subsequent policy-making reinforced and justified the new institutional linkages. Lastly, policy-makers increasingly turned to multilateral cooperation on regional and global levels to combat human trafficking.

A flurry of meetings, fora, new laws and modified enforcement resulted. But it was unclear how they affected human trafficking. One report contended that additional restrictive measures in the EU created greater demand for trafficker services (Morrison, 1998). A number of analysts attributed the deaths of 58 Asians being transported to Dover, UK, in 2000 to restrictive laws. A consequence, which was almost certainly unintended, of stricter laws and bolstered enforcement was a growing demand for smugglers and traffickers, and the clientele often included persons, such as Kurds fleeing Iraq, who might have valid claims for refugee status (Kyle and Koslowski, 2001: 349).

A number of national governments, including the US, instituted new procedures to aid the combating of human trafficking. A new category of visas was established for victims of trafficking who aided US authorities in the prosecution of traffickers. The USA also began to monitor anti-trafficking efforts by governments around the world and issued an annual report on the status of such efforts. The 2002 report estimated that 700 000 to 4 million people worldwide were victimized and placed 19 governments, including a number of key US allies like Saudi Arabia and Turkey (Purdum, 2002), in the category of states that made no real effort to stem human trafficking. The US law which obligated the US Department of State to issue the report did not mandate counter-measures against the 19 states. The 2001 report had listed even more states, a total of 23.

Some countries, such as the People's Republic of China, instituted very harsh measures to counter human trafficking, including capital punishment (Chin, 1999: 200). Studies of smuggled and trafficked Chinese revealed complex global networks which were difficult to dismantle through law enforcement. Lower echelon 'snakeheads' might be apprehended and punished but higher echelon criminals were more elusive. Koslowski assessed the overall picture of the effectiveness of anti-smuggling and human trafficking counter-measures as 'quite dim' (Kyle and Koslowski, 2001: 353). He noted, however, that fear of organized crime might 'galvanize serious international co-operation to curb human smuggling'.

Summing up: restriction and 'root causes'

Illegal migration to industrial countries increased after 1973. Together with the upsurge in refugee and asylum-seeker entries from the mid-1980s, it became a focus for aggressive campaigns from the extreme right. This contributed to the politicization of migration issues, and helped increase the pressure for migration control. Current political initiatives take two forms: one is a further tightening of restrictive measures, the other is the attempt to address what is often referred to as the 'root cause' of mass migration: the South–North divide.

Since the early 1990s, most OECD countries have changed laws and entry procedures, introducing such measures as stricter border controls, visa requirements, penalties for airlines which bring in inadequately-documented passengers, identity checks, workplace inspections, techniques for the detection of falsified documents and more severe penalties for those caught infringing regulations. Controls were devised to prevent people obtaining the papers needed to get work, social security benefits, schooling or medical services (for changes in various countries see OECD, 1997: 52–7). To give a few examples: the Netherlands reinforced rules for the detention and deportation of illegal residents, and introduced a new Aliens Employment Act in 1995. In Norway a central bureau for the fight against illegal immigration was established to centralize information provided by the police, national and foreign governments, airline companies and embassies. French authorities required employers to notify them prior to hiring a foreign worker in order to verify employment eligibility. The USA and Canada took steps to make it harder for immigrants with low income levels to bring in relatives.

The collapse of the Soviet Bloc made regulation of migration even more urgent. The 1991 European Convention on Security and Cooperation was intended to create a zone in which basic human rights and minimum living standards were respected, so that individuals from this area would have no basis to apply for asylum. The model was Poland, from which so many asylum seekers had come in the 1980s. By the 1990s, it was considered that requests by Poles for asylum could be uniformly denied. Further agreements were made in 1991–2 with a number of Eastern European countries on preventing use of their countries for transit of illegal migrants to Western Europe. These countries also signed the United Nations Refugee Convention and themselves began receiving asylum seekers. The European Convention on Security and Cooperation was supposed to prevent developments which would lead to mass population movements. The war in former Yugoslavia, which produced millions of refugees, questioned the efficacy of this mechanism to ensure security.

This general climate of restrictiveness led some observers to speak of a 'Fortress Europe' building walls to keep out impoverished masses from the South and East. Yet in the international debates on mass migration there has been growing agreement that entry restrictions could have only limited success. The amount of control and surveillance needed to make borders impenetrable is inconsistent with the trend towards increased interchange and communication. This is why sanctions against unauthorized employment have become crucial to control strategies. Some scholars believe that the disparities between economic and social conditions in the South and the North are such that illegal migration is likely to grow whatever the barriers (Cornelius *et al.*, 1994). Others are more circumspect and point out that some Western countries have developed a considerable capacity to deter illegal migration over the last quarter of a century (Miller, 1994; Messina, 1996). Nonetheless, there is an emerging consensus about the need to address the 'root causes' of mass migration by supporting efforts to improve conditions in the countries of origin.

Measures to reduce migration mean not just development aid, but also foreign and trade policy initiatives, designed to bring about sustainable development, and to improve political stability and human rights. This has long been understood in the USA. Yet during the Reagan presidency from 1980 to 1988, US foreign policy was hostile to the North–South dialogue (M. J. Miller, 1991: 36). US involvement in various insurgencies and civil wars at times exacerbated political instability, generating additional migrants. Ironically, US intervention, specifically in Central America, was partly justified by what was termed 'the fear of brown bodies'. In other words, if the USA did not quell insurgencies, they might succeed and the result would be millions of additional refugees from the new regimes: the Cuba scenario.

One indication of a changing approach was the conclusion drawn by a US federal study commission in 1990 that 'development and the availability of new and better jobs at home is the only way to diminish migratory pressures over time' (CSIMCED, 1990: xiii). But the report also found that development would increase international migration to the USA over the short to medium term. The persistence of illegal migration to the USA despite the legalization and the imposition of employer sanctions was taken by many as proof that a new strategy of 'abatement' was needed to replace or complement a strategy of deterrence. Such considerations played a major part in the discussions which led up to NAFTA in 1993. NAFTA did in fact spur investment and job creation in Mexico, especially along the border with the USA. But there is no evidence that it has yet had any appreciable effect in reducing the economic and demographic pressures which cause Mexicans to emigrate.

In Western Europe, too, immigration concerns loomed much more prominently in foreign policy debates than they had in the past. There was concern in Eastern Europe and North Africa that the Single European Market would create greater barriers to external trade, thus keeping out products from less-developed areas. In 1996, the EU established a customs union with Turkey and reached separate agreements with Morocco, Algeria and Tunisia to remove barriers to trade. As with NAFTA, however, it was far from clear whether the measures would reduce 'unwanted' migration in the long run. It was feared, for instance, that trade liberalization would endanger many medium-sized North African firms, while barriers to North African agricultural exports to the EU remained.

Three types of international cooperation that might abate migration to Europe have been suggested (Böhning, 1991b: 14-15). The first is trade liberalization. The problem is that the Western European industries that would be most affected by trade liberalization, such as agriculture and textiles, are precisely those which have been highly protected. Therefore it would appear that Western European governments possess only a limited capacity to liberalize trade, a conclusion confirmed by the trade dispute between the USA and the EC during the Uruguay Round of the General Agreement on Tariffs and Trade (GATT) negotiations in the early 1990s. The second means of international cooperation is direct foreign investment. But governments exercise little control over investors, who follow considerations of profit. Then there is foreign aid, which could abate international migration if designed to improve rapidly the economic and social conditions for people otherwise likely to depart. However, in the past, foreign aid has generally done little to improve average living conditions. If it is to help reduce migration, future aid policy will have to be much more concerned with social and demographic issues. Above all, a halt to military aid is essential.

The terrorist attacks of 11 September 2001 did little to alter the overall prognosis. UN Secretary General Kofi Annan convened a UN summit on global poverty in Monterrey, Mexico, in 2002 at which he urged donor states to increase their assistance to the world's poorest states, in part to address root causes of violence, instability and terrorism. He clarified that terrorism's roots are broader than poverty but 'where massive and systematic political, economic and social inequalities are found, and where no legitimate means of addressing them exists, an environment is created in which peaceful solutions all too often lose out against extreme and violent alternatives' (Crossette, 2002a).

To the surprise of many, the Bush Administration reversed itself and pledged to increase US foreign assistance by 14 per cent. The USA had spent .01 per cent of gross domestic product on foreign assistance previously (Blustein, 2002). However, only a few months later, the US president signed into law a new farm bill which massively subsidized US agriculture to the detriment of agricultural exports from around the world. The new US government largesse towards US agriculture reflected the usual pattern of lobbying and calculation of possible electoral gain that repeatedly stymies efforts to address root causes of growing illegal migration and human trafficking in addition to terrorism. While the months following the 11 September attacks witnessed tightening up of various elements of migration controls, such as issuing of visas and border enforcement, there was little evidence of the sea change required to make the quest for control more coherent and credible.

Guide to further reading

Concerning state capacity to prevent illegal migration and employment, see Cornelius *et al.* (1994 and 2003), Harris (1995), Bernstein and Weiner (1999), Martin and Miller (2000b), OECD (2000) and Zolberg and Benda (2001). On migration and regional integration in Europe and North America, see OECD (1998a), Geddes (2000) and Koslowski (2000). On human smuggling and trafficking, see Kyle and Koslowski (2001) and IOM (2000a). On legalization policies, see Miller, M. J. (2002), Bernstein and Weiner (1999) and OECD (2000).

Chapter 6

The Next Waves: The Globalization of International Migration

The North–South gap – the differentials in life expectancy, demography, economic structure, social conditions and political stability between the industrial democracies and most of the rest of the world – looms as a major barrier to the creation of a peaceful and prosperous global society. International migration is a major consequence of the North–South gap. However, the world can no longer be simply divided up between rich and poor nations. Long before the end of the Cold War, new poles of financial, manufacturing and technological power had emerged in the oil-rich Arab states and in East Asia. Oil-producing areas outside the Arab region, such as Nigeria, Venezuela and Brunei, have also become important areas of immigration. A wide range of industries attracts migrant workers: agriculture, construction, manufacturing, domestic services and more. Economic and social dislocation, political unrest, authoritarian rule and technological backwardness create conditions conducive to emigration.

The focus of this chapter is on current trends in international migration to, from and within the Arab, African and Latin American regions. The next chapter will deal with the Asia-Pacific region, which is home to more than half the world's people. The next waves of migrants will mainly come from these areas. Much of the international migration will continue to be intra-regional, but many migrants will also desire to go to Europe, Australia or North America. Understanding migration within the South is an essential precondition for formulating the future policies of the developed countries.

The Arab region

We use the term 'the Arab region' to include not only the Arab countries of North Africa and Western Asia, but also the non-Arab states of Turkey, Iran and Israel. The term refers to a geographical, rather than a political or ethnic area. The term 'the Arab region' is not entirely satisfactory, but seems better than the possible alternatives: the term 'Middle East' is Eurocentric and excludes North Africa; the term 'Western Asia' also excludes North Africa.

The area from the Atlantic beaches of Morocco to the western borders of Afghanistan and Pakistan is one of enormous diversity. Within it there are four key migrant labour subsystems: emigration from the Mediterranean littoral to Western Europe, Arab labour migration to oil-producing states, migration to non-oil producing Arab states, and East and South Asian labour migration to oil-producing states. The first three movements will be dealt with here, while Asian migration to the oil states will be discussed in Chapter 7. There are also large refugee flows, especially to Iran and Turkey, and mass immigration for settlement to Israel. On the whole, the situation of migrants in the Arab region is characterized by extraordinary deprivation of basic rights. Forced migration is prevalent and linked to the failure of political systems (Shami, 1994: 2).

There is no direct causal link between rapid population growth and international migration (Kritz, 2001). However, the swath from Morocco to Turkey is one of the world's most demographically fertile areas. There is enormous population growth and most of the population is young. Areas like Beirut, Gaza and the lower Nile Valley are very densely populated. Population density and the gap between job creation and the entry of new cohorts into the labour market help propel emigration. Nearby are lightly populated desert wastelands and zones of rapid economic growth possible only through massive recruitment of foreign labour. A 2002 United Nations Development Programme Report on the Arab states noted that average income of Arab citizens was 14 per cent of average income in the OECD area. It linked the anaemic economic growth of the preceding two decades, an average of 0.5 per cent, to authoritarian governance. Half of Arab youths expressed a desire to emigrate (*The Wall Street Journal*, 8 July 2002: A22).

North Africa and Turkey: still Western Europe's labour reservoir?

By 2002, Morocco and Turkey had the largest populations of expatriates in the EU, with 2 and 3 million of their citizens respectively residing as third country nationals (Belguendouz, 2001: 4; OECD, 2001: 253). Large Moroccan communities lived in Italy, Spain and the Netherlands in addition to France, where just over 500 000 Moroccans resided in 1999 (Lebon, 2000: 11). The vast majority of Algerians and Tunisians, 500 000 and 150 000 respectively, lived in France. The colonial legacy shaped all North African flows, but less completely in the Moroccan case.

The greater dispersion of the Moroccan population was linked to the French decision more or less to suspend further recruitment of foreign labour in 1974. In 1973, Algeria unilaterally suspended further emigration of its citizens for employment in France following a wave of violence against Arabs. Suddenly deprived of legal access to the French labour market, except for a small contingent of seasonal workers, Moroccans sought employment in Italy and Spain. Family reunion-related Maghrebi immigration to France continued when a French government effort to halt further family reunification was thwarted by a legal ruling.

Franco-Algerian tensions over immigration intensified during the presidency of Valery Giscard d'Estaing (1974–81), who sought to deny renewal of residency and employment authorization to several hundred thousand Algerians. It was in this context that the Algerian President Houari Boumédienne declared that nothing could stop the movement of persons from places like Algeria northward to Europe. His warning stoked French fears of uncontrollable migration from the South that were later amplified through books like Jean Raspail's influential *The Camp of the Saints* and films like *The March* (Zolberg and Benda, 2001: 1–15). The Algerian president then died unexpectedly and Algeria soon began its long descent into civil war, a conflict that took 120 000 lives between 1991 and 2001 (Spencer, 2002: 44).

But Boumedienne's apocalyptic vision did not prove prophetic. The election of President Mitterrand and the Socialists in 1981 eased Franco-Algerian tensions as the French government abandoned the effort to induce Algerian repatriation through non-renewal of permits. A decade later, the intensification of civil war in Algeria found the French government bracing for a huge influx of Algerians. But few Algerians were able to leave Algeria despite the wishes of many to do so, particularly among the young.

France severely restricted issuance of visas to Algerians, making it very difficult for Algerians to leave even if they had family members residing in France. Morocco and Tunisia closed their borders with Algeria to prevent the spread of the conflict and were quite successful in doing so. Some Algerians were able to emigrate and applied for asylum in Western Europe or North America, but they were the exceptions, not the rule. When considered along with the evidence concerning the feared *Võlkserwanderung* from Eastern Europe after 1989, which also did not materialize on the scale feared, the evolution of Algerian emigration in the 1990s disproves the categorical contention that states cannot regulate international migration. If that were true, there would have been much more Algerian emigration.

Migration did play a role in the civil war, as returning Afghanis, Algerian volunteers who fought against Soviet troops and allies in Afghanistan during the 1980s, returned to Algeria in the wake of the Soviet evacuation. They contributed importantly to the intensification of hostilities around 1990 (A. Miller, 2002: 75). And, as considered in detail in Chapter 11, the fighting in Algeria spilled over to France in the mid-1990s. Many Algerians and other North Africans were arrested for suspected ties to Al-Qaida or other radical Muslim organizations in the aftermath of 11 September 2001. But they comprised a tiny minority of North African-origin populations in the transatlantic area.

In December 2001, Algeria signed an accord with the EU, following in the footsteps of Morocco and Tunisia (Mohsen-Finan, 2002: 94). The free trade agreements call for free movement of goods, services, firms and capital and adopt certain competition rules of the EU. How the evolving relationship with the EU would affect population movements was unclear. The short- to medium-term prospect was that the accords would intensify North African emigration (White, 1999: 839–54).

Although the level of violence subsided in Algeria by 2000, enormous political and socio-economic problems remained. Thirty per cent of the economically-active population were unemployed as were over 60 per cent of Algerians under age 30 (Mohsen-Finan, 2002: 28). The democratic legitimacy of President Bouteflika was dubious and the armed forces remained dominant in government. A regional initiative begun in 1995 in Barcelona, Spain, had yielded modest results. The overall track record of international cooperation on migration matters, when viewed from a North African perspective, already was largely negative by 1990 (Boudahrain, 1991). Little had changed a decade later except for a broader recognition that North African migration questions centrally affected Western Mediterranean and transatlantic security.

The conferral of EU partnership status was part of an EU strategy to stabilize the Western Mediterranean and reduce population movements towards the EU. Tunisia, Algeria and Morocco came under pressure to cooperate. Algeria obliged by stepping up border enforcement along its long and porous borders with Libya and Niger. Migrants from sub-Saharan Africa travelled through Niger to Libya, and sought to transit through Algeria and Morocco to Europe. Others continued on to Europe from Libya. The dimensions of these movements are unknown, but they are very large and have resulted in many deaths, as African migrants perish in vast deserts. In one such incident near Dirkou in the north of Niger in 2001, at least 23 migrants died but the toll may have been much higher (Bensaâd, 2002: 15). Many other sub-Saharan Africans and North Africans perish at sea in attempting to reach Europe. In 2001, the Association of Moroccan Immigrant Workers in Spain estimated that nearly 4000 Moroccans alone had died trying to cross the Straits of Gibraltar over the preceding five years (Belguendouz, 2001: 5).

Some measure of the dimensions of sub-Saharan African flows to Morocco could be discerned through Moroccan immigration enforcement statistics which were published for the first time in 1999. In 1995, 444 sub-Saharan Africans were detained for attempting illegal migration. In 2000, 10 000 were detained. Overall, Moroccan security services arrested 19 037 persons attempting to emigrate illegally in 1999 and 25 613 persons in 2000.

In the first eight months of 2001, a total of 20 995 persons had been arrested, of whom 11 716 were Moroccans (Belguendouz, 2001: 14).

Moroccan relations with Spain and the EU deteriorated after signature of the partnership accord. Disagreements over population movements figured centrally, but there were also disputes over fishing rights, the former Spanish Sahara and Spanish possession of a small island just off the coast of Morocco. The migration-related tensions were highly significant because Morocco had long cooperated closely with France and the EU on regulation of international migration. Moroccans and sub-Saharan Africans had long transited through Spain to points in Europe, mainly to France prior to 1973 and afterwards Italy. But Spain's adhesion to the Schengen Agreement in 1991 led to the imposition of visas upon Moroccan citizens. This development roughly coincided with the first *pateras* carrying migrants to Spain (Belguendouz, 2001: 12).

Moroccans were the major beneficiaries of Spanish legalization policies but a new Spanish Conservative government declared that it would no longer legalize aliens. The Spanish-backed plan adopted at the Tampere summit of EU heads of state and government in 1999 rankled Moroccan authorities (Belguendouz, 2001: 7). Instead of dialogue, the Spanish government convoked the Moroccan ambassador in 2001 to complain about illegal migration and the insufficiency of Moroccan efforts to stem it. Morocco subsequently recalled its ambassador.

The EU plan expected Morocco to impose visa obligations on sub-Saharan African states where visa issuance was not obligatory. Morocco objected that this would adversely affect its relations with these states and would thereby affect its position concerning the Spanish Sahara which Morocco unilaterally annexed in two stages in 1976 and 1978. Moreover, Morocco was to sign readmission agreements with African states to facilitate repatriation of African migrants detained in Morocco. Nevertheless, Spain and Morocco did sign an agreement in 2001, similar to accords signed by Spain with Colombia and Ecuador, to authorize recruitment of 10–20 000 Moroccan workers annually (Belguendouz, 2001: 5).

Morocco wanted EU-Morocco migration dialogue to include measures against racism and discrimination as Moroccans were the principal victims of violence in Spain. The Moroccans view migration principally as a socio-economic phenomenon, not as a security matter, and the economic effects of Moroccan emigration were substantial. In 2000, Morocco received 21 billion dirhams in remittances from its citizens abroad (nearly US$2 billion). This total represented the equivalent of one-third of the total value of Moroccan exports and double the value of direct foreign investment received. Bank savings of Moroccans abroad amounted to over 50 billion dirhams (US$4.7 billion), about 40 per cent of all savings in Moroccan banks (Belguendouz, 2001: 4).

Migrant remittances similarly had an important effect on the economies of Tunisia and Algeria, although many cash flows to Algeria go unrecorded. Already by 1985, total transfers resulting from North African emigration to Europe stood at US$4–5 billion (Simon, 1990: 29).

The situation of Turkey both resembled and differed from the Maghrebi cases. Like Morocco, Turkey had a large expatriate population in the EU. It had received possibly as many as 4 million refugees and migrants fleeing political instability and economic distress in its region. In 1995, Turkey graduated from EU association status by signing a treaty with the EU that created a customs union between Turkey and the 15 EU member states. But Turkey continued its quest to achieve full membership in the EU. Of all candidate states, Turkey's prospects seemed bleakest, suggesting that the relationship between labour migration and regional integration was more complex than frequently assumed. In Turkey's case, extensive labour emigration may have become an impediment to its entry into the EU. In the 1990s, Turkey's asylum policy was heavily criticized by Europeans before reforms were undertaken to align Turkey's policies with those of the EU.

A key difference between Turkey and the North African states arose from Turkey's Ottoman heritage. Traditionally, the Turkish Republic's immigration policy bore some resemblance to Israel's law of return or Germany's policy towards ethnic Germans in Eastern Europe. As the Ottoman Empire expanded, the government ordered Muslim subjects to settle in recently acquired areas, a process termed 'sürgün' (Tekeli, 1994: 204–6). As the empire contracted, these settlers and their descendants frequently become victims. Between 1821, the start of the Greek War of Independence; and 1922, when the Turkish Republic emerged, an estimated 5 million Ottoman Muslims were killed and over 5 million were driven from their homes, mainly to the Anatolian Peninsula (McCarthy, 1995). Henceforward, the population of the Turkish Republic has always had large numbers of expellees and their descendants. In 1934, a Law of Resettlement was promulgated which authorized ethnic Turks from areas formerly comprising the Ottoman Empire to emigrate to and settle in the Turkish Republic (Tekeli, 1994: 217).

As recently as the 1980s, 300 000 ethnic Turks from Bulgaria fled to Turkey to avoid persecution. Many of them later returned to Bulgaria. Some Muslims displaced by conflicts in the Western Balkans in the 1990s also found refuge in Turkey. By 2000, Turkey had rewritten its immigration law so as to extend asylum beyond the traditional preference given to ethnic Turks from abroad.

This legal change reflected the new regional migration dynamics which made Turkey into a land of immigration as well as an important site for transit migration towards Europe. In the 1980s and 1990s, Turkey received several million Iranians, Afghanis and Iraqis who were not formally recognized as refugees but whose residency was tolerated by the Turkish government. Turkey's bid for full membership in the EU would necessitate further reform of Turkish immigration and refugee policies, perhaps ending the informal policy of toleration. Such a policy change would have significant implications for the entire region. Proposed legislation aimed at curbing illegal migration through measures like enactment of employer sanctions (OECD, 2001: 254).

The unstable situation in Iraq was of particular concern to Turkish authorities. The implications of the Kurdish situation for international migration are considered in Box 6.2.

The interdependency woven by decades of international migration between Mediterranean littoral states and Western Europe meant that they had forged a common future. However, it seems likely that a high degree of trans-Mediterranean tensions over international migration will persist. Only time will tell whether these tensions, which are at the frontline of the global North–South divide, will be attenuated through skilful government or exacerbated by the absence thereof. The conduct of the war against terrorism appeared fated greatly to affect the geo-strategic implications of population movements between the two zones.

Arab migration to oil-rich Arab states

Movements, mainly of male workers, from the poorer to the richer Arab states have taken on enormous political significance in this volatile region. In the 1970s and 1980s, Libya admitted large numbers of Egyptians and Tunisians. But when Egypt-Libya relations soured as Egyptian President Anwar al-Sadat reoriented foreign policy towards the West, thousands of Egyptian migrants were expelled. The Egyptian government received 18 000 complaints from ex-migrants after a 1985 crisis and agreed to compensate 6000 migrants for the financial losses they suffered (Farrag, 1999: 74). A similar fate befell Tunisian workers during a period of tensions, exacerbated by a foray by armed Tunisian insurgents from Libya that was crushed by the Tunisian government with French military assistance. After Yasser Arafat signed the Oslo Accords in 1993, thousands of Palestinians were ordered to leave. Many were stranded for months at the Egyptian-Libyan border as Israel refused to accept their 'repatriation' to the area controlled by the Palestinian Authority.

Since 1989, citizens of the four other Maghreb Union states (Morocco, Tunisia, Mauritania and Algeria) have theoretically been able to enter Libya freely, under the terms of the Treaty of Marrakesh which created the Union of Arab Maghreb. But the regional integration framework has had little effect (Safir, 1999: 89). Libya enjoyed the highest per capita income in Africa, estimated at US$5410 in 1990 (Farrag, 1999: 81). Migration to Libya by sub-Saharan Africans became more prevalent in the 1990s as

Libya's influence in African politics and diplomacy grew, in counterpoint to its poor relations with the USA and the UK. The influx of sub-Saharan Africans was largely tolerated by the Libyan government, which had launched a project to green the desert. The project created demand for labour in very difficult conditions (Bensaâd, 2002: 15). Concurrently, Arab migrant workers, especially from the Sudan and Egypt, had become a base of support for the anti-Qaddafi, radical Islamist opposition. (Silvestri, 1999: 167).

Sporadically, as had been the case earlier with Egyptians and Tunisians, violence erupted. Hundreds of Africans were killed in a massacre in 2000 (Bensaâd, 2002: 19). The killings were followed by a massive exodus of Africans, some of whom later returned to Libya. Libyan authorities detained thousands of African migrants in detention camps in the south of Libya, where some migrants were executed for rebellion or attempting to flee (Bensaâd, 2002: 19). Nevertheless, the influx continued as desperate Africans risked everything in the hopes of finding a job in Libya or going on from there to Europe.

Libya provides an extreme example of the interconnection between international migration and foreign policy issues. The mass expulsions also testify to the disregard of Libyan authorities for Arab League and ILO standards. As the Moroccan scholar Abdullah Boudahrain has argued, disregard for the rights of migrants is commonplace in the Arab world despite the existence of treaties designed to ensure protection (Boudahrain, 1985: 103–64).

In Iraq, the ruling Ba'ath Party regards freedom of entry, residence and employment for non-Iraqi Arabs as consistent with the ideal of pan-Arab unity and nationhood, notions also embraced by the Libyan government. In 1975, Iraq signed an agreement with Egypt to encourage settlement of Egyptian farmers in Iraq. But Iraq unilaterally renounced the agreement in 1977 after Egyptian President Sadat's historic visit to Israel (El Sohl, 1994: 123). Nevertheless, Iraq allowed several million mainly unskilled Egyptians to take up employment and residency in the 1980s. The secretary of state for Egyptians abroad estimated that there were 1.25 million Egyptians working in Iraq in 1983 (Roussillon, 1985: 642). The Iraqi government grew increasingly critical of the growing proportion of non-Arab workers employed in nearby states, which it regarded as a threat to the Arab character of the Gulf (Roussillon, 1985: 650–5). By the late 1970s, the number of Asian migrants in Iraq was declining in favour of Arab migrants, while the opposite trend was occurring in the oil-rich Gulf states to the south. The openness of Iraq to migrants from other Arab states helps explain the sympathy felt for Iraq during the Gulf crisis virtually throughout the Arab world. The significance of migration of Arabs to Iraq increased during the long and terrible Iraq–Iran war. In the late 1980s, reports of tensions between Egyptian migrants and the indigenous population became frequent. Moreover, Ba'athist rhetoric against non-Arab migrants intensified.

The consequences of the Gulf War of 1990–1 are examined in Box 6.1. Iraq was transformed from a regionally significant land of immigration into a zone of emigration as millions of migrants and Iraqi citizens fled.

Arab migration to the oil-rich states of the Arabian Peninsula was even larger. Some areas, like Kuwait, had already had immigrant labour policies under British rule, recruiting workers from British possessions in South Asia, particularly present-day India and Pakistan. Significant East Asian migration to the Gulf also began long before 1975 (Seccombe and Lawless, 1986: 548–74). Those areas under American domination, particularly Saudi Arabia, developed quite different foreign labour policies. In the 1950s and 1960s, Westerners and Palestinian refugees often provided the skilled labour required for oil production. In the wake of the 1967 and 1973 Arab–Israeli wars, labour migration skyrocketed as the rising price of oil financed ambitious development projects. Between 1970 and 1980, oil revenue in the Arab states belonging to OPEC (the Gulf states plus Iraq

Box 6.1 The Gulf War

After the oil-price rise of 1973, the oil-rich states of the Persian Gulf recruited masses of foreign workers from both Arab and Asian countries for construction and industrialization. At first most were men; later many female domestic servants were recruited from the Philippines and Sri Lanka. Resentments over the status accorded to various categories of aliens in Kuwait became a major factor in Iraq-Kuwait tensions. At the beginning of the Gulf crisis in 1990 there were 1.1 million foreigners in Iraq, of whom 900 000 were Egyptians and 100 000 Sudanese. Kuwait had 1.5 million foreigners: two-thirds of the total population. The main countries of origin were Jordan/Palestine (510 000 people), Egypt (215 000), India (172 000), Sri Lanka (100 000), Pakistan (90 000) and Bangladesh (75 000).

The Iraqi occupation of Kuwait and the subsequent war led to mass departures of foreign workers. Most Egyptians left Iraq, hundreds of thousands of Palestinians and other migrants fled Kuwait, and perhaps a million Yemenis were forced out of Saudi Arabia when their government sided with Iraq. An estimated 5 million persons were displaced, resulting in enormous losses in remittances and income for states from South-East Asia to North Africa.

The Gulf War suggested, as perhaps never before, the centrality of migration in contemporary international relations. Migrants were viewed as potentially subversive – a fifth column – by the major Arab protagonists, and became scapegoats for domestic and international tensions. Hundreds of migrants were killed in the outbreaks of violence. The political realignments occasioned by the conflict had major repercussions upon society and politics in the Arab region and beyond.

and Libya) increased from US$5 billion to US$200 billion. Saudi revenues alone increased from US$1 billion to US$100 billion (Fergany, 1985: 587).

From the mid-1960s to mid-1970s, most international migrants to the Gulf states were Arabs, mainly Egyptians, Yemenis, Palestinians, Jordanians, Lebanese and Sudanese. During the 1970s, however, the Gulf monarchies grew increasingly worried about the possible political repercussions. Palestinians, in particular, were viewed as politically subversive. They were involved in efforts to organize strikes in Saudi oil fields and in civil strife in Jordan and Lebanon. Yemenis were implicated in various anti-regime activities (Halliday, 1985: 674). Foreign Arabs were involved in the bloody 1979 attack on Mecca which was subdued only after the intervention of French troops. One result was increased recruitment of workers from South and South-East Asia, who were seen as less likely to get involved in politics and easier to control (see Chapter 7).

By the mid-1980s, the price of oil had plummeted and some observers, like the Central Intelligence Agency (CIA), concluded that the epoch of massive migration to the Arabian Peninsula had come to an end (Miller, 1985). Hundreds of thousands of Arab, South and East Asian workers did lose their jobs and return home. But the conclusion that massive labour migration to the oil-rich states had ended was premature. Migrant labour had become an irreplaceable component of the labour force (Birks *et al.*, 1986: 799–814). Despite government efforts to reduce dependency upon foreign labour, including mass expulsions of illegal aliens, foreigners continued to constitute the bulk of the Kuwait labour force on the eve of the Iraqi invasion

After the 1991 war, the Kuwaiti government announced plans to reduce its dependency on foreign labour, yet Kuwait could not rebuild without massive recourse to migrant labour. In Saudi Arabia, Egyptians began to take the place of politically suspect populations like the Yemenis and Palestinians expelled during the crisis (J. Miller, 1991). There was an increased proportion of South and East Asian workers in the migrant workforces of the Gulf oil states. Despite the military victory over Iraq, the Gulf monarchies had further eroded their legitimacy in most of the Arab world. This made even their key ally, Egypt, increasingly unattractive as a source of migrant labour.

By 2002, the sense of looming crisis deepened. The most compelling estimate held that non-nationals comprised roughly one-third of the total population of the states belonging to the Gulf Cooperation Council (GCC) (Saudi Arabia, Kuwait, Oman, Qatar, Bahrain and the United Arab Emirates) in 1995. But non-nationals vastly outnumbered nationals in the workforce by 5 232 000 to 2 378 000 (Evans and Papps, 1999: 208). National populations were growing rapidly while governmental revenues declined and deficits soared. Unemployment of young nationals grew, as governments were hard-pressed to create additional jobs in the public

Map 6.1 *Return migration from the Gulf during the 1990–1 Kuwait crisis*

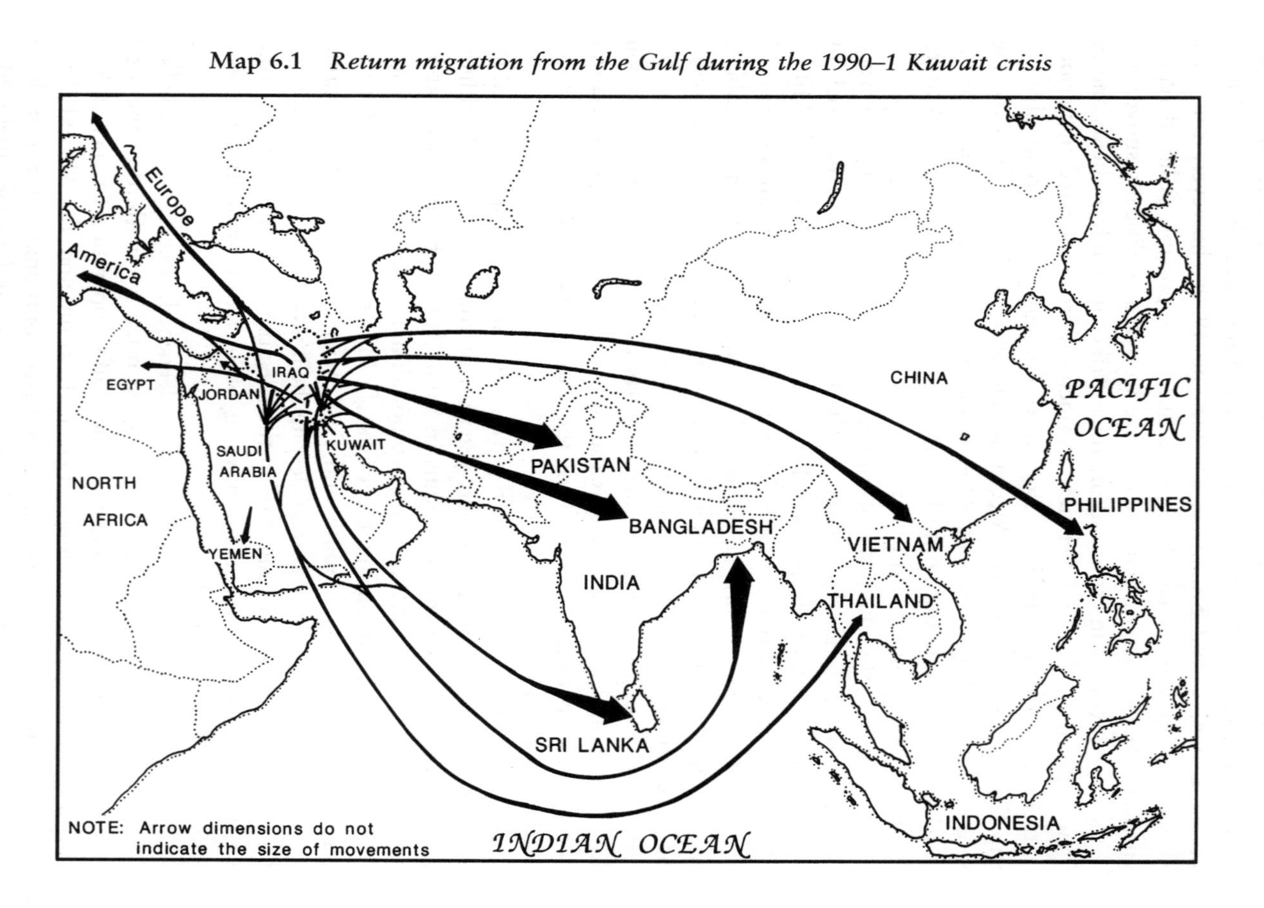

sector, the traditional means of ensuring loyalty to the state. Education and training patterns tended to exacerbate barriers to employment of young nationals, which in turn were related to long-term dependence upon foreign labour with cumulative distorting effects upon labour markets. This situation threw into question the wisdom of reliance upon cheap foreign labour in the development strategies of the GCC states (Evans and Papps, 1999: 227–33).

With a population of 70 million, Egypt is by far the most populous of Arab states and it has been the most affected by intra-regional labour migration. The evolution of Egyptian labour migration correlated not only with the ups and downs of oil revenues in nearby oil-producing states but also with changes in Egyptian domestic and foreign policies. Remittance of wages from Egyptians working abroad became a crucial economic concern as entire villages and regions depended on them for consumption and investment (Fadil, 1985).

Emigration greatly affected the fabric of life as peasants, craftsmen and highly-skilled professionals were lured away by wages many times higher than they could expect to earn at home (Singaby, 1985: 523–32). Labour emigration undoubtedly relieved chronic unemployment and underemployment, but it also stripped Egypt of much-needed skilled workers and disrupted, for better or worse, traditional village and family structures. Among the many significant effects of the massive emigration was growing Egyptian dependency upon regional political and economic developments: there were returning waves of migrants from Libya at the height of Egyptian differences with Colonel Qaddafi, during the oil price drop of the mid-1980s and during the Gulf crisis.

By 2002, Egypt needed to create 500 000 jobs per year to employ population cohorts entering the job market. It could not do so and therefore it signed 11 bilateral agreements with nearby states between 1974 and 1993 to facilitate emigration of Egyptians. In the mid-1990s, Egyptian statistics indicated that about 70 per cent of all migrants went to Saudi Arabia (Farrag, 1999: 56–7).

Arab migration to non-oil-producing states

Arab migration to non-oil-producing states within the Arab region is quantitatively and geopolitically less significant than migration to the oil-producing states, but is nonetheless important. It is often hard to differentiate between labour and refugee flows. Jordan is a prime example. By the mid-1970s, perhaps 40 per cent of the domestic workforce was employed abroad, primarily in the Gulf (Seccombe, 1986: 378). This outflow prompted replacement migration: the arrival of foreign workers who substituted for Jordanians and Palestinian residents of Jordan who emigrated abroad. However, much of the Jordanian labour that went abroad was skilled. Much of the labour that Jordan receives is skilled as well, but there is also a big inflow of unskilled Egyptians and Syrians. In the 1980s, this inflow is thought to have contributed to growing unemployment among Jordanian citizens and resident aliens. Wages in industries heavily affected by foreign workers have also declined (Seccombe, 1986: 384–5). The mass expulsion of Palestinians by Kuwait in 1991 greatly affected Jordan which received most of the influx (Shami, 1999: 151, 179–95). In the 1990s, 50 000–200 000 Iraqis resided in Jordan, many of whom were undocumented. Nearly 7000 applied for asylum in Jordan in 2000 (USCR, 2001: 185).

Another important migrant labour pattern involved Palestinian Arab residents of the territories occupied by Israel in the 1967 war. The Israeli labour market was opened up to workers from Gaza and the West Bank. This was part of an Israeli strategy aimed at integrating the occupied territories into the Israeli economy (Aronson, 1990). Most of the workers had to commute daily to work in Israel and were required to leave each evening. Palestinians found jobs primarily in construction, agriculture, hotels and restaurants and domestic services (Semyonov and Lewin-Epstein, 1987). Illegal employment of Palestinians from the territories was fairly widespread (Binur, 1990). In 1984, some 87 000 workers from the occupied territories were employed in Israel, about 36 per cent of the total workforce of the occupied territories.

By 1991, Soviet Jewish immigration was affecting employment opportunities for Arabs. The Israeli government clearly preferred to see Soviet Jews employed in construction or agriculture, rather than Palestinians, yet its efforts to employ Soviet Jews met with little or no success. Either the Soviet Jewish immigrants desired different jobs or the pay and working conditions were unsatisfactory (Bartram, 1999: 157–61). It was difficult to measure the displacement of Palestinians because other factors were also at work. The Gulf War heightened animosities and there was a wave of attacks by Arabs from the occupied territories on Jews in Israel. Israeli authorities introduced more restrictive regulations and admission procedures, aimed at weakening the Intifada, as well as ensuring greater security. A combination of all these elements resulted in a sharp decline in employment of Palestinian workers after 1991. Increasingly, foreign workers from Romania, the Philippines and Thailand were recruited to replace Palestinian Arab labour from the West Bank and Gaza. Concurrently, closure of Gulf state labour markets to Palestinians, which had long served as a safety valve for the population of the occupied territories, worsened the economic plight of Palestinians. This distress threatened the leadership of the Palestinian Authority and the entire regional peace process.

Israel received about 800 000 new immigrants from the former Soviet Union in the 1990s. Overall, Israel's population grew from 800 000 in 1947 to 6 million in 1998: net immigration accounted for 40 per cent of the total population growth (Kop and Litan, 2002: 23–5). Meanwhile, its foreign worker population grew to 250 000 (Kop and Litan, 2002: 107). Estimates of Israel's foreign worker population were shaky as the Interior Ministry in 1996 concluded that about 100 000 people had overstayed their visas in the previous decade. Other estimates placed the total of illegal foreign workers at 250 000 without any factual basis (Bartram, 1999: 164–5). In 2002, the Israeli government declared war against illegal employment of foreigners, but measures like employer sanctions and deportation appeared to have little deterrent effect, in part because fines were small. It seemed likely that some 'temporary' workers would become a permanent part of the complex fabric of Israeli society (Bartram, 1999: 167). The sharp contrast between the governmental generosity afforded Jewish immigrants and the lot of foreign workers in Israel prompted soul-searching and calls for a phase-out of foreign worker recruitment (Kop and Litan, 2002: 108).

Refugees and internally displaced persons in the Arab region

As of 2002, there were some 3.8 million Palestinian refugees scattered around the region and the world. The Palestinian-Israeli peace accords had done very little to alter their plight, although thousands of Palestine Liberation Organization officials and military or police personnel had been authorized to return to the area of the Palestinian Authority, consisting of parts of the Gaza Strip, Jericho, Hebron and several other urban areas of the West Bank. Negotiations concerning refugees, repatriation, compensation, reparations and access to the territory of the Palestinian Authority loomed as the most difficult aspect of the peace process. Israeli and Palestinian viewpoints and positions differed enormously, starting with the enumeration of refugees. With the Palestinian population of the West Bank and Gaza in dire economic straits, and high unemployment compounded by the size of new cohorts coming up to working age, prospects appeared bleak for mass repatriation of Palestinian refugees from Lebanon or Syria. The average woman in Gaza had ten children, one of the highest rates of fertility in the world. There were other significant refugee and internally displaced person populations in the region which similarly greatly affected regional politics (see Boxes 6.1 and 6.2).

Iran became the world's most important haven for refugees by the early 1990s. It succeeded in repatriating large numbers of Afghan refugees and announced that all Afghan refugees had to leave in 1997 (USCR, 1996: 111). However, 1.5 million Afghan refugees remained in 2000 and their numbers increased during the US-led invasion of Afghanistan in 2001

(USCR, 2001: 174). Meanwhile, hundreds of thousands of Iranian citizens had fled Iran for Turkey, Iraq and the West. Iranian dissidents occasionally attacked targets in Iran from bases in Iraq, and this brought retaliatory strikes. The Iranian government struck against Iranian political opponents abroad. Murders of Iranian opposition leaders in Western Europe severely strained Iran's relations with Germany and France. Iran also received many Azeris fleeing advancing Armenian forces in Azerbaijan. To stem the

Box 6.2 Kurdish refugees in regional conflicts

The Kurds constitute an important ethnic minority in the area, totalling between 20 and 25 million. About half of all Kurds reside in Turkey where they comprise about one-quarter of the total population. Kurds comprise about one-quarter of Iraq's population, about 12 per cent of Iran's and 10 per cent of Syria's (Gurr and Harf, 1994: 30–2). Kurdish hopes for an independent state were dashed in the wake of the First World War. Since then, Kurdish aspirations for independence or autonomy have fostered conflict.

Iraqi Kurdish politics have been dominated for many years by the Barzani family and the Kurdish Democratic Party (KDP). In 1976, the KDP split and a rival party emerged, the Patriotic Union of Kurdistan (PUK) led by Jalal Talabani. In 1975, the Shah of Iran, who, along with Israel, had been backing the Kurdish insurgency in Iraq, made a deal with the Iraqi government. In return for Iraq renouncing its claim to the Shatt-al-Arab waterway, Iran would close its border and stop aiding the Iraqi Kurdish insurgents. Mustafa Barzani, the legendary KDP leader, stopped fighting and many of his followers fled to Iran. Talabani and the PUK rejected Barzani's decision, and fighting between PUK and KDP forces ensued.

By 1987, PUK forces controlled much of Iraq's Kurdish area. However, the Iraqi army used poison gas and indiscriminate bombing of civilian centres to reassert Iraqi control. Tens of thousands of Kurds were killed and millions fled to Turkey and Iran. The Kurdish refugee influx proved very expensive for the Turkish government, which received little external aid to cope with the refugees. Meanwhile, a Kurdish insurgency in Turkey was growing. It was led by the Kurdish Workers Party (PKK), a Marxist-Leninist group with important bases of support in Syria, Lebanon and in Europe amongst Turkish guestworker populations.

In the 1990s, millions of Kurds were internally displaced by conflicts in Turkey and Iraq. Millions more found refuge in predominantly Kurdish regions of nearby states. In 1991, after the Gulf War, Kurdish rebels were crushed by Iraqi troops. Millions fled towards Turkey and Iran to escape retaliation. To prevent another mass influx, Turkey cooperated in a UN-authorized operation to create a protected area in Northern Iraq. Most of the Kurds returned home and an autonomous 'federated state' was proclaimed. A parliament was elected and an autonomous administration installed by 1993. However, KDP–PUK hostilities began again. The situation was further

→

tide, Iran set up refugee reception centres near its border with Azerbaijan. Iran's refugee policy evolved during the 1990s from one of reception of refugees and integration to more active intervention to prevent inflows or to contain them at the border. The huge cost of caring for refugees played a role in this changing orientation. However, core Iranian interests were at stake in conflicts like that in Azerbaijan as Azeris comprise a major ethnic minority in northern Iran.

→

complicated by entry into the area of PKK guerrillas. Increasingly, Turkish forces launched attacks on northern Iraq to destroy PKK bases, and the Turkish government declared its intent to establish a security zone along its border with Iraq.

The stakes involved in Kurdish uprisings and their suppression for regional politics rose steadily. The USA pursued a dual containment policy against Iraq and Iran. Israel and Turkey began to cooperate extensively in national security matters. Syria backed the PKK, and allowed it to use training camps in Lebanon's Bekaa Valley. The PKK's leader branded the USA, Israel and Germany as enemies of the Kurds and possible targets for PKK guerrillas (PKK activities in Germany are analysed in Chapter 11). By 1997, Syria, Iran and Iraq were attempting to improve relations to counter the Israel-Turkey axis backed by the USA.

Turkey finally succeeded in capturing the leader of the PKK and the Kurdish insurgency in south-eastern Turkey receded. But ships full of Kurds arrived in France, Australia, Greece and Italy. An alarmed EU developed a plan to stabilize the Kurdish area but it collided with the extension of the 'War on Terrorism' to the 'Axis of Evil' (Iraq, Iran and North Korea) in President Bush's State of the Union address of January 2002. US plans to invade Iraq ran the risk of setting in motion further mass displacement of Kurds and their neighbours. Concern over potential population movements figured importantly in European objections to US sabre-rattling.

Displacement, eviction, flight, refugee or non-citizen status has been the lot of millions of Kurds in recent decades. Their dispersed and transnational condition is increasingly characteristic of politics around the world. States advance their foreign policy and national security goals by mobilizing, training and backing Kurdish refugee populations. States strike against bases and havens used by armed opponents even if, by doing so, they violate the territorial sovereignty of another state. Perhaps most importantly, Kurdish refugee movements led to a multinational military operation aimed at preventing mass outflows. UN-authorized humanitarian intervention in northern Iraq in 1991 was necessitated by Turkey's refusal to become a haven for Iraqi Kurds, as it did in 1987. Caring for those refugees had cost too much and had further complicated Turkish internal politics and regional relations. In 1991, Turkey felt vital national security interests were at stake. All around the Arab region, states were coming to similar conclusions.

Illustration 6.1 *Displaced Iraqi Kurds return home 1991 (Photo: UNHCR/A. Roulet)*

Turkish citizens of Chechen and Circassian background have given support to insurrectionary movements abroad, principally in the Russian Federation. The Chechen population had been subjugated, slaughtered and dispersed by the Russian imperial rulers before 1917 and then deported during the Second World War by order of Josef Stalin. The support given by Turkish citizens of Chechen background to the Chechen revolt was typical of the transnational nature of many political conflicts in the post-Cold War era. Meanwhile, Russia defined the Chechen revolt as terrorism and noted that Al-Qaida was extensively involved. Forced migration in the past was an important grievance for Chechens, Palestinians, Jews, many Turks of Balkan background and others, just as it was a key factor in conflicts in Eastern Europe. The prevalence of non-voluntary population displacements in the Arab region was probably related to the extreme paucity of scholarship about migration in the area, especially in the wake of the Gulf War (Shami, 1994: 4–6).

Sub-Saharan Africa

Some social scientists believe that Africa, with one-quarter of the world's land mass and one-tenth of its population, is the continent with the most mobile population (Curtin, 1997: 63–4). Comparative data analysis could not at present be used to verify or disprove this contention, in part because credible statistics are so deficient, particularly concerning population movements between African states. However, in 1990 there were estimated to be 30 million voluntary international migrants in sub-Saharan Africa, about 3.5 per cent of the total population. The 1990s witnessed a major increase in involuntary or forced migration and, by the middle of the decade, refugees and internally displaced persons in some countries outnumbered voluntary international migrants by a ratio of more than two to one. By 1997, there were almost 17 million forced migrants, inclusive of nearly 4 million refugees (Findlay, 2001: 275–8).

Africa includes a large share of the world's poorest states. Migration is often a way to escape crushing poverty, or even death due to malnourishment. Some of the states have never had a census. The paucity of elementary population information, the frequent absence of identity documents and the habit of some individuals of declaring themselves to be nationals of one state when, in fact, they are citizens of another, makes analysis of international migration in sub-Saharan Africa particularly difficult.

Sub-Saharan Africa generates significant outflows of intercontinental migrants, mainly to Western Europe but also to North America and the Arab region. Many of these outflows were traditionally directed primarily to former colonial powers: for example, Congolese emigrating to Belgium, Senegalese to France or Nigerians to the United Kingdom. Many emigrants are college educated and the loss of scarce human capital though the 'brain drain' has been a long-standing African concern. Intercontinental migration has diversified, however, and increasingly includes poorly-educated labour migrants. Sub-Saharan Africans are emigrating in significant numbers to countries such as Spain, Portugal, Italy, Canada and the USA. But the vast bulk of international migration from sub-Saharan countries stays within the continent. A key question for the future is: Will this pattern hold? Some analysts contend that widening North–South socio-economic and demographic disparities are already shifting the pattern (Ouedraogo, 1994).

As in the Arab region, the legacy of colonialism still strongly influences migratory patterns. The European presence shifted the locus of economic activity and trade to coastal areas, producing migrations from the interior that have persisted after independence. The colonial powers carved up the continent into politico-administrative entities (which later became independent states) with little regard for the congruence of ethnic and territorial boundaries. Members of an ethnic group are often citizens of two or more adjoining or nearby states, while many states include members of several ethnic groups. This leads to confusion over legal status or national identity as well as to traditions of movement across international boundaries that are often poorly demarcated and controlled.

Map 6.2 *Migrations within and from Africa (c. 1970–2000)*

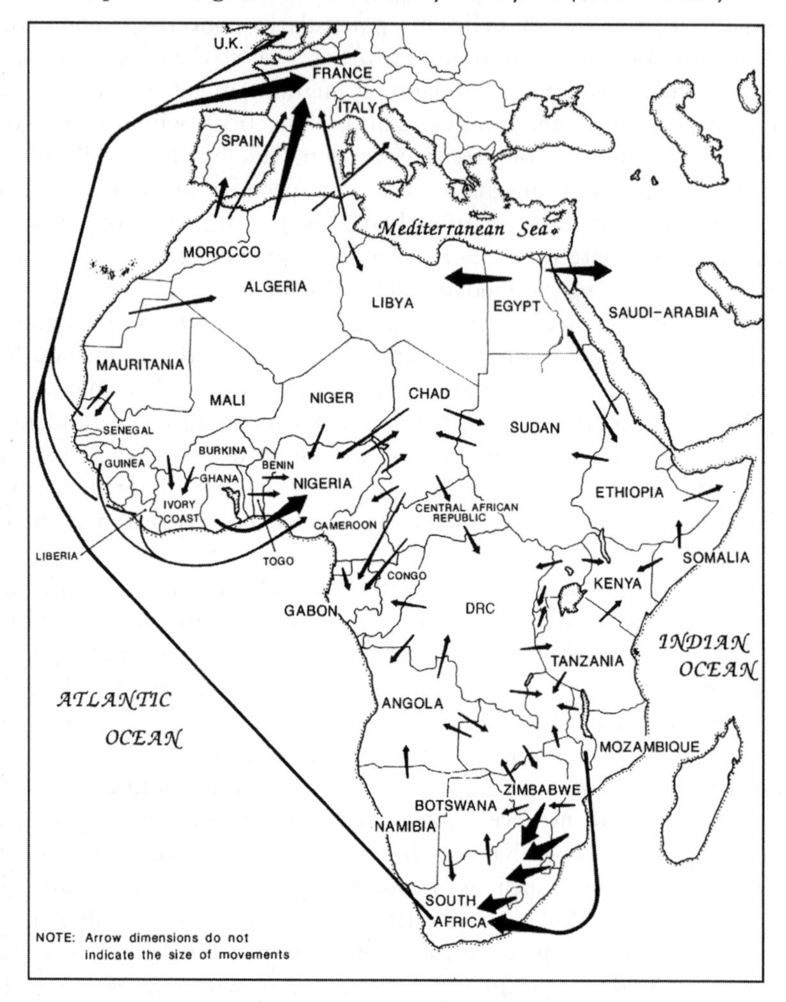

The colonial period brought not only European administrators and farmers, but also Syro-Lebanese merchants to West Africa, as well as merchants and workers from the Indian subcontinent to East and Southern Africa. In the post-independence period, these populations generally became privileged but vulnerable minorities. European-origin settler populations often departed en masse when independence was granted, with disastrous economic consequences as they had played key roles in agriculture, business and government. The ripple effects of the colonial period were felt long after it was over.

In Kenya, for instance, prime lands were appropriated and allocated to British farmers. When most of the European settlers fled, these lands reverted to the Kenyan government which allocated them to Kikuyus and other ethnic groups then dominant in Kenyan government. By the 1990s, power within Kenyan government had shifted and Kenyan President Moi aided and abetted attacks by pastoralist tribe members like the Masai who laid claim to the lands usurped by the British. Sporadic violence killed 1500 and produced as many as 300 000 internally displaced Kenyans. Donor country pressure upon the Kenyan government led to its grudging approval of a United Nations Development Programme initiative to resettle those displaced. Many lost their lands and were squatters or homeless. Some were able to farm their lands by day but left by night for fear of renewed violence. Many had lost everything and subsisted as day labourers or on charity. Such sequels of colonialism help explain why Africa encompassed 16 of the world's 41 major populations of IDPs in 2000 (USCR, 2001: 6). Sudan alone was estimated to have 4 million IDPs.

In the post-Cold War period, large-scale repatriations of refugees and resettlements of IDPs have occurred. In the early 1990s, there were an estimated 5.7 million uprooted Mozambicans, including 1.7 million refugees and 4 million internally displaced. By 1996, most had returned home (USCR, 1996: 12). From 1990 to 1996, some 4 million African refugees repatriated, mainly to Ethiopia, Eritrea, Mozambique, Zimbabwe, Namibia, the post-apartheid Republic of South Africa (RSA) and Uganda. In late 1996 and early 1997, tens of thousands of Rwandans also repatriated from Tanzania, when they were ordered to leave by the government. Many returned from Zaire, to escape fighting between insurgents, elements of the Rwandan and Zairian armies and Rwandan Hutu militia and ex-soldiers entrenched in the refugee camps. Some analysts pointed to these large-scale repatriations and resettlements in the post-Cold War era to refute the 'chaos theory' which viewed Africa as doomed to political disintegration, mass misery and wholesale uprooting of populations (USCR, 1996: 12). This more upbeat regional appraisal was challenged by unfolding tragedies in Sierra Leone and Liberia in the late 1990s as well as by events in the Great Lakes region of Central Africa (see Box 1.3).

Sub-Saharan Africa has witnessed the proclamation of numerous international organizations for the purpose of removing barriers to trade and the free movement of goods, capital and people. Generally these agreements have been poorly implemented or contradicted by policies and practices in member states (Ricca, 1990: 108–34; Adepoju, 2001). Despite the existence of many zones in which there is nominally freedom of movement by nationals of signatories to these agreements, there is nonetheless a great deal of illegal migration.

Illegal migration within sub-Saharan Africa is varied and complex. It is often tolerated in periods of good relations and economic prosperity, only to be repressed during economic downturns or periods of international tensions. The mass expulsions from Nigeria in 1983 and 1985 were the most significant in terms of persons uprooted – as many as 2 million – but

Box 6.3 Labour migration to the Republic of South Africa

Foreign worker recruitment to the Republic of South Africa (RSA) illustrates the connection between labour migration and economic and political dependency characteristic of much of Africa. The roots of recruitment go back to the colonial period. Most of the workers recruited during the apartheid period from Mozambique, Botswana, Lesotho, Swaziland and Malawi worked in gold mines, with less than 10 per cent in agriculture and other non-mining industries. Lesotho and Swaziland are landlocked states which border on the RSA (Lesotho is in fact completely surrounded by the RSA). Their populations can barely eke out a living from agriculture. The absence of economic opportunities made employment in RSA mines the only possibility for many, despite the rigours of mine work and the high risk of injury or death.

Recruitment to the RSA was highly organized. Candidates were subjected to a battery of physical and aptitude tests, and many were rejected. The successful ones were transported by air, rail or bus to the mines where they lived in hostels. Virtually only males were hired and most were young. They were given contracts which required them to return home after one or two years of work. In 1960, there were about 600 000 foreign workers in the RSA. The number declined to 378 000 by 1986. In 1973, the share of foreigners in the black miner workforce stood at 79 per cent. By 1985, it had been reduced to only 40 per cent. There was a shift away from recruitment of foreign workers from independent states, in favour of increased recruitment from the newly fashioned 'black homelands' due to the South African regime's fear of being deprived of foreign labour as a result of anti-apartheid policies (Ricca, 1990: 226–8).

With the end of the apartheid system, a new era unfolded in the RSA. The 'internalization' of employment in the mines continued and some foreign miners became legal residents. With the former 'homeland' areas facing

→

they are part of a much broader pattern. A clash between Mauritanian pastoralists and Senegalese farmers in the Senegal River valley in 1989 took 250 lives and resulted in the repatriation of 70 000 Senegalese and 170 000 Mauritanians. The conflict was rooted in an ecological crisis arising from a long drought but exacerbated by ethnic enmities (Kharoufi, 1994: 140–4). In 1991, many Zairians living in Congo were expelled. The director of Congo's air and border police said that three-quarters of the million or so Zairians in Congo would be expelled (Noble, 1991). Tensions over the unregulated arrival of Zairians and other aliens had been building for some time. Aliens were seen as contributing to rapid population growth in sprawling suburban areas and as overtaxing resources. Zairians were singled out by one specialist as contributing to vice (Loutete-Dangui, 1988: 224–6).

→

crushing poverty, there was no shortage of potential workers. The eclipse of apartheid allowed normalization of long-disrupted diplomatic and economic relations between the RSA and its neighbours. One of the key problems facing the post-apartheid government was unauthorized migration from abroad. There had been considerable illegal migration from neighbouring countries such as Mozambique during the apartheid era. Security measures, including an electrified fence on the border, made illegal entry quite dangerous. Since the collapse of apartheid, unauthorized entry grew enormously. Africans from as far away as Ghana flocked to the South African 'Eldorado'. Meanwhile substantial repatriation of South African refugees occurred, while South Africans who had been forced to relocate to 'homelands' sought to return home. Widespread unemployment and lawlessness further complicated the picture.

The uneasy partners in the post-apartheid government often had contrasting views on illegal migration. Certain trade union and African National Congress (ANC) leaders favoured policies which reflected the international solidarity that had been so important during the long struggle against apartheid. Other factions favoured more draconian policies to deter further unauthorized entry and to expel illegal immigrants. By 1996, the government had begun a legalization programme for aliens meeting certain criteria. It also planned enforcement of employer sanctions. The RSA continued to recruit foreign labour from nearby states. A total of nearly 51 000 mineworkers and 200 000 people from Southern African Development Community (SADC) member states applied and 103 000 applications were approved by 1997. The 1996 census did not provide accurate information about illegal migration to the RSA because census questionnaires did not ask about status. Hence, wildly conflicting estimates continued to be made of illegal immigrants, ranging from 2.5 to 8 million persons (Bernstein *et al.*, 1999). Deportations rose from 158 000 in 1995 to 181 000 in 1996 (Adepoju, 2001: 63).

This situation contrasted with the Nigerian case where technically illegal employment of aliens primarily from the Economic Community of West African States (ECOWAS) was looked upon benignly by the Nigerian government during a period of economic expansion. In the mid-1970s, many Ghanaians entered and found work in construction and the services (Adepoju, 1988: 77). A downturn in the economy, coupled with Nigerian governmental instability and with deteriorating relations between Nigeria and Ghana, prompted a new, stricter policy and mass expulsions. One scholar enumerated 23 mass expulsions conducted by 16 different states between 1958 and 1996 (Adepoju, 2001: 60–1).

Taking Africa as a whole, there are reasons for deep pessimism concerning the future of migration. Standards of living have fallen and political instability appears endemic in many areas. On the other hand, there were some notable economic success stories in the 1990s, such as Ghana, and an overall continental trend to democratization (Chazan, 1994). The track record of the seven regional organizations and groupings on regulation of international migration did not appear fated to improve, because wide socio-economic disparities between member states spurred lopsided, often unidirectional flows to richer member states (Adepoju, 2001: 61–4). Trafficking of women and children appeared to be increasing. Cases of young boys stowing away in the landing gear of aircraft and then being found crushed and frozen in Europe were emblematic of hopelessness and distress generating growing population movements.

The high numbers of refugees and IDPs in Africa is a symptom of the nation building and state formation process (Zolberg *et al.*, 1989), which can be compared to similar processes in Europe from the sixteenth to the twentieth centuries. There, too, ethnic and religious minorities faced persecution, while war and economic dislocation were rampant. Western European states took centuries to resolve basic issues of national identity and political legitimacy. Sub-Saharan Africa, like much of the former colonized world, has had to confront a broad spectrum of modernization issues in the decades since independence. This is the underlying cause of the proliferation of refugees and internally displaced persons.

Thus far, however, relatively few African refugees have left the continent. In view of the considerable resources – both financial and cultural – needed to move to developed countries, and the considerable barriers erected by potential receiving areas, the likelihood that large numbers of African refugees could leave in the future appears small. Unfortunately, this may help explain why the international community responded so belatedly and inadequately to the mass slaughter in Rwanda and related tragedies in Zaire, now the DRC.

Latin America and the Caribbean: transition from an immigration to an emigration region

The vast and highly diverse area to the south of the USA is sometimes portrayed as consisting of four principal areas:

1. The Southern Cone comprises Brazil, Argentina, Chile, Uruguay and Paraguay, which are all societies in which a majority of the population is of European origin. This was an area of massive immigrant settlement from Europe. There were also inflows from elsewhere: for example, Brazil received African slaves up to the nineteenth century and Japanese workers from the late nineteenth century until the 1950s.
2. The Andean area to the north and west differs in that Indians and *mestizos* (persons of mixed European-Indian background) comprise the

bulk of the population. Immigration from Europe during the nineteenth and twentieth centuries was less significant.
3. Central America where societies largely comprise persons of Indian and *mestizo* background, although there are exceptions, such as Costa Rica.
4. The Caribbean, made up predominantly of people of African origin but also people of Asian and European descent.

Quite a number of countries do not fit neatly into these four areas, but the categorization serves to underscore how immigration since 1492 has differentially affected the area as a whole and how many of these societies were forged by immigration.

De Lattes and de Lattes (1991) estimate that Latin America and the Caribbean received about 21 million immigrants from 1800 to 1970. The single largest migration was the estimated 3 million Italians who went to Argentina. The bulk of immigrants came from Spain, Italy and Portugal, and most of them went to the Southern Cone. States like Argentina and Uruguay encouraged immigration until the inter-war period. The economic depression of the 1930s brought significant changes in immigration policies. Apart from the Italian influx from 1947 to 1955, mass immigration from Europe had become a thing of the past by the 1930s (Barlán, 1988: 6–7). A significant exception to this general pattern was Venezuela, which had received very few European-origin immigrants until the rule of Pérez Jiménez, from 1950 to 1959. About 332 000 persons, mainly of Italian origin, settled in Venezuela under his regime. However, the so-called open door policy stopped with the overthrow of the military government in 1958 (Picquet *et al.*, 1986: 25–9).

As intercontinental inflows from Europe waned, intra-continental (or intra-regional) migrations developed. As in the Caribbean Basin back in the nineteenth century, labour migration predominated. The end of the Chaco War between Paraguay and Bolivia in 1935, for example, brought significant numbers of Bolivian army deserters into north-western Argentina. Some of them took jobs in agriculture. This marked the beginning of a seasonal labour migration from Bolivia to Argentina that lasted for over three decades, until mechanization reduced the need for labour. This labour flow was largely unregulated until 1958 when a bilateral agreement was signed to help protect the Bolivian migrants (Barlán, 1988: 8–9).

Similarly Paraguayan and Chilean labour migrants began to find employment in north-eastern Argentina and in Patagonia respectively in the 1950s and 1960s. Foreign workers spread from agricultural areas to major urban centres. Single, mainly male, immigrants were soon joined by families, creating neighbourhoods of illegal immigrants in some cities. Their entry and employment appear to have been tolerated, as long as they were seen as contributing to economic growth and prosperity, a view

Map 6.3 *Migrations within and from Latin America (c. 1970–2000)*

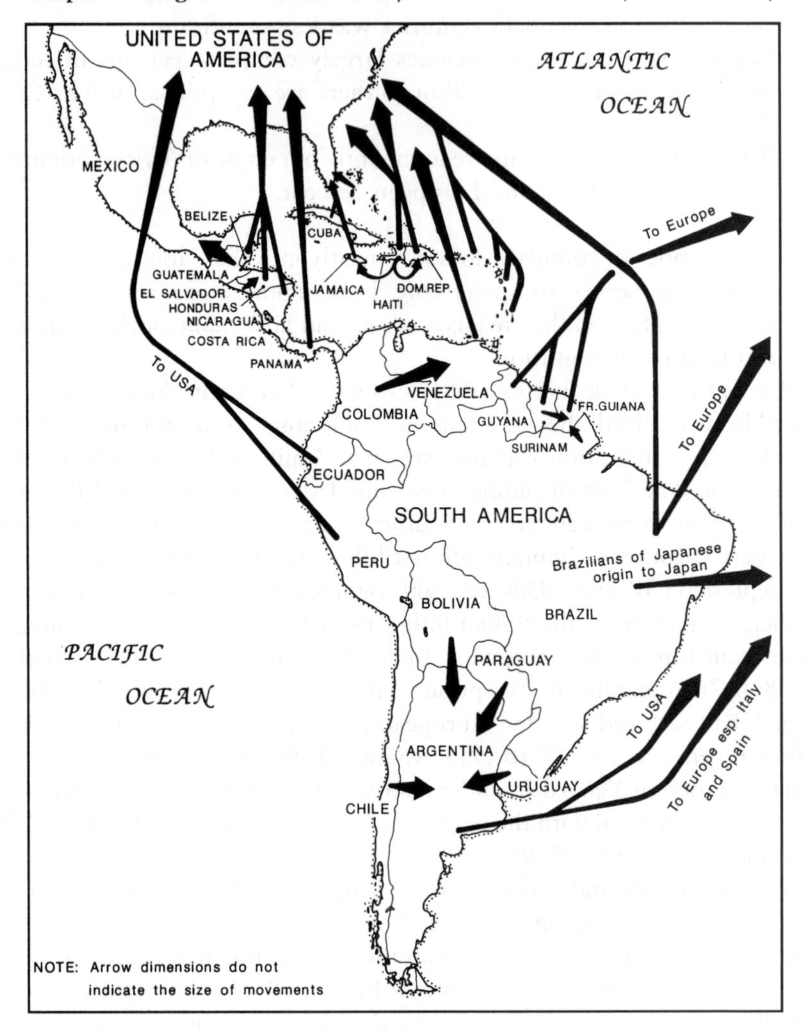

challenged only in the 1970s (Sanz, 1989: 233–48). Beginning in 1948, the Argentine government adjusted laws and policies to enable illegal foreign workers to rectify their status. Irregular or illegal migration is the predominant form of migration in Latin America, but this was not viewed as a problem until the late 1960s (Lohrmann, 1987: 258).

Venezuela is another country where legalization was deemed necessary. With the slowing of immigration from Europe and with oil-related economic growth, millions of Colombians flocked to Venezuela. Many arrived via the *caminos verdes* – the green highways – over which a guide would steer them across the frontiers (Mann, 1979). Other Colombians arrived as tourists and overstayed. By 1995, 2 million persons were thought to be residing illegally in Venezuela, most of them Colombian (Kratochwil, 1995: 33). Not only the oil industry but also agriculture, construction and a host of other industries attracted migrants. Declining incomes in Colombia and the attraction of the stronger Venezuelan currency were significant factors making work in Venezuela attractive in the 1980s (Martinez, 1989: 203–5). Many Colombians migrated to areas close to the Venezuelan-Colombian frontier and many were short-term migrants (Pelligrino, 1984: 748–66). However, most of the illegally-resident alien population of about 10 per cent of Venezuela's total population of 20 million lived in major cities (Kratochwil, 1995: 33). Another 2 million aliens resided legally (Dávila, 1998: 18).

Colombian seasonal workers traditionally helped harvest the coffee crop in Venezuela, and bilateral labour accords between the two countries were signed in 1951 and 1952. The Treaty of Tonchala in 1959 obliged the two governments to legalize illegally-employed nationals from the other country if legal employment could be found. In 1979, the Andean Pact was signed, obliging member states to legalize illegally-resident nationals from other member states (Picquet *et al.*, 1986: 30). This led to the Venezuelan legalization of 1980. Despite estimates ranging from 1.2 to 3.5 million illegal residents out of a total Venezuelan population of some 13.5 million, only some 280–350 000 aliens were legalized (Meissner *et al.*, 1987: 11). Either the estimates were much too high, which seems likely, or the legalization programme did not succeed in transforming the status of many illegal residents.

In the 1990s, drug-related violence and political turmoil in Colombia, where the government faced leftist insurgencies, drove tens of thousands of Colombians into Venezuela (Kratochwil, 1995: 15). An economic downturn and austerity measures in Venezuela sparked an attempted *coup d'état* and growing political unrest, contributing to significant outflows of Venezuelan citizens. Hundreds of Venezuelans, for instance, applied for asylum in Canada in 1995, prompting Canadian authorities to reinstate visa requirements for Venezuelan tourists (Kratochwil, 1995: 33). The Venezuelan government threatened to deport the illegally-resident alien population en masse, but it was uncertain whether it had the intent and means to carry out the threat.

The legalization policies implemented in Argentina and Venezuela testified to the changing character of migration within Latin America. Intra-regional labour migrations had supplanted immigration from Europe. According to a 1993 report, based on analysis of 1980s' census information, some 2 million Latin Americans and Caribbeans lived within the region outside their country of birth. Although foreign Latin Americans and Caribbeans did not exceed 10 per cent of the total population of any country in the region 'there has been an increase in the last decade, both of the overall magnitude of mobility within Latin America, and of the relative importance of Latin Americans and Caribbeans in migration between regions' (Maguid, 1993: 41).

The post-Cold War period in Latin America and the Caribbean was marked by efforts to reinvigorate and expand the many regional integration instruments like MERCOSUR and the Andean Group (CAN) (Derisbourg, 2002). The former includes Argentina, Brazil, Paraguay and Uruguay with a total population of 210 million. The latter consists of Bolivia, Colombia, Peru and Venezuela with a total population of 113 million. Movements of persons across national borders within these regional blocs were an important concern. However, coordination and cooperation was stymied by inadequate information (Maguid, 1993). Analysing earlier efforts within the Andean Group with regard to labour migration, Kratochwil concluded that 'the significant amount of work has been ultimately ineffective and the administrative agencies have collapsed erratically' (Kratochwil, 1995: 17). As in the Arab region and sub-Saharan Africa, Latin American and Caribbean regional integration projects had slender records of accomplishment in management of international migration.

A second significant feature of the post-Cold War period in Latin America and the Caribbean also echoed 1990 developments elsewhere. There were significant repatriations of refugees subsequent to peace accords in some countries but the eruption of new conflicts in the region produced new refugee flows. The most significant peace accords were reached in Central America where fighting in El Salvador, Nicaragua and Guatemala abated. In the 1980s, about 2 million Central Americans were uprooted, but only some 150 000 of these were recognized by the UNHCR as refugees (Gallagher and Diller, 1990: 3). There were significant repatriations of Guatemalans from Mexico, Nicaraguans from the USA and Costa Rica and of Salvadorans.

However, the political situation in all three countries continued to be tenuous. There were several reports of killings of returning Guatemalans, many of whom are Indians. Guatemalan migrants continued to come to the USA. Their presence was increasingly evident in labour-intensive agriculture and the poultry-processing industry. Most Guatemalans, Salvadorans and Nicaraguans in the USA did not repatriate despite the peace accords. Between 1984 and 1994, over 440 000 Central Americans applied for asylum in the USA. Most applications were denied, but most applicants stayed on (Martin and Widgren, 1996: 35). When the USA adopted a new law in 1986 to curb illegal immigration, the then president of El Salvador, Napoleon Duarte, wrote to the US president complaining that the law threatened El Salvador's stability because remittances from Salvadorans in the USA were vital to the economy. Other Latin American leaders voiced similar concerns, but they largely proved exaggerated (Mitchell, 1992: 120–3). In a similar vein, while there was some repatriation of Nicaraguans from Costa Rica, many others stayed on. In 1993, Nicaragua and Costa Rica signed a bilateral labour agreement concerning employment of Nicaraguans in Costa Rican agriculture (Maguid, 1993: 88). Illegal employment of Nicaraguans was widespread in Costa Rica and an increasingly salient question in bilateral relations between the two countries.

The Haitian outflow to the USA was part of a broader shift in the Latin American and Caribbean countries. By the 1970s, the region was a net exporter of people. The underlying reasons for this historic change are many, and the transition did not occur overnight. Since the colonial period, Caribbean migrants had been arriving on the eastern and southern shores of what is now the USA. These northward flows were accentuated during the Second World War, when Caribbean workers were recruited for defence-related employment in US Caribbean possessions, specifically the Virgin Islands, and for agricultural work on the US mainland. The origins of the British West Indies' Temporary Foreign Worker Programme, which recruited thousands of workers annually for employment in US agriculture, and which continued into the 1990s as the so-called H-2A programme, were not unlike the far larger temporary foreign worker programme established between Mexico and the USA.

Temporary labour recruitment helped set in motion the massive northward flows of legal and illegal immigrants from Latin America and the Caribbean to the USA and Canada after 1970. But the causes of the shift are to be found in other factors as well: the declining economic fortunes of the region, its demographic explosion, rural–urban migration, political instability and warfare. Many of these additional factors cannot be viewed as strictly internal. Policies pursued by the USA, such as its intervention in Central America, clearly played a role in the sea change that saw the area become a net area of emigration. The linkage was clearest in the case of the Dominican Republic, where US involvement in the assassination of the Dominican president Trujillo in 1961 led to mass issuance of visas to Dominicans to forestall a Cuban-style revolution (Mitchell, 1992: 96–101).

Up to 1990, the single most important factor behind the rise in emigration to the USA, Canada and certain Western European countries was the declining level of economic performance. Gross domestic product (GDP) per capita declined sharply in the 1980s, what some called the 'decade lost to debt' (Fregosi, 2002: 443). Democratic renewal and a trend toward liberalization of Latin American economies in the early and mid-1990s briefly buoyed Latin American economies before a succession of economic crises ravaged the area. By 2000, an estimated 78 million out of a total Latin American population of 480 million lived below the poverty line. Liberalization policies increased already severe inequality in countries like Mexico and Argentina. In the latter, the Gini index measuring inequality increased from 34.5 in 1974 to 50.1 in 1998. One consequence

Box 6.4 Haitian *braceros* in the Dominican Republic

One of most notorious migrations in the Caribbean sub-region is the employment of Haitian *braceros* (strong armed ones) in the Dominican Republic's sugar cane harvest. This migration is between two historically antagonistic states whose combined populations comprised almost half of the total population of the Caribbean region in 2002.

Every year between November and May, Haitians enter for harvest-related employment, both legally and illegally. These workers were predominantly men, but some families have followed, leading to settlement. In the early 1980s, the sugar crop represented only 12 per cent of cultivated land in the Dominican Republic but half of all exports and one-fifth of the revenue received by the government. Despite high unemployment in the Dominican Republic, practically all the sugar crop was harvested by Haitians. One reason for the rejection of such work by Dominicans was the horrific working conditions and pay of Haitian *braceros*. In 1937, some 15 000 Haitians were massacred on the Dominican side of the island of Hispaniola. In 1979, the London-based Anti-Slavery Society described the Haitian sugar cane workers' plight as slavery (Péan, 1982: 10). Every year, the government of the Dominican Republic would make a payment to the Haitian government for the provision of *braceros*. In 1980–1, US$2.9 million were paid for 16 000 *braceros* (French: 1990). This arrangement lapsed only in 1986 when the Haitian dictator 'Baby Doc' Duvalier was forced into his sumptuous exile in France.

Subsequently, sugar interests in the Dominican Republic relied increasingly on recruiters to find an estimated 40 000 workers needed for the harvest. In 1991, following democratic elections in Haiti and growing international

→

was a growing desire by Argentinians to emigrate. According to a 2001 poll, 21 per cent of all Argentinians wanted to emigrate and one-third of persons between 18 and 24 years old (Fregosi, 2002: 436).

Notable trends in early twenty-first century Latin American migration included growing emigration to Western Europe, especially to Spain, which had signed bilateral labour recruitment agreements with several South American countries. Ecuadorians figured prominently in protests in Spain for legalization. The growth in Brazilian and Argentine migration to Europe and North America also was noteworthy. Most of the former was illegal and went to the north-east region of the USA whereas Argentine applicants for dual citizenship with Spain surged as did applications for visas to the USA between 1999 and 2000 (Fregosi, 2002: 436). Growing trafficking of persons was in evidence throughout Latin American and many countries served as transit points for trafficking, mainly to the USA and Canada.

→

criticism of the plight of Haitian workers, the Dominican government ordered a mass expulsion of Haitians. Many of the more than 10 000 individuals expelled were persons of Haitian extraction who had long resided in the Dominican Republic or who had been born there (French, 1991: 15). This mass expulsion helped destabilize the fragile Haitian democracy. The overthrow of President Aristide in September 1991 led to a renewed outflow of Haitian emigrants to the USA. Most of them were intercepted by the US Coast Guard and detained at the US naval installation at Guantanamo Bay in Cuba before being repatriated.

In 1994, US military intervention in Haiti restored Aristide to power. However, the socio-economic and political crisis worsened. Haitian employment in the Dominican Republic's sugar cane industry declined to 20 000 but grew in other sectors like construction. In 2001, authoritative sources placed the population of Haitians and descendants of Haitians in the Dominican Republic at 500 000 (Alexandre, 2000: 18), about 6 per cent of the total population. About half of the Haitians were long-term residents or born in the Dominican Republic but not regarded as citizens (Segura, 2000: 4).

Large-scale deportations of Haitians continued. Between August 2000 and January 2001, the Dominican National Guard reported a total of 45 000 deportations (Segura, 2002: 5). IOM-initiated negotiations between the Dominican Republic and Haiti began in 1996 aimed at facilitating bilateral cooperation on humanitarian deportations, registration of Haitians and a possible legalization programme. But violence against Haitians continued and corrupt members of the National Guard facilitated trafficking of Haitian migrants (Alexandre, 2001: 51–3). Some Dominican political leaders feared that the potential implosion of Haiti would lead to a regional migration crisis, and viewed Haitian migration as a security threat.

Combating irregular migration was a major goal of the Puebla Process, formally the Regional Conference on Migration, begun in 1996. Eleven North American and Latin American states had become participants by 2000 and five other states were observers. Of the eleven regional consultative processes monitored by the IOM, the Puebla Process was regarded as one of the most successful (Klekowski von Koppenfels, 2001: 34–8). However, bilateral and regional cooperation of many issues related to irregular migration remained very problematic.

The government of Mexico decried the mounting toll of migrant deaths at the US–Mexico border. It estimated that 2000 migrants had died since 1994, and, in the most recent two-year period, an average of one per day (Nieves, 2002: A12). Most observers agreed that the increase in border-crossing deaths was associated with beefed-up enforcement at the US–Mexico border. US border patrol measures, like Operation Gatekeeper, begun in 1994, included deployment of supplementary border-monitoring personnel, physical barriers and enhanced surveillance equipment. This led migrants to rely more on traffickers who often attempted to cross into the USA through remote, dangerous areas (Andreas, 2001).

After the 'honeymoon' in US–Mexico relations came to an end by September 2001 (see Box 1.2 and Chapter 5) the future of the US–Mexico migration relationship, the world's single most important bilateral migratory nexus, remained unclear. The US$9 billion in remittances received yearly by Mexico from migrants abroad had become a mainstay of Mexico's economy. Roughly half of the 8 or 9 million Mexicans living in the USA had legal status and millions had become US citizens. A reform enabled US citizens of Mexican ancestry to become dual citizens of Mexico and President Fox and his administration clearly viewed the Mexico-origin population in the USA as an ally in US–Mexico relations. Indeed, the Mexican government pursued a strategy of fostering transnationalism amongst Mexican-background persons in the USA, a state of affairs which illustrated that transnationalism was scarcely inimical to states (Smith, 2001).

Overall trends and patterns in Latin American migrations discernible in the 1990s seemed likely to endure. Most emigration will continue to go to the USA and to Canada and the scale of intra-regional migrations will pale in comparison.

Conclusions

It is customary to differentiate between different categories of migrants, and regions of migration. But it is important to realize that all the movements have common roots, and that they are closely interrelated. Western penetration triggered off profound changes in other societies, first through colonization, then through military involvement, political links, the Cold War, trade and investment. The upsurge in migration is due to rapid processes of economic, demographic, social, political, cultural and environmental change, which arise from decolonization, modernization and uneven development. These processes seem set to accelerate in the future, leading to even greater dislocations and changes in societies, and hence to even larger migrations.

Thus the entry of the countries of the South into the international migration arena may be seen as an inevitable consequence of the increasing integration of these areas into the world economy and into global systems of international relations and cultural interchange. These new migratory movements are a continuation of historical processes that began in the fifteenth century with the European colonial expansion, and the ensuing diffusion of new philosophical values and economic and cultural practices around the globe.

The first effect of foreign investment and development is rural–urban migration, and the growth of cities. Leaving behind traditional forms of production and social relationships to move into burgeoning cities is the first stage of fundamental social, psychological and cultural changes which create the predispositions for further migrations. To move from peasant agriculture into a city like Cairo, São Paulo or Lagos may be a bigger step for many than the subsequent move to a 'global city' like Paris or Los Angeles.

It is therefore inappropriate to analyse migration as an isolated phenomenon; it is just one facet of societal change and global development. The different forms of migration – permanent emigration, contract labour, professional transients, students and refugees – all arise from these broader changes. The categories are interdependent: for instance, a refugee movement can start a permanent migration, or suspension of legal worker recruitment can lead to illegal movements. Migrations arise from complex links between different societies, and help to create new links. The mobility of people will remain a key issue in development strategies in the less-developed world, as well as a major element in North–South relations.

Guide to further reading

ILO and IOM publications, especially the ILO's *International Migration Papers* series, are particularly helpful for the Arab region, sub-Saharan Africa and Latin America. The four-volume series edited by Appleyard on *Emigration Dynamics in Developing Countries* (1998–9) is invaluable, as are Stichter (1985), Appleyard (1988, 1991), Stahl (1988) and Stalker (1994, 2000) for global perspectives on migration. Harris (1995), Martin and Widgren (1996), Bernstein and Weiner (1999), Siddique (2001) and Zolberg and Benda (2001) are also illuminating.

For the Arab region, Shami (1994) is authoritative. Kerr and Yassin (1982) and Semyonov and Lewin-Epstein (1987) provide information on labour migrations relating to Arab countries and Israel. For sub-Saharan Africa, Ricca (1990) and Adepoju in Siddique (2001) are most valuable. Mitchell (1992) provides an excellent overview of the relationship between US foreign policy and Western Hemisphere migration.

Chapter 7

New Migrations in the Asia-Pacific Region

Over half the world's population lives in the Asia-Pacific region. In the 1970s and 1980s international migration from Asia grew dramatically. (Strictly speaking, Asia includes the Arab region and Turkey. However, these countries have already been dealt with, so this section will be concerned mainly with South Asia (the Indian subcontinent), East Asia and South-East Asia.) The main destinations were North America, Australia and the oil economies of the Middle East. Since the 1990s, the major growth has been in migration within Asia, particularly from less-developed countries with massive labour surpluses to fast-growing NICs. The international movements are often linked to internal migration. In China, massive flows from rural areas in the centre and west to the new industrial areas of the east (especially Beijing, Shanghai and the Pearl River Delta) have created a 'floating population' of over 100 million people. Indonesia's *transmigrasi* programme has shifted about 1.7 million families from densely populated Java to more sparsely populated islands like Sumatra, Sulawesi and Irian Jaya since 1969 (Tirtosudarmo, 2001: 211). Internal displacement is also a major problem: some 5 million Asians have become IDPs due to conflict, violence or human rights abuses (Deng, 2001). Millions more are displaced by development projects, such as large dams, while others flee environmental change and natural disasters, like volcanoes and floods. Internal migration will not be dealt with here, but it is important to realize that it is often the first step in a process that leads to international movement.

Immigration is strictly regulated in Asia and the Persian Gulf. Policy-makers encourage temporary labour migration, but prohibit family reunion and permanent settlement. They are determined not to repeat the Western European experience of the 1970s, in which guestworkers became permanent residents and formed new ethnic minorities (Weiner and Hanami, 1998). However, trends towards long-term stay are becoming evident in some places.

The development of Asian migration

Asian migration is not new: westward movements from Central Asia helped shape European history in the Middle Ages, while Chinese migration to South-East Asia goes back centuries. In the colonial period, millions of indentured workers were recruited, often by force (see Chapter 3). Chinese settlers in South-East Asian countries (Sinn, 1998) and South Asians in Africa became trading minorities with an important intermediary role for colonialism. This often led to hostility – and even mass expulsions – after independence. However, it also helped create the ethnic networks that encouraged more recent migrations (IOM, 2000b: 69). In the nineteenth century there was considerable migration from China and Japan to the USA, Canada and Australia. In all three countries, discriminatory legislation was enacted to prevent these movements.

Migration from Asia was low in the early part of the twentieth century owing to restrictive policies by immigration countries and colonial powers. However, movements within Asia continued, often connected with political struggles. Japan recruited 40 000 workers from its then colony, Korea, between 1921 and 1941. Japan also made extensive use of forced labour in the Second World War. Manchuria experienced mass migration from the late nineteenth century, while within the Indian subcontinent there were huge movements, especially after independence in 1947.

External movements started to grow from the 1960s. The reasons were complex (Fawcett and Cariño, 1987; Skeldon, 1992: 20–2). Discriminatory rules against Asian entries were repealed in Canada (1962 and 1976), the USA (1965) and Australia (1966 and 1973). Increased foreign investment and trade helped create the communicative networks needed for migration. The US military presence in Korea, Vietnam and other Asian countries forged transnational links, as well as directly stimulating movement in the shape of brides of US personnel. The Vietnam War caused large-scale refugee movements. The openness of the USA, Canada and Australia to family migration meant that primary movements, whatever their cause, gave rise to further entries of permanent settlers. The huge construction projects in the Gulf oil countries caused mass recruitment of temporary contract workers. Rapid economic growth in several Asian countries led to movements of both highly-skilled and unskilled workers.

Asia's massive entry onto the world migration stage in the mid-twentieth century can be seen as the result of the opening up of the continent to economic and political relationships with the industrialized countries in the post-colonial period. Western penetration through trade, aid and investment created the material means and the cultural capital necessary for migration. At the same time, the dislocation of existing forms of production and social structures through industrialization, the 'green revolution' and wars (often encouraged by major powers as part of the Cold War) forced people to leave the countryside in search of better conditions in the growing cities or overseas. Later on, the rapid industrial take-off of some areas and the continuing stagnation or decline of others created new pressures for migration.

Map 7.1 *Migrations within the Asia-Pacific region (c. 1970–2000)*

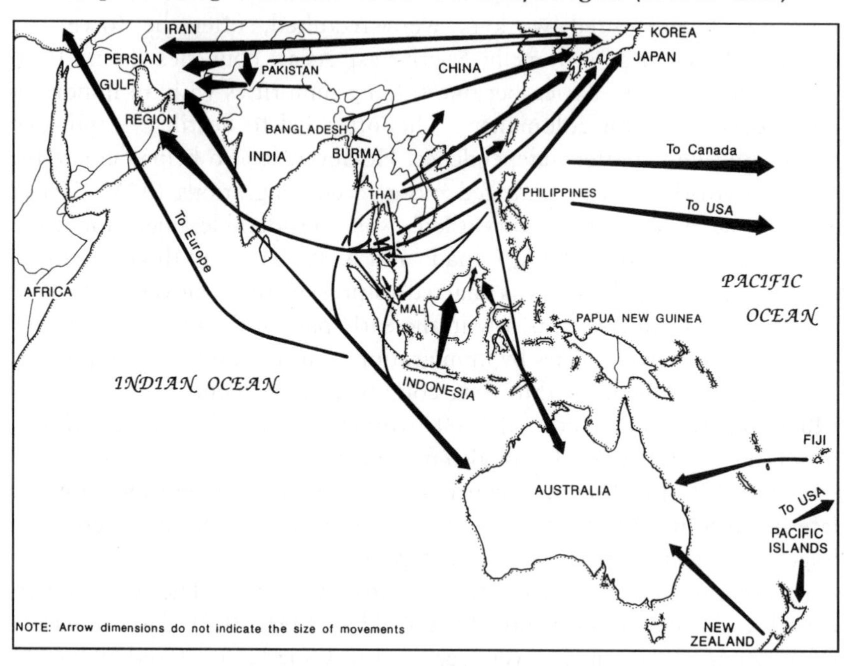

In recent years, social scientists have developed the notion of a 'migration transition'. Societies go through a number of fundamental changes in connection with economic development. The 'industrial transition' refers to the shift of economic activity and employment from agriculture to manufacturing, and then to the services. The 'demographic transition' involves falls in both mortality and fertility, leading to slower population growth and ageing populations. The 'migration transition' is seen as a result of all the preceding changes. At the beginning of the industrialization process, there is frequently an increase in emigration, due to population growth, a decline in rural employment and low wage levels. This was the case in early nineteenth-century Britain, just as it was in late nineteenth-century Japan, or Korea in the 1970s. As industrialization proceeds, labour supply declines and domestic wage levels rise; as a result emigration falls and labour immigration begins to take its place. Thus industrializing countries tend to move through an initial stage of emigration, followed by a stage of both in- and outflows, until finally they become predominantly countries of immigration (Martin *et al.*, 1996: 171–2). The early stage of increasing emigration linked to economic development has also been called the 'migration hump'. It has important policy implications: attempts to reduce migration by encouraging development may achieve the opposite in the short term, and only work in the long term (Martin and Taylor, 2001).

By 2000, there were estimated to be 6.2 million Asians employed outside their own countries within the Asian region, and another 5 million employed in the Middle East. Emigration for employment from the region grew at about 6 per cent a year from 1995 to 1999, despite the Asian financial crisis of 1997–9 (Abella, 2002). In addition, there are millions of refugees and family members. All countries experience both emigration and immigration, but it is possible to differentiate between mainly labour-importing countries (Japan, Singapore, Taiwan and Brunei), countries which import some types of labour but export others (Hong Kong, Thailand, Malaysia), and countries which are predominantly labour exporters (China, Philippines, India, Bangladesh, Pakistan, Sri Lanka, Indonesia). (Official names for some countries differ from customary usage. We use Taiwan for the country the UN refers to as Chinese Taipei, and Hong Kong for what has, since 1997, become the Hong Kong Special Administrative Region (SAR) of China. The Republic of Korea (South Korea) is called Korea, unless there is any risk of confusion with North Korea. We use Burma, rather than Myanmar.)

In this chapter, we will examine the main Asian migration systems: movement to western countries, contract labour to the Middle East, intra-Asian labour migration, movement of highly-skilled workers, student mobility and refugee movements. Most of these movements include substantial irregular migration. This often takes the form of tourist visa-holders overstaying their permits, but there is also a great deal of smuggling of undocumented workers. The number of irregular migrants in Japan, Korea, Taiwan, Malaysia and Thailand combined was estimated at 845 000 in 1997 (Scalabrini Migration Center, 2001), although this figure should be seen as very approximate.

Asian migration to Western Europe, North America and Australasia

Three European countries experienced large Asian migrations connected with decolonization: the Netherlands from the former Netherlands East Indies (Indonesia); France from Vietnam; and Britain from the Indian subcontinent and Hong Kong. There were also some smaller movements, like those from Goa, Macau and East Timor to Portugal. Such movements had declined considerably by the late 1970s. In the 1980s, Vietnamese workers were recruited by the Soviet Union and the German Democratic Republic. Although often called trainees, these migrants shared many of the characteristics of contract workers.

Since the 1990s, there has been an increase in labour migration from Asia to Europe, including recruitment of medical and information technology personnel. Female domestic workers from the Philippines and

China have moved in growing numbers to Italy. Irregular flows of manual workers to Britain and other destinations are growing. A recent trend is the growth of East Asian migration to Europe: in 1999, China was among the top ten source countries for Hungary, Italy and Finland, while Japan was in the top ten for France and the Netherlands (OECD, 2001: Chart 1.4).

The largest movement was that to the USA after the 1965 Immigration Act. The number of migrants from Asia increased from 17 000 in 1965 to an average of more than 250 000 annually in the 1980s (Arnold *et al.*, 1987) and over 350 000 per year in the early 1990s (OECD, 1995: 236). Most Asians came to the USA through the family reunion provisions of the 1965 Act, though refugee or skilled worker movements were often the first link in the migratory chain. Since 1992, Asia has been the source of about one-third of all immigrants, and by March 2000 there were over 7 million residents of Asian origin. In 1999, China was the second largest source of immigrants, with 37 000 (following Mexico with 132 000). India, Philippines, Vietnam and Korea were also among the top ten source countries (OECD, 2001: Chart 1.4).

Asian immigration to Australia developed after the repeal of the White Australia Policy in the late 1960s–early 1970s, with additional stimulus from the Indo-Chinese refugee movement at the end of the 1970s. By the beginning of the 1990s, about half of new immigrants came from Asia. Among the top ten source countries in 1999 were China (third after New Zealand and the UK), India, Philippines, Taiwan and Vietnam (OECD, 2001: Chart 1.4). Official estimates for 1999 put the Asia-born population at over 1 million, nearly a quarter of the immigrant population and about 5 per cent of the total population (OECD, 2001: Table B.1.4)

In Canada, it was the 1976 Immigration Act, with its non-discriminatory selection criteria and its emphasis on family and refugee entry, which opened the door to Asian migration. The 1981 Census showed the presence of 674 000 people of Asian ethnic origin (Kubat, 1987: 237). Since 1993, over half of all immigrants have come from Asia. Between 1995 and 1998, the six most important source countries were China, Taiwan, Hong Kong, India, Pakistan and the Philippines. By the 1996 Census, the roughly 1.6 million residents of Asian origin made up almost a third of the immigrant population (OECD, 2001: 66).

New Zealand is a traditional immigrant country which built up its population through immigration from the UK, with racially-selective entry policies to keep out non-Europeans. However, since the 1950s, economic and political links with nearby Pacific islands, such as Tonga and the Cook Islands, have given rise to new inflows (Trlin, 1987). From 1991, policies encouraged immigration of people with professional skills and capital for investment. Most of these came from Hong Kong, Taiwan, Korea and Japan (Lidgard, 1996: 6). In 1995, 22 000 of the total 56 000 permanent entrants were from Asia. By 1999, Asian entries had declined to 11 000 out of a total of 29 000 (IOM 2000b, 279). New Zealand's ethnic composition has become more complex: the Maori people have grown to over 10 per cent of the total population, Pacific Islanders make up about 5 per cent and Asians about 3 per cent (Pool and Bedford, 1996). This has led to heated public debates and electoral campaigns focusing on immigration policy (IOM, 2000b: 282–3).

The movements from Asia to the classical immigration countries of North America and Oceania have certain common features. Unexpectedly large movements have developed mainly through use of family reunion provisions. The countries of origin have become more diverse. Vietnamese and other Indo-Chinese refugees were a dominant flow in the 1970s and 1980s; Hong Kong became a major source in the run-up to incorporation into China in 1997. Movements from these countries continue and have been joined by flows from the Philippines, India, Japan and Korea. The most important trend is the growth in migration from China. In the last few years all these countries have changed their immigration rules to encourage entry of skilled and business migrants. A global labour market for highly-skilled personnel has emerged, with Asia as the main source.

Contract labour migration to the Middle East

Large-scale migrations from Asia to the Middle East developed rapidly after the oil price rises of 1973. Labour was imported at first from India and Pakistan, then from the Philippines, Indonesia, Thailand and Korea, and later from Bangladesh and Sri Lanka. By 1985, there were 3.2 million Asian workers in the Gulf states, of whom over 2 million were in Saudi Arabia. The Iraqi invasion of Kuwait and the Gulf War in 1990–1 led to the forced return of some 450 000 Asians to their countries of origin. After the war, recruitment of Asian workers increased again, partly due to reconstruction needs but also due to the replacement of 'politically unreliable' Palestinians in Kuwait and Yemenis in Saudi Arabia (Abella, 1995; see also Chapter 6 above). Israel began to recruit large numbers of Thais and Filipinos for agriculture, construction and domestic work, after security measures blocked entry of Palestinians from the West Bank and Gaza.

The largest migrant worker flows remain those from South Asia to the six countries of the Gulf Cooperation Council (GCC). In the late 1990s about 1 million contract workers left South Asia each year. In 1997, the total labour outflow was 416 000 from India (93 per cent to the Middle East), 231 000 from Bangladesh (about 75 per cent to the Middle East), 154 000 from Pakistan (96 per cent to the Middle East) and 150 000 from Sri Lanka (85 per cent to the Middle East) (IOM, 2000b: 110). The relatively small national labour forces of the GCC states work mainly in the public sector, leaving huge gaps in the private sector. The result is extreme dependence on foreign labour. Saudi Arabia, with a population of

20 million, had a foreign labour share of 28 per cent. The smaller GCC states had even higher foreign shares: Kuwait 65 per cent, Bahrain 37 per cent, Qatar 77 per cent, United Arab Emirates (UAE) 73 per cent and Oman 27 per cent (IOM, 2000b: 108).

In the 1970s, most migrants were male workers employed in the many construction projects funded by petro-dollars. Governments of sending countries like India, Pakistan and the Philippines actively marketed their labour abroad, and made labour-supply agreements with Gulf countries. Korean construction companies were encouraged to take on contracts in the Arab region, which included provision of labour. The Asian labour-sending countries also allowed private agencies to organize recruitment (Abella, 1995).

The temporary decline of the construction sector after 1985 encouraged more diverse employment of contract workers, particularly a shift into the services sector, such as hotels and personal services. There was an upsurge in demand for domestic servants, leading to a feminization of contract labour flows. Most women workers came from the Philippines, Indonesia, Thailand, Korea or Sri Lanka; neither Pakistan nor Bangladesh sent females abroad (Skeldon, 1992: 40–1). Many Filipinos and Koreans were skilled workers who worked as drivers, carpenters, mechanics or building tradesmen. Others were professionals or para-professionals (engineers, nurses and medical practitioners). Labour migrants were not part of the unemployed rural and urban poor, but rather educated people, whose departure could have a negative effect on the economy (Skeldon, 1992: 38). As flows became more diverse, undocumented migration grew sharply. The UAE expelled more than 160 000 unauthorized workers in a three-month period in 1996 and Saudi Arabia reportedly expels 350 000 to 450 000 per year. In the latter case, recent government attempts to encourage nationals to take private sector jobs have led to increased deportations (IOM, 2000b: 107–15).

Asians in Arab countries encounter difficult conditions, due both to the lack of worker rights and the very different cultural values, especially with regard to the position of women. Migration takes place within rigid contract labour frameworks: workers are not allowed to settle or bring in dependants, and are often segregated in barracks. They can be deported for misconduct and often have to work very long hours. Women domestic workers are often subjected to exploitation and sexual abuse. The big attraction for workers is the wages: unskilled workers from Sri Lanka earn eight times more in the Middle East than at home, while Bangladeshis earn 13 times more (IOM, 2000b: 119). Many migrant workers are exploited by agents and brokers, who take large fees (up to 25 per cent of their pay). Agents sometimes fail to keep their promises of providing work and transport, and wages and working conditions are often considerably inferior to those originally offered.

Labour migration within Asia

Since the mid-1980s, rapid economic growth and declining fertility have led to considerable demand for migrant labour in the new industrial economies of East and South-East Asia. Labour migration within Asia grew exponentially in the first half of the 1990s. There was some return migration during the Asian financial crisis of 1997–9, but in the meantime labour migration has resumed. In all the 'tiger economies', migrant workers are doing the '3D jobs' (dirty, dangerous and difficult – or just low skilled and poorly paid) that nationals can increasingly afford to reject. It is impossible to deal in detail with the complex migration experience of each Asian country. Instead we will discuss some general trends, look briefly at a number of immigration countries and present a few more detailed case studies

A key recent development has been the feminization of migration. As demand for service workers has grown, women have come to dominate certain flows. Two-thirds of Indonesian migrants from 1984 to 1994 were women (Amjad, 1996: 346–9). The female share among first-time migrant workers from the Philippines rose from 50 per cent in 1992 to 61 per cent in 1998. Intra-Asian movements were particularly female dominated, while men were still the majority in flows to Saudi Arabia (Go, 2002: 66). Most migrant women are concentrated in jobs regarded as 'typically female': domestic workers, entertainers (often a euphemism for prostitutes), restaurant and hotel staff and assembly-line workers in clothing and electronics. These jobs offer poor pay, conditions and status, and are associated with patriarchal stereotypes of female docility, obedience and willingness to give personal service. Female migration has considerable effects on family and community dynamics. Married women have to leave their children in the care of others, and long absences affect relationships and gender roles.

The increase in domestic service reflects the growth of dual career professional households in Asia's new industrial countries. Singapore is a good example. Due to attractive job opportunities for Singaporean women, employment of foreign domestic servants is very high: in 1993, 15 per cent of households had a live-in servant, and there were estimated to be 81 000 foreign domestic servants, of whom 50 000 were from the Philippines, 17 000 from Sri Lanka and 10 000 from Indonesia. Fees for recruitment (mainly through specialized agencies) were deducted from the maid's wages and could be up to S$2000 (Wong, 1996). Domestic service leads to isolation and vulnerability for young women migrants, who often have little protection against the demands of their employers (Lim and Oishi, 1996).

Another form of Asian female migration that developed in the 1980s was that of so-called 'mail order' brides to Europe and Australia (Cahill, 1990). In the 1990s, foreign brides were recruited increasingly by farmers in rural areas of Japan and Taiwan, due to the exodus of local women to more attractive urban settings. This is one of the few forms of permanent immigration permitted in Asia. The young women involved, who come from the Philippines, Vietnam and Thailand, can experience severe social isolation (IOM, 2000b: 65).

A further feature of Asian labour migration is the major role played by the 'migration industry'. Most recruitment of migrant workers both to the Gulf and within Asia is organized by migration agents and labour brokers. Governments and employers in receiving countries find it easier to rely on such intermediaries than to organize movements themselves. Authorities of labour-sending countries have found themselves powerless to control the activities of the industry. Martin (1996: 201) estimates that migrants typically pay fees equal to 20-30 per cent of their first year's earnings. For the whole of Asia, the labour broker industry could be worth US$2.2 billion per year.

While some agents carry out legitimate activities, others deceive and exploit workers. For instance, certain Thai agents dupe young rural women into going to Japan, ostensibly to work in restaurants or factories, then hand them over to *Yakuza* gangsters, who keep them in conditions of near-slavery as prostitutes (Okunishi, 1996: 229-30). Imprisonment, deportation and even death are the risks faced by illegal migrants, while the leaders of the smuggling gangs are rarely apprehended. On Christmas Day 1996, up to 280 Indians, Pakistanis and Sri Lankans drowned in the sea between Malta and Sicily. They were the victims of an international network of migrant smugglers, with tentacles in South Asia, the Middle East and Southern Europe. The migrants had paid up to US$9000 for the fatal voyage to a dream of prosperity in Europe (Ferguson, 1997: 29). Even when they arrive safely in North America or Europe, many illegal entrants are subject to 'debt bondage', working for years to pay off fees to the smugglers. Chinese workers smuggled from Fujian Province by 'snakehead' gangs pay up to US$30 000. It is believed that between 100 000 and 200 000 unauthorized migrants leave China each year with the help of smuggling gangs (IOM, 2000b: 68–70).

East Asia

The East Asian economic miracle has led to strong demand for labour, but governments have rejected recruitment of foreign workers for fear of bringing about cultural and social change to nations considered as homogeneous and monocultural. The combination of fertility decline, ageing populations and growing undocumented migration has led to serious contradictions, most evident in Japan, but also emerging in Korea, Hong Kong and Taiwan.

Korea exported labour to the Gulf in the 1970s and 1980s, but has now passed through the migration transition: by 1995, the GDP per capita was US$10 000 and labour departures had fallen sharply. In 2000 there were 312 000 foreign workers in Korea. The major source was China – 100 000 in 2000, of whom 57 000 were ethnic Koreans holding Chinese nationality. Other main sources were the Philippines, Bangladesh, Thailand and Mongolia. Official policy is fairly similar to Japan. Unskilled migrants are barred and many foreign workers (105 000 in 2000) are classified as 'trainees', but in fact carry out labouring jobs. The majority (189 000 in 2000) are undocumented workers, who are paid low wages and lack basic rights (Seol and Skrentny, 2003).

Hong Kong has been transformed from a labour-intensive industrial economy to a post-industrial economy based on trade, services and investment, leading to shortages of both skilled and unskilled workers. Highly-qualified expatriate workers from North America, Western Europe and India are recruited for well-paid jobs in finance, management and education. Unskilled workers from China have entered illegally in large numbers. Maids are recruited in the Philippines and elsewhere. In 1997, the situation was complicated by fears about the effects of reunification with China. Many highly-skilled Hong Kong workers emigrated to the USA, Canada and Australia, to seek a safe domicile (Skeldon, 1994). Some stayed in the receiving country only long enough to gain permanent resident status or citizenship, and then returned to Hong Kong to work, often commuting back and forth. This group are sometimes known as 'astronauts', while the children they leave in Canada or Australia are called 'parachute children' (Pe-Pua *et al.*, 1996). In the meantime, the political situation has stabilized, and labour entries continue to rise, with an estimated 509 000 foreign residents in 1998 (IOM, 2000b: 63).

Taiwan introduced a foreign labour policy in 1992, permitting recruitment of migrant workers for occupations with severe labour shortages. Duration of employment was limited to two years. In 2000, there were 380 000 legal foreign workers and an unknown number of illegals. Workers came mainly from Thailand, the Philippines, Malaysia and Indonesia. Most recruitment is carried out by labour brokers, who charge high fees to workers. Many workers stay on illegally after two years, or change jobs to get higher wages and to escape repayments to brokers (Lee and Wang, 1996).

South-East Asia

South-East Asia is characterized by enormous ethnic, cultural and religious diversity, as well as by considerable disparities in economic development. Governments of immigration countries are concerned about

Box 7.1 Dilemmas of an East Asian immigration country: Japan

Japan has been experiencing severe labour shortages since the mid-1980s. At first women were admitted, mainly from the Philippines and Thailand, to work as dancers, waitresses and hostesses. They were followed by men from these countries as well as Pakistan and Bangladesh, who worked – generally illegally – as factory or construction workers. The foreign population of Japan increased from 817 000 in 1983 to 1.6 million in 1999. About 41 per cent are permanent residents (OECD, 2001: 198), mainly Koreans, who were recruited as workers before and during the Second World War. Other foreign groups have also grown rapidly: the Chinese grew from 75 000 in 1985 to 294 000 in 1999; Brazilians (mainly *Nikkeijin*, descendants of earlier Japanese emigrants) increased from 2000 to 224 000, and Filipinos from 12 000 to 116 000.

The Japanese government is strongly opposed to immigration, due to its concern to preserve ethnic homogeneity. In 1989, revisions to the Immigration Control Act introduced severe penalties for illegal foreign workers, brokers and employers. However, recruitment of unskilled foreigners of Japanese origin was permitted, leading to a scramble to recruit *Nikkeijin* from Brazil and Peru. Other 'side doors' to Japan include recruitment of 'trainees' from developing countries, or employment of foreigners registered as students of Japanese language schools, who are permitted to work 20 hours per week. Trainees are often used as cheap labour (Oishi, 1995: 369). The 'back door' of irregular labour migration appears to be tolerated by the Japanese authorities, who probably have the institutional capacity to curtail it if they wanted to. Official estimates put the number of irregular immigrants in Japan in 1999 at 252 000 – a decline from the 1995 peak of 285 000 (OECD, 1997, 2001: 198).

Immigrants make up only 1.2 per cent of Japan's population of 126 million. However, low birth rates and population ageing make it likely that immigration will grow in future, despite current economic stagnation. Well-educated young Japanese are unwilling to take factory jobs. Government industry policy encourages investment in new technology to raise labour productivity, while many companies shift labour-intensive workplaces to

→

maintaining complex ethnic balances, and combating possible threats to security.

Singapore is heavily dependent on migrant workers from Malaysia, Thailand, Indonesia, the Philippines, Sri Lanka, India and China. There were about 590 000 in 2000 – 28 per cent of the labour force. Foreign employment grew threefold between 1993 and 2000, with the Asian Crisis of 1997–9 causing only a brief slowdown (Abella, 2002). Foreign men work in construction, shipbuilding, transport and services; women are mainly in domestic and other services. The government imposes a foreign worker levy to encourage employers to invest in new technology rather than hiring

→

low-wage countries. But there are limits to these approaches: it is hard to increase productivity or to relocate construction and services jobs, and many factory jobs, such as making car components, are part of complex supply chains which cannot easily be divided geographically. A topical issue is the need for aged-care workers to look after Japan's growing elderly population.

The key question is whether settlement is taking place. Recent research shows that immigrant workers were heavily concentrated in certain sectors or occupations, causing structural dependence (Mori, 1997: 155). This could encourage employer pressure to regularize undocumented workers. Differing employment patterns are linked to varying legal status: regular workers (especially *Nikkeijin*) find jobs in large enterprises, while irregular workers are mainly in small enterprises or informal-sector jobs. A study of Asian newcomers in the Shinjuku and Ikebukuro districts of Tokyo found some long-term settlement, as well as intermarriage with Japanese (Okuda, 2000). A study of *Nikkeijin* in Toyota City found high levels of concentration in certain apartment blocks, and frequent isolation from the Japanese population. Conflicts frequently arose around issues of daily life such as rubbish disposal, noise and traffic offences (Tsuzuki, 2000). Komai (1995) found tendencies to international marriages, family formation, residential concentration and the building of ethnic communities. Ethnic places of worship, businesses, associations and media were beginning to emerge.

Another significant trend is the gradually improving – though still weak – situation of immigrants with regard to civil, political and social rights (Kondo, 2001). Long-standing residents, mainly of Korean origins, may remain non-citizens even into the third or fourth generation due to restrictive naturalization laws (Esman, 1994). However, legal changes in 1992 led to a gradual rise in naturalizations: from 6794 in 1990 to 16 120 in 1999 (of whom 10 059 were Koreans) (OECD, 2001: 337). Mori (1997: 189–206) found that public authorities were gradually including foreign residents – even irregular workers – in health, education and welfare services. Social integration programmes have been introduced, including employment service centres for foreign workers and education for children of foreign nationals on equal terms with native Japanese (OECD, 1998b: 131). Many voluntary associations have been set up to work for improved rights for immigrants.

migrants. However, this has led to downward pressure on migrants' wages, rather than reductions in foreign employment. Unskilled workers are not permitted to settle or to bring in their families. Migrants usually work long hours, six days a week, and live in barracks. However, the government favours entry of skilled workers and professionals and gives them a privileged status. There were about 55 000 of these in 1997, about 12 per cent of the foreign workforce (IOM, 2000b: 82). Such migrants – especially those of Chinese ethnicity – are encouraged to settle permanently.

Thailand became a major exporter of workers to the Gulf in the 1980s. Fast economic growth in the 1990s initiated a migration transition, though

Box 7.2 Dilemmas for Asian immigration countries: Malaysia

Today, Malaysia has the largest foreign share in its population of any Asian immigration country – at least 5 per cent. The share in the employed labour force is probably double this. Malaysia experiences both emigration and immigration. Low-skilled Malays work in Singapore, while middle-class ethnic Chinese and Indians migrate to Australia and North America. But inflows far exceed departures: in 2000 there were 850 000 registered foreign workers in Malaysia (Abella, 2002). Nearly two-thirds were from Indonesia, with smaller numbers from Bangladesh, the Philippines and Thailand (IOM, 2000b: 85). In 1997 there were officially estimated to be at least 1 million undocumented workers. A more recent estimate is 200 000 in 2000 (Abella, 2002). Whether numbers have really declined, or whether it is simply a matter of unreliable statistics is hard to say. The East Malaysian island states of Sabah and Sarawak are even more dependent than Peninsular Malaysia on foreign workers, with up to 700 000 migrants in 2000 – mainly Indonesians and Filipinos. For centuries, these islands have been part of a geographic zone of free circulation between peoples linked by ethnicity and trading relationships. Immigrants play a major role in plantations and the informal economy (IOM, 2000b: 87).

Malaysia is a multi-ethnic, middle-income country. Its complex ethnic balance is a result of colonial labour import for the tin mines and rubber plantations. Today the citizen population is made up of 61.9 per cent Malays, 29.5 per cent Chinese and 8.6 per cent Indians (*Far Eastern Economic Review*, 2000: 161). Successful economic management has led to rapid economic growth and industrialization since the 1980s, making Malaysia intro a 'second-wave tiger economy' with severe labour shortages, especially in the plantation sector. Malaysia made the 'migration transition' from labour export to labour import in the mid-1980s, relatively early in its development process. Lim attributes this phenomenon to two special factors: the multi-ethnic population, which facilitated rapid reactivation of historical migration networks; and the open export-oriented economy, with high rates of foreign investment (Lim, 1996).

Government policies consist of a mixture of attempts at regulation of foreign labour, legalization campaigns and border control measures – such as a plan to build a 500-kilometre wall along the northern border with Thailand in 1996. In 1998, in response to the Asian Crisis, the government announced plans to reduce the foreign labour force by up to 1 million through deportation of undocumented workers and non-renewal of contracts of legal

→

some Thai workers still seek work abroad. Trafficking of Thai women for prostitution remains a major problem. Construction, agricultural and manufacturing jobs have attracted large numbers of workers from Burma, Laos, Bangladesh and India. As elsewhere, the Asian Crisis led to attempts to expel foreign workers, with some 300 000 being repatriated (IOM,

→

workers (Pillai, 1999). However, it soon became clear that the long coastline was impossible to control, and that undocumented workers within the country were hard to identify. Poor conditions in Indonesia led to increased migration pressures, while Malaysian employers sought to retain workers in industrial and plantation jobs which were not attractive to local workers. Estimates put actual repatriations in 1998 at around 200 000.

The failure to cut the foreign labour force significantly during the Crisis demonstrated the structural dependence of the Malaysian economy on labour import. Does this imply that long-term settlement is taking place? Research by Kassim (1998) in squatter settlements around Kuala Lumpur documented processes of community formation. Moreover, family migration is common in Sabah, while in Peninsular Malaysia more Indonesian and Filipina women are entering services such as domestic work and hotels (Pillai, 1999: 181–2). Increased female migration is conducive to family formation and long-term stay. However, in August 2002, the government introduced a new law to deter illegal migrants, through severe penalties including heavy fines, caning and up to five years in prison. Tens of thousands of Indonesians and Filipinos fled, with their home countries sending naval vessels to evacuate them. Human rights groups pointed out that deportees included asylum seekers such as Rohingyas from Burma and Achehnese from Indonesia, who face persecution at home (BBC, 2002).

Pillai points to a politicization of migration. Until 1995, it was not an important public topic, but since then it has become a key issue, with frequent media debates and statements by politicians (Pillai, 1999: 182–6). This development is linked to the realization that migration is not a temporary phenomenon, and may have unpredictable social and cultural consequences. By 1999, the government was under pressure from the Malaysian Agricultural Producers Association, the construction industry and some state governments to bring in more workers. The Malaysian Trade Unions Congress opposed labour recruitment due to its effects on jobs and wages for local workers, while Chinese political groupings feared that Indonesian immigration would alter the ethnic balance to their disadvantage. The government party, UMNO, and the main Islamic opposition party, PAS, both supported Indonesian entries as a potential boost to Malay and Islamic interests. There were frequent polemics against illegal immigrants as a threat to public order and health. However, a growing number of NGOs are supporting migrants. The trial of Irene Fernandez, leader of the women's rights organization Tenaganita, for exposing bad conditions in migrant detention centres became a major public issue (Jones, 2000).

2000b: 92). However, growth soon resumed: there were an estimated 665 000 foreign workers in 2000. Regulation is poor and only 103 000 were legal residents (Abella, 2002). It is hard to distinguish clearly between migrant workers and refugees, especially in the case of the Burmese – the largest group – and the Cambodians.

Countries of emigration

Just as the Mediterranean periphery fuelled Western European industrial expansion up to the 1970s, industrializing Asia has its own labour reserve areas: China, the South Asian countries, the Philippines and Indonesia have all become major labour providers for the region and indeed for the rest of the world. Most Asian sending-country governments have set up

Box 7.3 Dilemmas for emigration countries: the Philippines

The Philippines is today's the labour exporter par excellence (rather like Italy a generation ago) with nearly one-tenth of its people overseas. The government estimates that about 7 million Filipinos work abroad, and that they remitted US$7 billion in 1999 (IOM, 2000b: 96). Filipinos are to be found all over the world. Permanent settlement in the USA, Canada and Australia grew from the 1960s. Under the Marcos martial law regime of the 1970s, export of labour became a key element of economic policy. Since then, ever-increasing numbers of temporary workers or Overseas Contract Workers (OCWs) have been deployed overseas: first to the Gulf states and then to other Asian countries. Filipinos also have an increasing presence in Europe, particularly in Italy and Spain. Undocumented migration, often organized by agents, has grown, with an estimated 1.9 unauthorized migrants living abroad (IOM, 2000b: 96). Emigration has become part of normal life for millions of Filipinos and their communities.

Official annual deployments of OCWs grew from 300 000 in 1984 to 559 000 in 1997. In 1997, labourers made up 38 per cent of new hires, service workers 34 per cent, professionals 23 per cent, maids 21 per cent and entertainers 12 per cent (IOM, 2000b: 96–7). In addition, 188 469 Filipinos left home in 1997 as seafarers on foreign ships (Battistella and Assis, 1998: 234). Despite fears that the Asian Crisis would cause returns of up 100 000 workers, OCW deployments actually grew slightly in 1998 (IOM, 2000b: 98).

The Philippine government takes an active role in migration management. People who wish to work abroad have to register with the Philippine Overseas Employment Administration (POEA). The Overseas Workers' Welfare Administration (OWWA) has the function of assisting workers and protecting them from exploitation and abuse. Pre-departure orientation seminars are provided for entertainers, domestic workers and nurses. The Philippines has special officials at its consulates but their number is relatively low: in 1993 there were 31 labour attachés, 20 welfare officers and 20 coordinators to respond to the needs of 4.2 million migrant workers in 120 countries (Lim and Oishi, 1996: 120). Philippine officials often find themselves powerless against unscrupulous agents and abusive employers, who may have the backing of the police and other authorities in receiving countries.

The weakness of the Philippine government in protecting vulnerable workers led to a politicization of migration policy in 1995. On 17 March, a

→

special departments to manage recruitment and to protect workers, such as Bangladesh's Bureau of Manpower, Employment and Training (BMET) and India's Office of the Protector of Emigrants. The governments of labour-sending countries see migration as economically vital, partly because they hope it will reduce unemployment and provide training and industrial experience, but mainly because of the worker remittances (Appleyard, 1998b).

→

Filipina domestic worker, Flor Contemplacion, was hanged in Singapore, after being found guilty of murder. The case strained relations between the two countries and led to a heated debate in the Philippines. At that time there were over 60 000 Filipino workers in Singapore – the great majority were female domestic workers. Frequent cases of abuse had been reported, including non-payment of salaries, poor working conditions, ill-treatment and sexual harassment (Gonzalez, 1998: 5; Wong, 1996). Contemplacion was accused of the murder of a fellow Filipino and her employer's child. The case appeared as the culmination of a long series of humiliations suffered by OCWs in Singapore and elsewhere. There was large-scale mobilization by opposition parties, church associations, women's groups, labour unions and OCW organizations. A series of mass demonstrations was organized, culminating in the presence of more than 25 000 people at Contemplacion's funeral (Gonzalez, 1998: 6–7).

The Ramos Administration was forced to act. Migration of domestic workers to Singapore was suspended, albeit temporarily and ineffectively. In June 1995, the Philippine Parliament passed the Migrant Workers and Overseas Filipinos Act – the 'OCWs' Magna Carta'. This Act claimed to represent a shift in philosophy away from the primacy of economic goals, in favour of protecting the dignity and human rights of Filipinos. Specific policies included selective deployment favouring certain occupations and destinations; measures to improve information for prospective migrants; and a 'country-team approach' to improve cooperation between government agencies (Go, 1998).

These measure appear to have had little effect. There is no evidence of dramatically reduced migration of female entertainers or domestic workers, nor of substantial improvement in the conditions of Filipino OCWs. Nor does the Philippine government seem any more effective in providing legal protection to workers. A major difficulty was the unwillingness of labour-recruiting countries to cooperate by entering into bilateral agreements with the Philippines, or by adhering to multilateral instruments such as ILO Conventions and the 1990 UN Convention on the Rights of All Migrant Workers and Members of their Families. To enforce the 1995 Act fully, the Philippine government would have to stop most labour emigration. Since labour force growth remains rapid, while economic development is slow, the result might be mass unemployment and considerable discontent. Thus the Philippines cannot break its dependence on labour export, and the overwhelming market power remains with the labour-importing countries.

Remittances make a major contribution to the balance of payments of countries with severe trade deficits. Pakistani workers remitted US$1.4 billion in 1994, 17 per cent of the country's total revenue from export of goods and services. Indian workers remitted US$5 billion, 14 per cent of such revenue. The figure for Bangladeshis was US$1.1 billion (34 per cent), and for Sri Lankans US$0.7 billion (17 per cent) (IOM, 2000b: 123). (These figures are worldwide remittances, but the largest component is remittances from the Middle East.) Asian governments have introduced special policies to encourage repatriation of worker savings through state banks, to help make them available for development and investment purposes (Taylor, 1999: 71). However, workers also bring back earnings in the form of cash or consumer goods. Millions of families have become dependent on remittances, and have improved living standards because of them. However, the money is often spent on luxury goods, dowries and housing, rather than on productive investments. Since the migrants generally come from the middle strata rather than the poorest groups in the areas of origin, remittances often exacerbate social inequality, and lead to increased concentration of land ownership (Castles, 2000).

Highly-qualified migrants and students

Most Asian migration is of low-skilled workers, but there is also growing mobility of professionals, executives, technicians and other highly-skilled personnel (see Chapter 4 above). One form is the 'brain drain': university-trained people moving from less-developed to highly-developed countries. Europe, North America and Australia have obtained thousands of doctors and engineers from India, Malaysia, Hong Kong and similar countries. Britain recruits nurses from the Philippines for the National Health Service. Germany competes with other highly-developed countries to attract Indian information technology (IT) specialists. The USA obtains many highly-trained workers from Asia. According to the US State Department, 69 per cent of employer-sponsored H-1B visas from 1990 to 1997 were issued to persons from India, the Philippines, Japan and China (Abella, 2002).

Such labour mobility may be a drain on the resources of the poorer countries, leading to shortages of skilled personnel. There are reports of Philippine hospitals closing down operating theatres because all the trained staff have gone to the UK. On the other hand, many educated people cannot find jobs at home. It appears that opportunities for work abroad stimulate the growth of training facilities for IT personnel in India and medical personnel in the Philippines. The remittances of the skilled migrants may be beneficial, and many do return when opportunities become available, bringing with them new experience and sometimes additional training. Unfortunately, many highly-skilled migrants find their entry to appropriate employment in highly-developed countries restricted by difficulty in securing recognition of their qualifications, or by discrimination in hiring and promotion practices. If they fail to get skilled jobs, their migration is both a loss to their country of origin and a personal disaster.

Another form of highly-qualified migration is of executives, professionals and experts sent overseas by their companies to or by international organizations. Capital investment in less-developed countries may be seen as an alternative to low-skilled migration to developed countries, but it leads to movements of skilled personnel in the opposite direction. China had some 200 000 foreign specialists in 2000, while Malaysia had 32 000 and Vietnam about 30 000. They came from other Asian countries, but also from the USA, Europe and Australia (Abella, 2002). Short-term business visitors are also important. There were 2.6 million business travellers from Japan in 2000, of whom 1.6 million went to other Asian countries (Abella, 2002). Capital investment from overseas is a catalyst for socio-economic change and urbanization, while professional transients are not only agents of economic change, but also bearers of new cultural values. The links they create may encourage people from the developing country to move to the investing country in search of training or work. Lim (1996: 329) has shown that 'the three largest foreign investors in Malaysia – Taiwan, Japan and Singapore – are also the three main destinations of Malaysian emigrant workers'. The returning professional transients bring new experiences and values with them. Some Japanese observers see the stationing of highly-trained personnel overseas as part of the 'internationalization' of Japan, and a powerful factor for cultural change (Suzuki, 1988: 41).

Student mobility is often a precursor to skilled migration. By the late 1980s, there were 366 000 foreign students in the USA, of whom nearly half came from Asia (Skeldon, 1992: 35). Australia issued 86 277 student visas in 2000–1, with the great majority of students coming from Asia (DIMIA, 2001). There is considerable competition among developed countries to attract fee-paying students. Many Australian universities now have Asian campuses. In 1996, when racist speeches and attacks on Asian students took place in Australia, university vice-chancellors demanded action from the government to counter a feared decline in student enrolments.

Many former students stay on in developed countries upon graduation, especially those with PhDs. According to the National Science Foundation, there were 23 559 science and engineering faculty members of Asian origin in the USA in 1997 – 10.5 per cent of all academics in these fields (Abella, 2002). Australia changed its immigration rules in 1999: in the past, students had to leave Australia on graduation and wait at least two years before applying to migrate to Australia. Now they are allowed to remain in the country as they pursue their immigration application. Many of these full-fee overseas students come from Asia and are concentrated in the business and IT fields (Birrell, 2001). Student movements to developed countries may thus be part of the brain drain.

Refugees

At the end of 2000, the UNHCR recorded nearly 5 million refugees in Asia and the Pacific – 41 per cent of the global total of 12.1 million. Using the wider concept 'populations of concern to UNHCR' (which includes asylum seekers, returnees, some internally-displaced persons and others) the Asia and Pacific total came to 7 million – one-third of the global total (UNHCR 2000a: 21). The biggest source of refugees was Afghanistan, with some 4.5 million Afghans in other countries, mainly in Pakistan and Iran. Other major refugee sources included Burma, Iraq, Vietnam, China, Bhutan, East Timor, the Philippines and North Korea (USCR, 2001: Table 4).

Asia's two largest enduring exoduses since 1945 have been from Indo-China and Afghanistan. Over 3 million people fled from Vietnam, Laos and Cambodia following the end of the Vietnam War in 1975. Many left as 'boat people', sailing long distances in overcrowded small boats, at risk of shipwreck and pirate attacks. Over the next 20 years, 2.5 million found new homes elsewhere, while 0.5 million returned. Over a million were

Illustration 7.1 *Vietnamese boat people in Malaysia 1978 (Photo: UNHCR/K. Gaugler)*

resettled in the USA, with smaller numbers in Australia, Canada and Western Europe. China accepted about 300 000 refugees, mainly of ethnic Chinese origin. Other Asian countries were unwilling to accept settlers. In 1989, a 'Comprehensive Plan of Action' was adopted by all the countries concerned. People already in the camps were to be resettled, while any new asylum seekers were to be screened to see if they were really victims of persecution. Those found to be economic migrants were to be repatriated. In 1979 Vietnam introduced an 'Orderly Departure Programme' to permit legal emigration, particularly of people with relatives overseas, that was considerably stepped up in 1989. By 1995, most of the camps were closed and the emergency was considered over (UNHCR, 2000b: 79–103).

Up to a third of Afghanistan's 18 million people fled the country following the Soviet military intervention in 1979. The overwhelming majority found refuge in the neighbouring countries of Pakistan (3.3 million in 1990) and Iran (3.1 million) (UNHCR, 2000b: 119). There was hardly any official resettlement overseas. The Afghan emergency came just after the Indo-Chinese exodus, and there was little willingness in Western countries to provide homes for new waves of refugees. Moreover the *mujahedin* (Islamic armed resistance) leaders wanted to use the refugee camps as bases for recruitment and training. For political, humanitarian, religious and cultural reasons, Pakistan and Iran were willing to provide refuge for extended periods. Pakistan received substantial compensation from the USA in the form of military, economic and diplomatic support. Iran, on the other hand, received very little external assistance, despite being one of the world's principal havens for refugees (UNHCR 2000b: 118).

The different handling of the Vietnamese and Afghan cases is an example of the way refugee movements can become part of wider foreign policy considerations for major powers (Suhrke and Klink, 1987). With the end of the Soviet intervention in 1992, about 1.5 million Afghan refugees returned home. However, the outbreak of new conflicts, the seizure of power by the fundamentalist Taliban, a four-year drought and the devastated condition of the country delayed the return of the rest. In 2000, Afghans remained the world's largest refugee population. Some Afghan men went to work in the Gulf states, to help fund the costs of rebuilding their villages (UNHCR, 1995: 182–3). Increasing numbers of Afghans moved on to Western countries: between 1990 and 2000, 155 000 Afghans sought asylum in the EU (unpublished UNHCR data).

The events of 11 September 2001 made the world aware of the consequences of protracted situations of conflict and political anarchy. Afghanistan had become the centre of the global Al-Qaida terrorist network. It was also the leading global producer of heroin. The huge Afghan refugee diaspora came to be seen as one component of a threat to global security. The US-led invasion of Afghanistan was designed to destroy Al-Qaida and the Taliban, establish a legitimate government, and permit the return of the refugees. The initial attack in late 2001, however, was expected to precipitate major new flows of refugees and IDPs. The UNHCR and other humanitarian agencies were included in plans for the crisis, and provided with special funding. In the event, outflows of refugees were limited, partly because Pakistan and Iran closed their borders, and partly because the Taliban were quickly defeated.

In March 2002, the Afghan Transitional Authority and UNHCR started a mass return programme. By July, more than 1.3 million Afghans had returned, 1.2 million from Pakistan and 100 000 from Iran. This unexpectedly rapid repatriation put severe strain on UNHCR finances (UNHCR, 2002). The agency was forced to cut the assistance given to returning families. Western countries – willing to spend billions on armed intervention – were not ready to top up relief funds. Meanwhile, the governments of Australia, the UK and other Western countries also began sending back Afghan asylum seekers, even though it was far from clear that conditions were safe in Afghanistan.

Apart from these two huge refugee movements, there have been many exoduses smaller in number, but no less traumatic for those concerned. After the failure of the democracy movement in 1989, thousands of Chinese sought asylum overseas. Conflicts linked to the break-up of the former Soviet Union led to mass displacements in the 1990s affecting many new states, including Georgia, Chechnya, Armenia, Azerbaijan and Tajikistan. Around 2 million people were internally displaced or forced to flee across borders (UNHCR, 1995: 24–5). At least 50 000 North Koreans have fled to China. Other long-standing refugee populations include Tibetans and Bhutanese in India and Nepal, Burmese in Thailand and Bangladesh, and Fijian Indians in Australia and New Zealand. Muslims from Mindanao (southern Philippines) have fled to Malaysia to escape persistent internal conflict. The long civil war in Sri Lanka has led to mass internal displacement as well as refugee outflows. In 2001, an estimated 144 000 Sri Lankan Tamils were living in camps in India, while other Tamils were dispersed around the world. After the people of East Timor voted for independence in September 1999, violence by the pro-Indonesian militias supported by the Indonesian Army forced at least 250 000 people across the border into West Timor (still part of Indonesia). Another half million East Timorese (the majority of the population) were forced to flee into the mountains (USCR, 2001).

The Asian experience shows the complexity of refugee situations in less-developed countries: they are hardly ever a simple matter of individual political persecution. Almost invariably, economic and environmental pressures play a major part. Refugee movements, like mass labour migration, are the result of the massive social transformations currently taking place in Asia (Van Hear, 1998). Long-standing ethnic and religious differences exacerbate conflicts and often motivate high levels of violence. Resolution of refugee-producing situations and the return home of refugees is hampered by scarcity of economic resources and lack of guarantees for human rights in weak and despotic states. Western countries have often become involved in struggles about state and nation formation in Asia, including the Vietnam War, the conflicts in Afghanistan and many others. Responses to asylum seekers have often been conditioned by such experiences. One of the latest expressions of this malaise is Australia's 'Pacific solution' of pushing asylum seekers on to neighbouring islands like Nauru and Papua New Guinea (see next chapter).

Perspectives for Asian migration

Asian migration has grown rapidly since the 1970s. Statistics are poor, but it seems that by 2000 there were 5 million Asian workers in the Gulf oil countries and over 6 million working in the major Asian labour-importing countries. Millions more had migrated permanently to the USA and other Western countries. In addition there were 5 million refugees. Most Asian migrants came from just a few source areas, especially the Philippines, Indonesia, China, Thailand and South Asia, which have become labour reserves for the region and the world. The majority of the Asian migrants are low-skilled workers, but flows of highly-skilled personnel are on the increase.

Every migratory movement in Asia has its own special features, yet there are significant general trends. One is the lack of long-term planning: movements have been shaped not only by government labour policies, but also by the actions of employers, migrants and the migration industry. Illegal migration is very high, and agents and brokers play a major role. Official policies range from 'near denial' of the presence of foreign labour (Japan and Korea) to 'active management' (Singapore), with most countries somewhere in-between (Miller and Martin, 1996: 195).

Fairly general features of labour migration systems in the Middle East and Asia include rigid control of foreign workers, the prohibition of settlement and family reunion, and the denial of basic rights. Many of the governments concerned refer explicitly to the European experience, in which temporary guestworkers turned into settlers and new ethnic minorities. The strict regulatory systems are designed to prevent this. Will they succeed? Although most movements are temporary in intention, trends towards permanent settlement are beginning to emerge in some places, as the examples of Japan and Malaysia showed. However, these trends are limited, and not officially sanctioned. When Western Europeans tried to reduce foreign populations in the 1970s, they found it difficult for several reasons: their economies had become structurally dependent on foreign labour, employers wanted a stable labour force, immigrants were protected by a strong legal system, and the welfare state tended to include non-citizens. Do such pressure for settlement exist in Asia (Castles, 2001)?

There are certainly signs of increasing dependence on foreign workers for the '3D jobs', as labour force growth slows in industrializing countries and local workers reject menial tasks. The limited success of repatriation policies during the Asian Crisis was a clear indicator that migration cannot easily be reversed. In these circumstance employers seek to retain 'good workers', migrants prolong their stays, and family reunion or formation of new families in the receiving country takes place. Trends towards democratization and the rule of law also make it hard to ignore human rights. The growth of NGOs working for migrants' right in Japan and Malaysia indicates the growing strength of civil society in Asia's new democracies. It therefore seems reasonable to predict that settlement and increased cultural diversity will affect many Asian labour-importing countries; yet no Asian government has plans to deal with long-term effects of migration – even to discuss the matter is still almost taboo in many Asian countries.

Despite the rapid growth, movements are still quite small in comparison with Asia's vast population. Migrant workers make up a far smaller proportion of the labour force in countries like Japan and Korea than in European countries (although the proportion is large in Singapore and Malaysia). However, the potential for growth is obvious. The Indian subcontinent provides a vast labour reservoir. Economic and political reform in China could open the door for mass labour migration, while setbacks to reform could lead to refugee movements. Indonesia and the Philippines have considerable population growth, and regard labour export as a vital part of their economic strategies. The fast-growing economies of East and South-East Asia seem certain to pull in large numbers of migrant workers in the future. It is hard to believe that this will not lead to some degree of settlement, with far-reaching social and political consequences. The twenty-first century has been dubbed the 'Pacific century' in terms of economic and political development, but it will also be an epoch of rapidly-growing population mobility in the Asian region.

Guide to further reading

Literature on Asian migration has grown exponentially in the last few years. The publications of the Scalabrini Migration Center (Quezon City, Philippines) provide excellent ongoing documentation. They include the *Asian and Pacific Migration Journal* (APMJ), a magazine, *Asian Migrant*, a web atlas (www.scalabrini.asn.au/atlas/) and an electronic information service (www.scalabrini.asn.au/philsmc.htm). The Asia-Pacific Migration Research Network is also a good source of contacts and information (www.capstrans.edu.au/aprmn or through www.unesco.org). Appleyard (1998b) is good on emigration from South Asia. IOM (2000b) has useful summary chapters on Asia and the Pacific. On Japan, Komai (1995), Mori (1997) and Weiner and Hanami (1998) provide good studies in English. For most other countries, journal articles are still the best sources.

Chapter 8

Migrants and Minorities in the Labour Force

Employment of foreign workers is strongly influenced by broad macro-economic trends. This was evident in Western Europe when foreign labour employment generally stagnated or declined between 1975 and 1985 in a period of recession and restructuring. By 1997, however, a general pattern of upturn in international migration to the OECD area could be discerned (OECD, 2001). This was linked to factors like the spectacular growth of the US economy and related recovery of Western European economies and growing demand for highly-skilled labour in many OECD countries. Most legal immigration to OECD countries, however, continued to be authorized for family reunification rather than on economic grounds.

The events of 11 September 2001 contributed to a global recession. Historically, migrant workers are disproportionately adversely affected by such economic downturns and there was some evidence of this in the USA in the months following the attacks and in Malaysia where tens of thousands of mainly Indonesian migrants were deported in 2002 as the government sought to implement stricter immigration regulations. Despite apprehension over possibly untoward economic consequences of the war on terrorism, the short- to medium-term prospect for global migration for employment remained quite robust on the first anniversary of 11 September.

The OECD's 1986 Conference on the Future of Migration identified the underlying reasons for the long-term prospects for increasing employment of immigrants: the ageing of Western societies, demographic imbalances between developed and developing regions in close proximity to each other, the North–South gap, continuing employer demand for foreign labour and the growth of illegal migration (OECD, 1987). Furthermore the conference stressed the necessity of understanding immigration in its global context as something inextricably bound up with economic and foreign policies, developments in international trade and growing inter-dependence.

This book has shown how most post-1945 movements started as labour migration, often organized by employers and governments. The movements have changed in character over time, with increasing participation of non-economic migrants, including dependants and refugees. The economic migrants, too, have become differentiated, with increasing participation of highly-skilled personnel and entrepreneurs. The political economy-based theories of labour migration which developed in the 1960s and 1970s emphasized the crucial role of migrant workers in providing low-skilled labour for manufacturing industry and construction, and in restraining wage growth in these sectors. In the post-Cold War era, there developed a need to re-examine this political economy in the light of the shift from temporary labour to permanent settlement and the increasing economic differentiation of migrant workers. Key questions to be asked include the following:

1. What was the impact of economic restructuring since the 1970s on migrant workers?
2. Have the patterns of labour market segmentation by ethnic origin and gender which had emerged by the 1970s persisted, or have there been significant changes?
3. What variations are there in employment patterns according to such criteria as ethnic background, gender, recentness of arrival, type of migration, legal status, education and training?
4. What variations are to be found between immigration countries, especially pertaining to scope of the underground economy, and how are they to be explained?
5. What is the situation of second and subsequent generation immigrants in the labour market (is disadvantage passed on from generation to generation)?
6. Is institutional or informal discrimination a major determinant of employment and socio-economic status?
7. What strategies have migrants adopted to deal with labour market disadvantage (for example, self-employment, small business, mutual aid, finding 'ethnic niches')?

This chapter addresses the above questions by reviewing some of the major theoretical and empirical findings concerning immigrants and labour markets since the 1970s. The growing complexity of immigrant labour market effects is examined, along with material illustrating cross-national trends in labour market segmentation and the growing polarization of immigrant labour market characteristics. A case study of the evolution of foreign employment in the French motor and building industries is included to demonstrate the adverse effects of economic restructuring since the early 1970s on foreign labour in certain industries and to illustrate processes of labour market segmentation.

Migrants in the informal economy

Understanding of the key role played by the employment of many migrants in the underground economy has increased since the mid-1970s. Studies of legalized aliens, in particular, have provided deeper insight (OECD, 2000: 53–78). As governments sought to deter illegal employment of aliens, they required better understanding of labour market dynamics in sectors known to employ large numbers of migrants illegally. Such sectors characteristically included labour-intensive agriculture, building construction, gardening and lawn maintenance, the garment industry, hotels and restaurants, domestic services, janitorial and cleaning services, nursing and, in the USA, the meat-packing industry.

Employer demand for migrant workers in sectors like these often persisted despite recession and high unemployment of citizens. Employment of eligible aliens also persisted in the face of enforcement of employer sanctions and other measures intended to deter illegal employment (see Chapter 5). Indeed, some employment-eligible aliens and French citizens of immigrant background have claimed they had to pose as illegal aliens in order to get agricultural work in southern France.

As suggested in Chapter 4 with reference to Northern and Southern Europe, the dimensions and nature of the underground economy vary from country to country and from region to region. There are also important variations in the willingness and capacities of governments to regulate labour markets. Virtually all labour migration to Southern Europe in recent decades has been directed to employment in the informal sector (Reyneri, 2001).

Some recent scholarship focusing on illegal migration and alien smuggling views these processes as the unproblematic meeting of labour market demands and locates the problems elsewhere, in governmental efforts to regulate international migration (Harris, 1996). Such views are also expressed by labour-sending countries, such as Mexico, which view illegal migration of their nationals as driven by unmet demand for labour in the destination countries. Such perspectives sometimes portray illegal employment as heroic and emphasize that the migrant workers can improve their overall socio-economic welfare and that of their families as well as that of the host country through such employment. Using World Bank data which divides the world into 22 'high income' and 110 'middle and low income countries', Martin and Taylor calculated that the average person who moves from the latter to the former increases his or her income by ten to twenty times (Martin and Taylor, 2001: 98). The significance of migrant wage remittances to homelands was examined in Chapters 6 and 7.

Governments and societies in countries receiving large numbers of unauthorized migrant workers sometimes choose to ignore the inflow or to view it as benign. This was the case of France up to roughly 1970. But much more commonly, the existence of an underground economy and the role played by migrant workers in it is viewed as unlawful and as socially harmful. This is what motivated adoption of employer sanctions and other measures to punish illegal employment, not only of aliens but of citizens as well. What accounts for the persistence of the underground economy and illegal migrant employment in it?

In many instances, governments simply lack the administrative wherewithal or the political will to enforce their laws and regulations. The case of Mexican farmworkers in the USA is particularly illustrative. In 1970, according to Philip Martin, there were about 750 000 Mexican-born US residents; in 2002, there were over 9 million. Martin estimates that 95 per cent of new entrants to the seasonal farm workforce are foreign-born. Many Mexicans thus begin their sojourns in the USA as illegal farmworkers. Martin estimates that, of the 1.8 million farmworkers employed in crop production (as distinct from livestock rearing), over half were illegally employed in 2002 (Martin, 2002). Farmworkers constitute the poorest segment of American society and migration trends in the 1980s and 1990s were clearly linked to growing poverty in rural areas, for example in California (Taylor *et al.*, 1997).

While employers in labour-intensive agriculture frequently raise the spectre of labour shortages and crops rotting in fields, characteristically there is oversupply of labour. One consequence is wage depression. Farmworker wages stagnated in the 1980s and 1990s, mainly due to the influx of illegal immigrants (Taylor *et al.*, 1997: 13–14). A second is that employers have little incentive to improve working conditions or management techniques. Unionization of US farmworkers, particularly in California, progressed between 1965 and 1975 with salutory effects upon wages, but declined thereafter, in part due to increased farmer recourse to farm labour contractors (Taylor *et al.*, 1997: 14–16).

It is frequently claimed that US citizens shun farm work and such views are echoed by farmers in countries like Germany, France, Spain and Italy. However, the purported 'dependency' of labour-intensive agriculture upon migrant workers requires careful scrutiny. Farmers are often exempt from compliance with rules and regulations that apply in non-agricultural sectors. Moreover, acreage planted with labour-intensive crops, primarily fruits and vegetables, is often increased on the assumption that an ample supply of foreign workers will be available. As globalization has enabled the year-round supply of fruits and vegetables in more developed countries, there is reason to question whether expanded labour-intensive agricultural production in the more economically-advanced countries constitutes something that is desirable, particularly if such production competes with agricultural exports from less-developed countries with high rates of unemployment or underemployment and of emigration. From a public policy perspective, especially in view of policies directed against illegal migration, labour-intensive agricultural production would optimally occur in areas with a comparative advantage in overall production and marketing costs. The pervasiveness of illegal alien employment in labour-intensive agriculture in the richest countries is not only exploitative of migrant workers, who are usually badly paid and generally work in difficult conditions at the mercy of their employers, but foreign agricultural workers and their dependants are often forced to live in substandard housing and experience segregation and racism. The outbursts of anti-immigrant violence in Spain in 2001 and in southern France in the early 1970s were very much connected to social tensions surrounding largely illegal employment of foreign agricultural workers (see Chapters 4 and 5).

Other factors accounting for the persistence of employer demand for unauthorized workers include the growth of subcontracting in sectors like building construction, garments and janitorial services. The growth of service industries like lawn maintenance and gardening services in the USA and of domestic services virtually throughout the more developed countries, but also in many Asian and Middle Eastern countries, plays an important role.

The weakening of trade unions in the current phase of globalization also constitutes an important factor. Illegal employment is unusual in sectors, firms or industries with strong unions. In the USA in the 1990s, however, some of the most successful unionization drives involved illegally-employed workers. These occurred against an overall backdrop of declining unionization and helped prompt the change in leadership of the AFL-CIO and its policy towards illegal migration that set the stage for the US–Mexico migration initiative in 2001 (see Box 1.2).

As Claude-Valentin Marie has observed (2000), the illegally-employed alien worker in the informal sector is in many ways emblematic of the current era of globalization. The precariousness of such workers, their absence of rights and flexibility respond to the exigencies of firms in an era of intense globalization. In the most extreme exploitative circumstances, trafficked men, women and children become latter-day slaves whose numbers worldwide amount to millions (see Chapter 5).

Growing fragmentation and polarization of immigrant employment

Perhaps what is most distinctive about immigrant employment is clustering or concentration in particular jobs, industries and economic sectors. The sectoral nature of immigrant employment concentration varies from country to country due to historical factors and other variables, such as entrepreneurial and foreign worker strategies (OECD, 1994: 37). The pattern of immigrant employment concentration within a particular state and society often evolves through time. In France, a decline in alien employment in the motor and building industries since 1973 has been paralleled by new concentrations of aliens in the rapidly-growing services sector. A nine-country OECD study revealed:

> contrasting situations in the structure of foreign labour compared with national employment in each type of economic activity. The role of foreign labour differs in the countries covered by the study ... Despite these differences, concentrations of foreign workers persist in sectors often neglected by nationals, though at the same time there has been a spread of foreign labour throughout all areas of economic activity, especially services. (OECD, 1994: 37)

The persistence of labour market segmentation is a theme common to many studies on immigrants and labour markets. Castles and Kosack demonstrated a general pattern of labour market segmentation between native and immigrant workers in Western Europe in the 1970s (Castles and Kosack, 1973). Collins regards the 'impact of post-war immigration on the growth and fragmentation of the Australian working class' as 'one of the most salient aspects of the Australian immigration experience' (Collins, 1991: 87). A US Department of Labor report concluded:

> the most important current consequence of internationalization, industrial restructuring, and the increase in the national origins and legal status of new immigrants is the dramatic diversification of conditions under which newcomers participate in the US labor market. Newcomers arrive in the United States with increasingly diverse skills, resources and motivations. In addition, on an increasing scale, they are arriving with distinct legal statuses. In turn, this proliferation of legal statuses may become a new source of social and economic stratification. (US Department of Labor, 1989: 18)

The range and significance of immigrant labour market diversity is obscured by policy and analytical perspectives that stress the homogeneity of competitive labour markets or sharp contrasts between primary and secondary labour markets (US Department of Labor, 1989: 18). It is often meaningless to generalize about average earnings and other labour market effects of immigration, just as it is meaningless to assume a general interest in discussions of immigration policy. Immigration has extremely unequal effects upon different social strata. Some groups gain from policies facilitating large-scale expansion of foreign labour migration, while other groups lose (Borjas. 1999: 12–13). The winners are large investors and employers who favour expanded immigration as part of a strategy for deregulation of the labour market. The losers are many of the migrants themselves, who find themselves forced into insecure and exploitative jobs, with little chance of promotion. Among the losers are also some existing members of the workforce, whose employment and social condition might be worsened by such policies.

In the 1980s, awareness grew that immigrant workforces were becoming increasingly bipolar, with clustering at the upper and lower levels of the labour market. The head of ILO's migrant workers section termed Western Europe's growing number of professionals, technicians and kindred foreign workers the 'highly invisible' migrants and estimated that they comprised one-quarter of legally-resident aliens living in the former EC (Böhning, 1991a: 10). Americans, Canadians, Japanese and Europeans from nearby states that did not belong to the EC comprised most of the highly invisible migrants. However, resident alien populations, such as Turks in Germany, who are stereotypically seen as blue-collar workers, also included surprising numbers of professionals and entrepreneurs.

A bifurcation in the labour market characteristics of immigrants was apparent in the USA as well. Borjas found an overall pattern of declining skills in post-1965 immigrant cohorts as compared to earlier immigrants. This is a result of the 1965 changes in immigration law which opened up the USA to immigration from around the world (see Chapter 4). As entries from Western Europe declined in favour of those from Asia and Latin America, the differences in the prevailing socio-economic and educational standards between the regions were reflected in the declining skills and rising poverty of post-1965 immigrants (Borjas, 1990, 1999). The USA is far more attractive to poorer and less-privileged Mexicans than it is to the Mexican middle and upper classes, who are little inclined to emigrate from a society marked by extreme inequality in income distribution and life chances (Borjas, 1990: 126). Hence it was scarcely surprising that the Mexican farmworkers who were legalized after 1986 on average possessed only four years of schooling.

The growing bifurcation of immigrants to the USA was apparent in the sharply contrasting poverty rates of various national origin groups. The fraction of immigrants from Germany and Italy living in poverty was 8.2 per cent, whereas Chinese and Koreans had poverty rates of 12.5 and 13.5 per cent respectively, and immigrants from the Dominican Republic and Mexico suffered poverty rates of 33.7 and 26 per cent (Borjas, 1990: 148). Similarly, Borjas found a strong link between rising welfare utilization by immigrants and the changing character of immigration to the USA (Borjas, 1990: 150–62). These trends prompted Borjas to advocate changes in US immigration law to increase the skill levels of immigrants. The Immigration Act of 1990 aimed nearly to triple the number of visas reserved for qualified workers from 54 000 to 140 000 yearly. Moreover, 10 000 visas for investors were set aside annually.

As in Western Europe, labour market projections for the USA circa 1990 forecast growing shortages of highly-qualified personnel. The Immigration

Act of 1990 was designed to enhance US competitiveness in what was perceived as a global competition to attract highly skilled labour. Concurrently, one of the major future challenges facing the USA was deemed finding gainful employment for the existing and projected stocks of low and unskilled workers, many of whom are minorities. Nonetheless, advocacy of temporary foreign worker recruitment for industries such as restaurants and hotels, agriculture, and construction continued on both sides of the Atlantic, and many employers, such as in Germany and France, complained about labour shortages despite relatively high unemployment rates. The politics of the second generation of temporary foreign worker policies was examined in Chapter 5

A sharp pattern of labour market segmentation was also apparent in Australia (see Chapter 9). Collins identified four major groups: (1) men born in Australia, in English-speaking countries and Northern Europe, who were disproportionately found in white-collar, highly-skilled or supervisory jobs; (2) men from non-English speaking countries who were highly concentrated in manual manufacturing jobs; (3) women with an Australian or English-speaking background, found disproportionately in sales and services; and (4) women with a non-English-speaking background who tended to get the worst jobs with the poorest conditions (Collins, 1978). For Collins: 'perhaps the crucial point in understanding post-war Australian immigration is that English-speaking and non-English-speaking migrants have very different work experiences' (Collins, 1991: 87).

Significant labour market segmentation is thus evident in industrial democracies. Traditional gender divisions, which concentrated women in low-paid and low-status work, have been overlaid and reinforced by new divisions affecting immigrant workers of both sexes. As migration is globalized, there are widening gaps both between immigrants and non-immigrants, and among different immigrant categories. Future trends in the labour market will favour highly-skilled immigration, but the pool of aspiring low-skilled immigrants is enormous and will expand exponentially in coming years.

Labour market segmentation leads to long-term marginalization of certain groups, including many of the new immigrants from non-traditional sources. Generally there are not rigid divisions based on race, ethnicity or citizenship status. Instead, certain groups have become over-represented in certain disadvantaged positions. Some individual members of disadvantaged groups do well in the labour market, but most do not. The causes for this are not only found in specific factors like education, length of residence, prior labour market experience or discrimination. Much more complex explanations are usually required, which provide historical understanding of the processes of labour migration and settlement, along with their role in a changing world economy.

Global cities, ethnic entrepreneurs and immigrant women workers

Patterns of international migration are tightly bound up with capital flows, investment, international trade, direct and indirect foreign military intervention, diplomacy and cultural interaction. Pioneering work by Sassen (1988) stressed how patterns of foreign investment and displacement of certain US manufacturing jobs abroad have fostered new migratory streams to the USA (or have tended to expand pre-existing flows). Sassen underscores the significance of the emergence of global cities, like New York or Los Angeles, for understanding future patterns of migration. Linkages between global cities and distant hinterlands create paradoxes wherein enormous wealth and highly remunerated professional employment uneasily coexist with growing unskilled service industry employment and Third-World-like employment conditions in underground industries. The casualization of labour and growing illegal alien employment are characteristic of global cities. Considerable illegal employment of aliens often coincides with high unemployment of citizens and resident aliens. The latter are likely to belong to minorities and have often been victims of job losses in industries that have shifted manufacturing operations abroad.

As noted in previous chapters, some immigrant groups have traditionally played key economic roles as traders and entrepreneurs. Since the 1970s recession, a growing body of research has examined immigrant entrepreneurship and its effects. Across industrial democracies, growing numbers of immigrants are self-employed and owners of small businesses (Waldinger *et al.*, 1990). Most typical are ethnic restaurants, 'mom and pop' food stores and convenience stores. Immigrant-owned businesses frequently employ family members from the country of origin. Light and Bonacich, in their influential study *Immigrant Entrepreneurs* (1988), traced the origins of the Korean business community in Los Angeles to the Korean War, which led to the establishment of extensive transnational ties and eventually migration between the Republic of Korea and the USA.

Studies in France similarly stressed the complex historical genesis of immigrant entrepreneurship. Abdelmalek Sayad, the French sociologist, noted that 'sleep merchants' who supplied lodging for illegal aliens, usually compatriots, figured among the first North African businessmen in France (Vuddamalay, 1990: 13). In Germany, there were 150 000 foreigner-owned businesses by 1992, including 33 000 owned by Turks. The Turkish-owned businesses generated 700 000 jobs in 1991 and recorded sales of DM25 billion (about US$17 billion) and invested DM6 million in Germany (*This Week in Germany*, 18 September 1992: 4).

Immigrant entrepreneurship has been assessed divergently. Some scholars, such as Fix and Passel, stress the economic dynamism of immigrant entrepreneurs with their positive effects upon economic growth and quality of life for consumers:

> Another source of job creation is the entrepreneurial activities of immigrants themselves. In 1990 almost 1.3 million immigrants (7.2 percent) were self-employed, a rate marginally higher than natives (7.0 percent) ... During the 1980's, immigrant entrepreneurship increased dramatically. In 1980, 5.6 percent of immigrants living in the United States were self-employed but by 1990 the same group of pre-1980 immigrants (who had now been in this country for an additional decade) had a self-employment rate of 8.4 percent. (Fix and Passel, 1994: 53)

A more critical viewpoint stresses the human suffering entailed by intense competition, the long hours of work, exploitation of family labour and of illegally-employed aliens, resultant social problems and so on (Light and Bonacich, 1988: 425–36; Collins *et al.*, 1995). The Los Angeles riots of 1992 revealed an undercurrent of tension between blacks and Korean business people in Los Angeles. Tensions between urban black Americans and Korean entrepreneurs were manifested in other major US cities, frictions that were similar to anti-Jewish business sentiments when US ghettos boiled over in the 1960s. Such tensions again point to the need for a broad-gauged approach to apprehension of immigration. The downside of immigrant entrepreneurship was summarized in a 1997 report:

> The ethnic solidarity hypothesized to be conducive to immigrant business can be seen in another light, as exclusionary and clannish, impeding access to business and employment opportunities for the native-born ... The informal business transactions in immigrant communities that are normally regulated by gossip and ostracism can sometimes be enforced in ways that are distinctly illegal. To some of the relatives involved, the much-vaunted 'strong family ties' that keep a corner store open 24 hours a day may seem exploitative and unfair. There is even reason to suspect that migrant self-employment is more of a survival strategy than an indication of socio-economic success – more, that is, of a lifeboat than a ladder. (*Research Perspectives on Migration*, 1997: 11)

Research in the 1980s and 1990s shed a great deal of additional light on the labour market role of immigrant women. Houstoun *et al.* (1984) documented a female predominance in legal immigration to the USA since 1930. They concluded that deployment of US military forces abroad played a significant role in this. They noted that an estimated 200 000 Asian-born wives of US servicemen resided in the USA in the early 1980s. While working-age immigrant men reported a labour force participation rate (77.4 per cent) similar to US men, female immigrants were less likely to report an occupation than US women. The bifurcation pattern considered above was more pronounced with immigrant women. They were more concentrated in highly-skilled occupations (28.1 per cent) than US women but also more concentrated in low-status, white-collar clerical employment (18.0 per cent), semi-skilled blue-collar operation jobs (17.9 per cent) and in private household work (13.9 per cent) (Houstoun *et al.*, 1984).

Data on female immigrant employment in Australia revealed sharp segmentation. Collins and Castles used 1986 Census data to examine the representation of women in manufacturing industry. The index figure 100 indicates average representation. They found high degrees of over-representation for women born in Vietnam (494), Turkey (437), Yugoslavia (358) and Greece (315). Women born in the USA (63), Canada (68) and Australia (79) were underrepresented (Collins and Castles, 1991: 15). Female clustering in manufacturing industries undergoing restructuring rendered them disproportionately vulnerable to unemployment. Immigrant women of a non-English-speaking background were thought to be over-represented in outwork for industries such as textiles, footwear, electronics, packing and food and groceries. Collins and Castles considered these workers as perhaps the most exploited section of the Australian workforce (Collins and Castles, 1991: 19).

Morokvasic has argued that, in general, immigrant women from peripheral zones living in Western industrial democracies:

> represent a ready made labour supply which is, at once, the most vulnerable, the most flexible and, at least in the beginning, the least demanding work force. They have been incorporated into sexually segregated labour markets at the lowest stratum in high technology industries or at the 'cheapest' sectors in those industries which are labour intensive and employ the cheapest labour to remain competitive. (Morokvasic, 1984: 886)

Patterns of labour migration in the post-post-Cold War era continued this type of incorporation of women's labour, and extended it to new areas of immigration, such as Southern Europe and South-East Asia. The exploitation of women in human trafficking was examined in Chapter 5.

Foreign labour in France's car and building industries

In many highly-developed countries, migrant workers became highly-concentrated in the car and building industries. Employer recourse to foreign labour in these sectors has been particularly significant in France, both in quantitative and in political terms. At the height of labour immigration in the early 1970s, some 500 000 foreigners were employed in the building industry. About a quarter of all foreigners employed in France were in the building industry. In motor car construction, some 125 000 foreigners were employed, representing one out of every four car workers. Only the sanitation services industry had a higher ratio of foreign to French employees by 1980 (Miller, 1984).

The disproportionate effects of the 1970s' recession upon foreign workers in the car and building industries were incontrovertible. Although foreigners comprised one-third of building sector employees, they suffered nearly half of the total employment loss from 1973 to 1979, and declined to 17 per cent of the building industry workforce by 1989 (OECD, 1992: 24). In the car industry, total employment actually increased by 13 000 in the same period, yet foreign workers were hard hit by layoffs, their number falling by 29 000. During the 1980s, tens of thousands of additional jobs were lost, with aliens again being disproportionately affected.

A report compiled by the Fédération Nationale du Bâtiment, the main French building sector association, revealed that total employment in the building sector declined by 11.7 per cent from 1974 to 1981. But the reduction of the foreign employee component, some 150 000 jobs, represented a loss of 30 per cent of the 1974 foreign workforce, whereas the 45 000 decrease in the number of French workers employed represented only a 3.9 per cent decline from 1974 employment levels. In other words, three out of every four jobs lost in the building industry from 1974 to 1981 had been held by foreigners.

Foreign worker employment in the building and car industries reached its height in 1974 and then contracted sharply. Nonetheless, according to a Ministry of Labour survey, foreign workers still comprised 28 and 18.6 per cent of the building and car construction industries workforces respectively in 1979. This was all the more remarkable because, in addition to the halt in recruitment, the French government sought to reduce foreign worker employment through a programme offering a cash incentive for repatriation. There was also a *revalorisation du travail manuel* programme, which sought to substitute French for foreign workers through improving the conditions of manual jobs. Both the repatriation and *revalorisation* programmes fared poorly.

Prior to 1974, the car assembly industry was characterized by a high rate of foreign employee turnover. This pattern was profoundly altered by the 1974 recruitment ban. Major consequences of the stabilization of the foreign workforce in motor manufacturing were the ageing of the foreign workforce, its mounting unionization and socio-political cohesiveness as well as resentment of perceived discrimination against foreigners in terms of career opportunities. By the 1980s, most foreign car workers had been employed for at least five years by their company. At the Talbot-Poissy plant by 1982, for example, only one out of the 4400 Moroccan manual workers had worked there less than five years. Some 3200 of the

Moroccans had worked there for ten years or more (Croissandeau, 1984: 8–9).

Foreign workers often chose to join or to vote for various unions as a group, whether from a specific nationality or from a specific shop. Hence support could swing sharply from one union to another, depending on foreign workers' views of a union's specific programme on issues of concern to them. The volatility of ties to French unions stemmed in part from the parallel development of largely autonomous shop-floor organization among foreign workers. In many cases, shop-floor cohesion was based upon national or religious solidarity. By the 1980s, Islamic solidarity groups, whose loci of contact were Muslim prayer-rooms provided by management within the factories, had become an important force. In other instances, underground revolutionary groups affected the form of foreign worker integration into union structures.

The extraordinary sense of collective identity evidenced by foreign car workers by the 1980s stemmed from the stratification which bound together workers of similar ethnic and religious backgrounds in assembly line and other manual jobs. The striking concentration of foreign workers in unskilled or lowly-qualified jobs at Renault-Billancourt was typical of car plants which employed large numbers of foreign workers. Any explanation for the low certified skill levels of most foreign car workers must return to the recruitment process. Citroën, and to a lesser extent other French motor manufacturers, deliberately sought out physically able but poorly-educated foreigners to fill manual labour positions. It was felt that their low levels of education and general backwardness made them better suited for monotonous and often physically taxing jobs than Frenchmen. Hence many foreign car workers were illiterate.

With few hopes for professional advancement, many foreign car industry workers grew frustrated with their jobs. Their frustration and the difficulty of their work was reflected in rising absenteeism and generally less-disciplined work habits (Willard, 1984). Whereas employers once prized foreign workers for their industry and discipline, they began to complain of production and quality control problems. Employer misgivings over hiring of foreign labour were crystallized by a wave of strikes of primarily foreign workers which plagued the industry in the 1970s before rocking its very foundations in the 1980s.

The car workers' strikes hastened plans to restructure and modernize the French motor manufacturing industry. Both Peugeot and Renault, the two major automobile firms (Peugeot having acquired Citroën and Chrysler Europe in the late 1970s), announced plans to automate production through the use of industrial robots. Unrest in French car factories continued sporadically into the early 1990s, but would never again reach dimensions comparable to those of the 1973–83 period. The building industry, with its weaker unionization rate, rampant illegal alien employment, widespread subcontracting and predominance of small and medium-sized employers, did not experience parallel unrest. However, economic restructuring, as seen through the window of these two French industries, had disproportionately affected immigrant employment, with far-reaching political consequences.

In other French industries, however, immigrant employment grew over the 1973–93 period. This was particularly true of services and the apparel industry. In other countries, similar seemingly contradictory developments have been documented. Migrants are disproportionately vulnerable to job loss during recessions and periods of economic restructuring in declining industries, but not in others. Tapinos and de Rugy suggest that 'immigrant workers, more sensitive to fluctuating demand, would appear to be more popular than nationals in sectors subject to strong cyclical swings, but also more at risk during a recession' (OECD, 1994: 168).

The process of labour market segmentation

The French car and building industries were typical of the situation in all highly-developed countries, in that they exhibited a pattern of foreign worker concentration in less desirable jobs. These jobs were frequently unhealthy, physically taxing, dangerous, monotonous or socially unattractive. This state of affairs was shaped by many factors. In both industries, employment of foreign and colonial workers had already become traditional before the Second World War. In the post-1945 period, both industries faced a serious shortfall of labour, a problem solved by recourse to aliens. The legal foreign worker recruitment system aided employers by making employment and residence contingent on employment in a certain firm or industry – usually within one city or region – for a period of several years. Many foreign workers only gradually earned freedom of employment and residential mobility.

The recruitment system funnelled foreign workers into less attractive jobs. Employers might have had to improve working conditions and wages if it had not been for the availability of foreign labour, or they might have been unable to stay in business. Illegal alien employment was rare in the car industry: the size of firms and the presence of strong unions made it difficult. Illegal employment was common in the building industry, where it adversely affected wages and working conditions. This had the paradoxical effect of making the industry all the more dependent on foreign labour. As employment in the industry became socially devalued, employers could often find only foreigners to work for them. Similar processes affected female foreign workers, who became highly concentrated in certain sectors of manufacturing, such as clothing and food processing, and in service occupations such as cleaning, catering and unskilled health service work. Undocumented employment of women was even more common than for men, since ideologies about foreign women as mothers and housewives made it easy to conceal their role in the labour force.

There was little direct displacement of French workers by foreigners. Certain types of jobs became socially defined as jobs for foreign labour, and were increasingly shunned by French workers who, during the long period of post-war expansion, could generally find more attractive employment elsewhere. Indeed, massive foreign worker employment enabled the upward mobility of many French workers. This general process prevailed until the late 1970s or early 1980s, when France went into a prolonged recession and unemployment grew.

Employer recruitment strategies also contributed to labour market segmentation between French and alien workers. Some building industry employers preferred to hire illegal aliens because they could increase profits, through non-payment of bonuses and payroll taxes for instance, and they ran little risk of legal sanctions until the 1980s. Some motor industry employers deliberately sought to hire poorly-educated peasants without industrial experience in order to frustrate left-wing unionization efforts. This strategy had the effect of making assembly-line work even less attractive to French workers. In the same way, clothing industry employers found it particularly easy to pressure foreign women into undocumented and poorly-paid outwork; again a situation to be found in virtually all industrial countries (Phizacklea, 1990). In France, between 1983 and 1991, overall employment in the clothing industry fell by 45 per cent, but foreign worker employment rose by 53 per cent (OECD, 1994: 40).

Eventually the pattern of ethnic stratification within French car plants became a major factor in labour unrest. The strategy of divide and rule practised by some employers ultimately boomeranged when foreign car workers struck for dignity in the late 1970s and the early 1980s. The ethnic solidarity produced by the process of labour market segmentation in many French car factories was a key factor in the prolonged unrest. Again, parallels can be found in migrant worker movements in other countries (for Australia, for instance, see Lever-Tracy and Quinlan, 1988).

The process of labour market segmentation usually results from a combination of institutional racism and more diffuse attitudinal racism. This applies particularly in countries which recruit 'guestworkers' under legal and administrative rules which restrict their rights in a discriminatory way. The legally-vulnerable status of many foreign workers in turn fosters resentment against them on the part of citizen workers, who fear that their wages and conditions will be undermined. This may be combined with resentment of foreign workers for social and cultural reasons, leading to a dangerous spiral of racism. Such factors have profoundly affected trade unions and labour relations in most countries which have experienced labour immigration since 1945.

Immigration, minorities and the labour market needs of the future

The plight of laid-off Moroccan car workers in France was emblematic of a host of critical problems facing many industrial democracies. Even in the early 1980s, a Paris-area car plant typically finished painting cars by hand. Teams of immigrant workers generally did the work and, in many cases, it was done by Moroccans. One-quarter of all Moroccans employed in France in 1979 were employed by the car industry alone. The Moroccans were recruited because they were eager to work, recruitment networks were in place and because they were reputed to be physically apt and hardworking people. By 1990, most of the painting teams had been replaced by robots. Many of the workers were unemployed and, owing to their lack of educational background, there was little hope of retraining them to take jobs requiring more advanced educational backgrounds. Their only hope for re-employment lay in finding another relatively low-skilled manual labour job, but such jobs were disappearing.

Throughout Western Europe, economic restructuring led to alarmingly high unemployment rates for foreign residents by the mid-1980s. In 2002, their unemployment rates generally remained well above those for the population as a whole. All indications are that there will continue to be an aggregate surplus of manual workers over employment opportunities for the foreseeable future. Job opportunities will be found primarily in the highly-skilled sector where shortages are already apparent and will continue into the future, or in the informal economy.

The labour market difficulties of laid-off foreign workers were compounded by several other worrisome trends. Immigrant children comprised a growing share of the school-age population but were disproportionately likely to do poorly in school, to be early school leavers or to enter the labour force without the kind of educational and vocational credentials increasingly required for gainful employment (Castles *et al.*, 1984: Chapter 6). The worst scenario for the French socialists involved the sons and daughters of the laid-off Moroccan car workers leaving school early and facing bleak employment prospects. The fear was of a US-style ghetto syndrome in which successive generations of an ethnically distinctive population would become entrapped in a vicious cycle of unemployment leading to educational failure and then socio-economic discrimination, and finally housing problems.

France faced an uphill struggle to ensure that the most vulnerable members of its society enjoyed a reasonable measure of equality of opportunity. Immigrants and their descendants comprised a large share of the at-risk population. This was the major motivation behind Western European efforts to curb illegal immigration. It was generally felt that the population that was the most adversely affected by competition from illegal aliens on labour markets was existing minority populations. The overall economic effects of immigration are thought to be marginally positive (US Department of Labor, 1989; Borjas, 1999: 12-13). But labour market effects of immigration, and particularly of illegal immigration, are uneven and spatially concentrated. In the USA, it was thought by some specialists that Afro-Americans and Hispanic citizens were the two groups most affected by illegal migration. These conclusions were disputed, however, and many Hispanic advocacy groups in the USA viewed illegal immigration as a benign, if not positive, inflow, since it provided much-needed workers and helped in family reunion and community formation processes.

Conclusions

This chapter has argued that the economic restructuring since the 1970s has given rise to new immigration flows and new patterns of immigrant employment. One major result has been increasing diversification of immigrants' work situations and of their effects on labour markets. A major review of the literature on macro-economic impacts of immigration since the mid-1970s found that studies converge 'in concluding that immigration causes no crowding-out on the labour market and does not depress the income of nationals ... This is probably the most important contribution economists have made toward clarifying the issues involved' (OECD, 1994: 164).

Opponents of immigration often argue that it harms low level workers by taking away jobs, and that it may damage the economy of the receiving country by harming the balance of payments, causing inflation and reducing the incentive for productivity improvement and technological progress. Economists in long-standing immigration countries like the USA and Australia have done a great deal of empirical research and econometric analysis on these topics. In Europe, by contrast, such research is in its infancy. A recent authoritative report by a National Research Council (NRC) panel of leading US economists and other social scientists found that the aggregate impact of immigration on the US economy was quite small. However, they did find that immigration 'produces net economic gains for domestic residents for several reasons. At the most basic level, immigrants increase the supply of labour and help produce new goods and services. But since they are paid less than the total value of these new goods and services, domestic workers as a group must gain' (Smith and Edmonston 1997: 4). However, the report goes on to warn:

> Even when the economy as a whole gains, however, there may be losers as well as gainers among different groups of US residents. Along with immigrants themselves, the gainers are the owners of productive factors that are complementary with the labour of immigrants – that is

> domestic, higher-skilled workers, and perhaps owners of capital – whose incomes will rise. Those who buy goods and services produced by immigrant labour will also benefit. The losers may be the less-skilled domestic workers who compete with immigrants and whose wages will fall. (Smith and Edmonston, 1997: 5).

This finding is not unexpected, but nonetheless very important. Competing groups of local workers may be genuinely threatened by immigration, which explains the readiness of some working-class people to support anti-immigration parties. However, the econometric studies carried out by the NRC panel revealed that 'immigration has had a relatively small adverse impact on the wages and employment opportunities of competing native groups'. The minor impact was evidently due to the dispersal effect of migration – where wages fall, workers tend to move to areas where the wages are better (Smith and Edmonston 1997: 7).

Australian economists have been studying immigration for many years, as it has been the motor of economic growth in Australia since the 1940s (Wooden 1994; Castles *et al.*, 1998; Foster 1996). A recent authoritative study by economist Will Foster concludes

> that immigration impacts on both demand and supply sides of the economy. Immigrants create jobs as well as fill them; they pay taxes as well as make demands of government; and they bring funds from overseas and contribute to higher exports as well as to imports. ... But beyond their mere presence, the research evidence shows that the demand- and supply-side effects in fact balance each other so closely that no more than marginal impacts can be detected for any of the key economic indicators.... To the extent that any of the usual measures of economic health have been significantly affected, the evidence is that immigration has been generally beneficial for the Australian economy and for the employment prospects and incomes of Australian residents. (Castles *et al.*, 1998: Chapter 3).

Patterns of labour market segmentation by ethnic origin and gender which had emerged by the 1970s have generally persisted and, in many ways, become even more pronounced in the 1990s. However, the growth of illegal migration, continuing deficiencies in statistics, and the growing transnational interdependence of which international migration is an integral part make it difficult to generalize about the labour market effects of immigrants. Writing of immigrant women, Morokvasic observed that 'it is probably illusory to make any generalizations based on these findings in different parts of the world ... They can only be interpreted within the specific socioeconomic and cultural context in which these changes are observed' (Morokvasic, 1984: 895). There are tremendous variations in immigrant employment patterns according to ethnic and national background, gender, recentness of arrival, legal status, education and training.

Varying economic structures, governmental policies, patterns of discrimination and legal traditions further complicate matters.

In Western Europe, an authoritative study (Commission of the European Communities, 1990) documented the continuing pattern of employment, educational and housing disadvantages encountered by immigrants. Discrimination endured despite the integration policies of many governments. Disadvantage is often intergenerational and poses a grave challenge to Western European social democratic traditions. In the USA, the passage of time has generally witnessed intergenerational upward mobility for European-origin immigrants. The quintessential question asked about immigrants to the USA is: will the Mexican or Dominican immigrants be like the Irish and Italian immigrants of the nineteenth and early twentieth centuries? It seems too early to answer this question, but the intergenerational mobility evinced by earlier immigrant waves to the USA has created a more optimistic context and expectation than prevails in Western Europe. Much the same could be said for Australia and Canada. However, Borjas marshalled strong evidence that intergenerational disadvantage for poorly-educated poor migrants and their offspring in the USA was emerging as a pattern (Borjas, 1999).

Institutional and informal discrimination has clearly contributed to immigrant disadvantage. In Western Europe, the discrimination inherent in the employment and residential restrictions characteristic of guest-worker policies funnelled immigrants into specific economic sectors and types of jobs. The analysis of foreign worker employment in the French motor manufacture and building industries demonstrated the disproportionate effects of job losses through economic restructuring since the 1970s upon foreign workers. However, in the 1980s, immigrant employment in France grew sharply in the expanding services sector. Legally-resident aliens enjoyed more secure legal status and more extensive rights than in the past. This enabled many foreigners to adjust to restructuring. Some migrants have developed their own strategies to cope with labour market disadvantage. The unionization of foreign employees in Western Europe and strike movements like those witnessed in the French car industry were forms of adaptation. The proliferation of immigrant entrepreneurs was another.

Labour market segmentation is a central element in the process which leads to formation of ethnic minorities. Labour market segmentation has complex links with other factors that lead to marginalization of immigrant groups (see Chapters 2, 9 and 10). Low-status work, high unemployment, bad working conditions and lack of opportunities for promotion are both causes and results of the other determinants of minority status: legal disabilities, insecure residency status, residential concentration in disadvantaged areas, poor educational prospects and racism.

Some sociologists argue that, in the 1990s, the conflict between labour and capital is no longer the major social issue in advanced societies. It has been replaced by the problem of the *exclusion* of certain groups from the mainstream of society. These groups are economically marginalized through insecure work, low pay and frequent unemployment; socially marginalized through poor education and exposure to crime, addiction and family breakdown; and politically marginalized through lack of power to influence decision-making at any level of government. All these factors join to produce spatial marginalization: concentration in certain urban and suburban areas, where minorities of various kinds are thrown together, virtually cut off from and forgotten by the rest of society (Dubet and Lapeyronnie, 1992). Certain immigrant groups have a very high propensity to suffer social exclusion. These immigrants are doubly disadvantaged: they are not only among the most disadvantaged groups in contemporary society, but they are also frequently labelled as the cause of the problems. Thus immigrants experience a rising tide of racism, which isolates them even more. This process of ethnic minority formation will be discussed in the following chapters.

Guide to further reading

Böhning (1984) provides comparative perspectives on migrants in the labour market. Sassen's work (1988) is significant for this topic, too. Borjas (1990) and Portes and Rumbaut (1996) examine the US situation. Piore's earlier work (1979) is still useful. Lever-Tracy and Quinlan (1988) and Collins (1991) give good analyses for Australia. Waldinger *et al.* (1990) is excellent on ethnic small business, while Phizacklea (1990) looks at the links between gender, racism and class, through a case study of the fashion industry.

In addition to its annual report, now entitled *Trends in International Migration*, the OECD has produced a significant stream of reports and publications pertinent to the theme. Many result from international conferences including *The Changing Course of International Migration* (1993) and its sequel, *Migration and Development* (1994). *Combating the Illegal Employment of Foreign Workers* (2000) offers important insights into illegal migrant employment.

Major works on globalization and migrant employment include Stalker (1994, 2000) and Castles (2000). Key recent books on the economics of international migration include Portes (1995), Borjas (1999) and his edited volume (2000) and Rotte and Stein (2002). Philip L. Martin's many books on the economics of migration (1991–2002) are important. On Australia, see Castles *et al.* (1998). On the UK, see Glover *et al.* (2001). For an influential viewpoint of the USA, see Committee for Economic Development (2001).

Chapter 9

The Migratory Process: A Comparison of Australia and Germany

This chapter presents comparative case studies of the migratory process in two countries with very different traditions and institutional frameworks. Despite these differences, there are significant parallels in the development of migration and ethnic diversity, as will become apparent. This leads to the conjecture that the dynamics of the migratory process (as discussed theoretically in Chapter 2) can be powerful enough to override political structures, government policies and the intentions of the migrants. It does not mean, however, that these factors are unimportant: although settlement and ethnic group formation have taken place in both cases, they have done so under very different conditions. There have also been differing outcomes, which can be characterized as the formation of ethnic *communities* in the Australian case, as against ethnic *minorities* in Germany. The examples gain additional interest through the major changes in attitudes and policies towards migration in both countries since the mid-1990s.

Australia and Germany: two opposing cases?

Australia and the Federal Republic of Germany (FRG) have both experienced mass population movements since 1945. In both cases, foreign immigration started through the official recruitment of migrant workers. Early on, the areas of origin of migrants were partly the same. However, there the similarities seem to end, and the two countries are often seen as opposite poles on the migration spectrum.

Australia is considered one of the classical countries of immigration: a new nation which has been built through colonization and immigration over the last two centuries. Like the USA and Canada, it is a sparsely populated country which has been open to settlement from Europe and also, more recently, from other continents. Since 1947, there has been a continuous policy of planned immigration, designed both to build population and to bring about economic growth. Immigration has been mainly a permanent family movement of future citizens and has made Australia into a country of great ethnic diversity, with policies of multiculturalism. After 1996, this model was challenged and important changes have occurred.

Germany, by contrast, is generally seen as a 'historical nation', with roots that go back many centuries, even though unification as a state was not achieved until 1871. Post-1945 policies emphasized the recruitment of temporary 'guestworkers', although there have also been large influxes of refugees and 'ethnic Germans' from Eastern Europe. Despite the zero immigration policy pursued since 1973, the end of the Cold War led to massive new population movements. Until the late 1990s, leaders claimed that Germany was 'not a country of immigration'. Since then, however, there have been important changes in attitudes and policies on immigration and integration.

In comparing the two countries, we shall look at the way the migratory process is shaped by a number of factors: the origins and developments of migratory flows; labour market incorporation; the development of immigrant communities; the evolution of legal frameworks and government policies; and the immigrants' various forms of interaction with the society of the receiving country.

Tables 9.1 and 9.2 overleaf give figures for the immigrant populations in the two countries. The figures for Australia are birthplace figures as many overseas-born people have become citizens. About 22 per cent of the total population in 2001 were born overseas. In addition, about 27 per cent of those born in Australia had at least one immigrant parent. Thus nearly half the Australian population were either born overseas or had one or both parents born overseas. There were also about 400 000 Aboriginal people and Torres Strait Islanders (2.2 per cent of the total population) who are the only true 'non-immigrants' in Australia.

In 1999, Germany had 7.3 million foreign residents, making up 8.9 per cent of the total population. In 1990, prior to reunification, there were 5.3 million foreign residents in the old *Länder* (states) of the Federal Republic. These figures omit foreigners who have become citizens. However, this is a fairly small group, since naturalization rates are low.

Origins and development of the migratory movements

Migratory movements to Australia and Germany were described in Chapters 3 and 4. Here, we will discuss differences and similarities in the post-1945 immigration experiences.

In 1947, the Australian government started a large-scale immigration programme, designed to increase the population for both strategic and economic reasons. The original aim was to attract British settlers, but areas of origin quickly became more diverse. In the late 1940s, many immigrants came from Eastern and Central Europe, while in the 1950s and

Table 9.1 ***Australia: Immigrant population by birthplace (thousands)***

Country of birth	*1971*	*1981*	*1991*	*1996*	*2001*
Europe	2197	2234	2299	2217	2133
UK and Ireland	1088	1133	1175	1124	1086
Italy	290	276	255	238	219
Former Yugoslavia	130	149	161	n.a.	n.a.
Greece	160	147	136	127	116
Germany	111	111	115	110	108
Other Europe	418	418	457	618	604
Asia (incl. Middle East)	167	372	822	1007	1155
New Zealand	81	177	276	291	356
Africa	62	90	132	147	183
America	56	96	147	151	161
Total	2563	2969	3676	3813	3988

n.a. = not available.

Sources: Australian censuses.

Table 9.2 ***Foreign residents in Germany (thousands)***

Country	*1980*	*1990*	*1995*	*1999*
Turkey	1462	1695	2014	2054
Former Yugoslavia	632	662	798	737
Italy	618	552	586	616
Greece	298	320	360	364
Poland		242	277	292
Croatia			185	214
Austria	173	183	185	186
Bosnia-Herzogovina			316	168
Spain	180	136	132	130
Netherlands		111	113	111
Portugal	112	85	125	133
Other countries	978	1355	2084	2340
Total	4453	5343	7174	7343
of which EC/EU		1632	1812	1856

Notes: Figures refer to 31 December and apply to the old *Länder* of the pre-reunification Federal Republic of Germany up to 1990, to the whole of Germany for 1995 and 1999. 'Former Yugoslavia' includes Bosnia-Herzegovina and Croatia in 1980 and 1990, but these countries are shown separately in 1995 and 1999. EC/EU refers to the European Union of 15 member states.

Sources: OECD (1992: Table 10; 2001: Table B.1.5).

1960s Southern Europeans predominated. Many Eastern Europeans were selected for settlement from displaced persons camps, while Southern Europeans were recruited through bilateral agreements with the Italian, Greek and Maltese governments. Since the Australian government wanted both workers and settlers, initial recruitment led to processes of chain migration, through which early migrants helped relatives, friends and fellow villagers to come and join them.

Germany has had several major migratory movements since 1945. The first and largest was that of over 8 million expellees (*Heimatvertriebene*) from the lost eastern parts of the *Reich* and 3 million refugees (*Flüchtlinge*) who came to the FRG from the German Democratic Republic (GDR) up to 1961. These people were of German ethnicity, and immediately became citizens of the FRG. Despite initial strains, they were absorbed into the population, providing a willing source of labour for Germany's 'economic miracle' (Kindleberger, 1967).

The next major movement – that of 'guestworkers' from the Mediterranean area – was to be the one that did most to turn Germany into a multi-ethnic society. As in Australia, government labour recruitment was the driving force, but there was a major difference: 'guestworkers' were not meant to settle permanently (see Box 4.1). The 'guestworker' system was designed to recruit manual workers (both male and female) to work in factories and other low-skilled jobs in Germany. Foreign workers had a special legal status which restricted family reunion, limited labour market and social rights, and gave little chance of becoming citizens. But the ending of labour recruitment in 1973 led to trends towards family reunion and permanent settlement. Many Southern Europeans did leave, but those who stayed were mainly from the more distant and culturally-different sending countries, especially Turkey and Yugoslavia.

The Turkish case illustrates how temporary migration became transformed into settlement and community formation. Turkey had no tradition of international labour migration: the initial movement was a result of German recruitment policies. The Turkish government hoped to relieve domestic unemployment and to obtain foreign exchange through worker remittances. The migrants themselves sought an escape from poverty, unemployment and dependence on semi-feudal landowners. It was hoped that money earned and skills gained abroad would encourage economic development at home. Thus the Turkish participants initially shared the German expectation of temporary migration. However, when Germany stopped recruitment in 1973, many Turkish workers stayed on, and family reunion continued. Migrants realized that economic conditions at home were bad, and that there would be no opportunity to re-migrate to Germany later. In 1974, there were just over 1 million Turkish residents out of a total foreign population of 4.1 million. Their number grew to 1.6 million by 1982, and to 2 million by 1995. Family reunion was not the only form of continued migration: political unrest and ethnic conflict in Turkey generated waves of asylum seekers, who found shelter in Turkish and Kurdish communities abroad. German government policies were ineffective in preventing further immigration and settlement. Mass deportation, though debated, was never a real option for a democratic state, committed to a wide range of international agreements.

By the mid-1970s, both Australia and Germany had large permanent settler populations. Australian authorities accepted settlement and were beginning to seek ways of managing cultural diversity. Germany, by contrast, officially denied the reality of settlement. From the late 1970s, both countries experienced new forms of immigration, leading to even greater diversity.

Australia's immigration intakes became global in scope. The White Australia Policy was formally abandoned in 1973, and large-scale Asian immigration began in the late 1970s with the arrival of Indo-Chinese refugees. By the mid-1980s, Asia was the source of 40–50 per cent of entries. Australia also attracted Latin Americans (both workers and refugees) and Africans (in fairly small numbers). Immigration of New Zealanders (who can enter freely) also grew. In the 1990s, economic and political crises brought about new inflows from the former Soviet Union, former Yugoslavia, the Middle East and South Africa. Both skilled migration and refugee entry led to entry of dependants. The changing composition of immigrant intakes reflects these changes. In 1962–3, 84 per cent of entrants came from eight European countries, led by the UK (44 per cent). (Australian migration statistics relate to financial years, from July to June.) The rest were mainly from Italy, Greece, Yugoslavia, Spain, Malta, Germany and the Netherlands. New Zealand (1.3 per cent) and Egypt (1.1 per cent) were the only non-European countries in the top ten (Shu *et al.*, 1994). In 1998–9 by contrast, only one of the top ten was European – the UK with 10.4 per cent. New Zealand came first with 22 per cent. Then came the UK, followed by China, South Africa, Philippines, Yugoslavia, India, Indonesia, Vietnam and Hong Kong (DIMIA, 2001).

Germany has had three main types of immigration in recent years: first, entries of asylum seekers from Eastern Europe and from non-European countries; second, entries of economic migrants (both skilled and unskilled, documented and undocumented) from Europe and outside; and, third, immigration of *Aussiedler* or 'ethnic Germans' from the former Soviet Union, Poland, Romania and other Eastern European countries.

Article 16 of Germany's 1948 Basic Law laid down a right for victims of persecution to seek asylum. Until 1993, anybody who claimed to be an asylum seeker was permitted to stay pending an official decision on refugee status, which often took several years. Asylum-seeker movement became significant in the late 1970s, and grew fast as the Soviet Bloc imploded, reaching 100 000 in 1986, 193 000 in 1990, and 438 000 in 1992 (OECD,

1995: 195). Fears of mass East–West and South–North migrations of desperate and impoverished people were seized on by the extreme right, and there was an upsurge in racist incidents. After a lengthy and emotional debate, Paragraph 16 of the Basic Law was amended in 1993, allowing German frontier police to reject asylum seekers for a variety of reasons. Measures were also taken to speed up the processing of applications. As a result, asylum applications fell to 322 600 in 1993,116 400 in 1996 and 95 100 in 1999 (OECD, 2001, 170).

In 1992, there was an upsurge in asylum seekers from former Yugoslavia. At first, many were gypsies, as were the majority of asylum seekers from Romania. This ethnic group became the main target of racist violence in mid-1992. Asylum-seeker flows grew sharply during the war in Bosnia-Herzogovina, and by 1995 Germany was host to 345 000 displaced persons from former Yugoslavia (UNHCR, 2000b: 239). After 1995, Germany made agreements with the Bosnian and Serbian governments to facilitate repatriation. People fleeing from Kosovo in 1999 received temporary protection, but most were repatriated quickly after the crisis. In 1999, refugees and asylum seekers made up 1.2 million of Germany's 7.3 million foreign residents. Over 400 000 of these were 'de facto refugees' – persons denied refugee status, but unable or unwilling to leave for political reasons or fear of persecution (OECD, 2001: 173).

Germany ended recruitment of migrant labour in 1973, but reintroduced temporary labour schemes for workers from Poland and other Central and Eastern European countries in the 1990s. Some 40 000 contract workers were admitted in 1999. Seasonal workers are also admitted for up to three months per year for work in agriculture, hotels and catering. There were 223 400 seasonal workers in 1999. Under the new 'Green Card' system set up in 2000 to attract highly-skilled workers (especially for the information technology sector), about 5000 persons had been admitted by February 2001. This was far less than the planned 20 000 recruits (OECD, 2001: 174–5). The shortfall appears to have been due to the limited duration of the work permits, and restrictions on family reunion. There is also some undocumented labour migration to Germany, although no official figures are available. In addition, many EU citizens live and work in Germany. Since they do not require work permits, the numbers are not known.

The *Aussiedler* are people of German ancestry whose ancestors have lived in Russia and Eastern Europe for centuries. Like the post-1945 expellees, they have the right to enter Germany and to claim citizenship. *Aussiedler* are generally of rural origin, and may have considerable problems of social adaptation and labour market entry. They are provided with a range of services and benefits to facilitate settlement, which act as a powerful attracting factor. *Aussiedler* arrivals rose from 86 000 in 1987 to a peak of 397 000 in 1990. In the latter year, 148 000 were from the Soviet Union, 134 000 from Poland and 111 150 from Romania. At a time of social stress and growing unemployment, the influx of *Aussiedler* became unpopular. Despite the principle of free admission, the German government introduced an annual quota. Entries of *Aussiedler* declined to an average of around 220 000 per year from 1991 to 1995. Then quotas were cut further, with just 105 000 entries in 1999 (OECD, 2001: 170).

Australia maintains a regular Migration Program planned by the government on the basis of economic, social and humanitarian considerations. Permanent arrivals averaged 110 000 per year in the 1980s and 90 000 in the 1990s. The current government has increased skilled migration and limited family entries. The 2001–2 Migration Program (85 000) represents a considerable increase in planned skilled-worker intakes (45 500), while family entries are projected at 21 200. This does not include entrants from New Zealand, who numbered 31 600 in 1999–2000. In addition, the Humanitarian Program has averaged 12 000 entrants per year since the early 1990s. Australia remains one of only about

Box 9.1 Australia's immigration panic and the '*Tampa* Affair'

Due to its inaccessibility, Australia has had little illegal immigration in the past. Immigration has always been a political issue, but the level of conflict has been limited. This situation has changed dramatically since 1996 due to two factors: the rise of the One Nation Party and the increase in arrivals of boat people. Matters came to a head in 2001 with the '*Tampa* Affair' and a Federal Election fought largely on immigration.

During the March 1996 Federal Election, a Liberal Party candidate in Queensland, Pauline Hanson, attacked services for Aboriginal people in such an extreme way that she was disendorsed as a candidate by her party. Yet she won the seat as an Independent, and quickly set up the One Nation Party, which sought to build on anti-minority feelings. In her inaugural speech in the Federal Parliament, Hanson attacked Aboriginal people, called for the stopping of immigration and the abolition of multiculturalism, and warned of 'the Asianization' of Australia. Both the ruling Liberal-National coalition and the Australian Labor Party (ALP) were slow to condemn Hanson's politics. The trend towards racialization of politics had immediate effects. The tightening of immigration policy was targeted at categories which were claimed to be hurting national interests: family reunion and asylum seekers. The result was a much more hostile climate towards immigration and multiculturalism.

The situation was exacerbated by the increase in boat-people arrivals in northern Australia. These fell into two main groups: Chinese people being smuggled in to work, and asylum seekers from Iraq, Afghanistan and other countries brought in from Indonesia, usually on fishing boats chartered by people smugglers. Numbers were not high by international standards, never

→

ten countries in the world that have programmes to resettle refugees from countries of first asylum in collaboration with the UNHCR. However, since the late 1990s a growing number of asylum seekers have arrived illegally by boat (see Box 9.1). (For detailed sources and analysis of recent Australian immigration, see Castles and Vasta (2003).)

Australia's integration into global and regional economic networks has led to increased temporary migration. In 2000–1, 160 157 temporary residence permits were issued, of which 45 669 were for skilled workers. Overseas Student Visas are also a growing category: 86 277 were issued in 2000–1. Most students come from Asia (DIMIA, 2001). Emigration by Australians is also increasing. Working abroad has become an important part of professional and personal experience, and it is believed that as many as 800 000 Australian citizens currently live overseas.

Australia has received 6 million immigrants since 1947. Although settlement was always planned, it has had unforeseen consequences: the

→

going much above 4000 in a year, but they provoked hostile media campaigns and popular outrage. Immigration Minister Ruddock attacked the asylum seekers as 'queue jumpers' claiming that they took places from 'genuine' refugees allocated by UNHCR. He declared that boat people were a threat to Australian sovereignty. In 1999 the government introduced the three-year Temporary Protection Visa (TPV), which confers no right to permanent settlement or family reunion. Another deterrent has been to stop boat people from landing on Australian shores, and to try to send them back to Indonesia. Those who do land are detained in isolated and remote camps, where they are isolated from lawyers, the media and supporters. As a result, they can languish in mandatory detention for up to three years. Hunger strikes, riots, self-inflicted injuries and even suicide have become commonplace. The government also introduced legal measures to limit the power of the courts in asylum matters (Crock and Saul, 2002: Chapter 5).

Matters got even worse in August 2001, when the Norwegian freighter MV *Tampa* picked up over 400 asylum seekers (mainly from Afghanistan and Iraq) from a sinking boat off Northern Australia. The government refused the captain permission to land, and the *Tampa* anchored near the Australian territory of Christmas Island. This was the start of a saga involving international diplomacy, heated public debates in Australia, and feverish political activity. In the 'Pacific Solution', Australia tried to export the asylum seekers to its neighbours, Nauru and New Guinea – and was willing to spend vast sums of money in doing so. Asylum became the central issue in the November Election, giving victory to Liberal-National Prime Minister Howard. Before the *Tampa* affair, a Labour victory had been predicted. The 2002–3 Federal Budget included A$2.8 billion for border control measures – an increase of A$1.2 billion over the previous year. Even stricter border control legislation was introduced in 2002 (Castles and Vasta, 2003).

ethnic composition of the population has changed in a way that was never intended by the architects of the migration programme. This has been partly because the need for labour during expansionary phases has dictated changes in recruitment policies. It has also been due to the way in which chain migration has led to self-sustaining migratory processes. Most of Germany's foreign population is the legacy of guestworker recruitment in the 1960s and 1970s. Altogether over 20 million people have migrated into Germany since 1945. Over 60 per cent of the foreign residents have lived in Germany over eight years (OECD, 2001: 170). The largest group is the 2 million Turks, of whom 750 000 were actually born in Germany (Oezcan, 2002). Current movements are complex and unpredictable.

Labour market incorporation

Up to about 1973, both Australian and German immigration policies were concerned with the recruitment of a manual labour force. Non-British migrants who received assisted passages to Australia were directed into jobs on construction sites such as the Snowy Mountains Hydroelectric Scheme, in heavy industry, or in factories (Collins, 1991). Similarly, the German Federal Labour Office channelled foreign workers into unskilled and semi-skilled jobs on building sites and in factories, and used restrictionary labour permit rules to keep them there as long as possible.

The pre-1973 movements were mainly rural to urban migration: Mediterranean farmers and rural workers emigrated because of poverty, breakdown of social structures through war, and decline of local industries. Many intended to work temporarily in industrial economies, in order to use their earnings to improve their farms or set up small businesses upon return. Even in Australia, many Southern European workers expected to return home, and indeed did so in the 1960s and 1970s, as conditions improved in their countries of origin. However, in time, many migrant workers' intentions changed when it became clear that it would not be possible to achieve their objectives in the home country as quickly as originally expected. The result was an increasing orientation towards long-term stay and occupational mobility in the immigration country (a trend which went hand in hand with family reunion: Piore, 1979: 50ff).

Workers found that, having entered the labour market at the bottom, it was hard to gain promotion. Migrants' qualifications were often not recognized, forcing skilled workers into dead-end jobs. Typical workplaces for men were car assembly lines, construction sites and foundry work; and, for women, clothing, textiles and food processing. Service occupations such as catering, refuse collection, office cleaning and unskilled jobs in public utilities also became known as 'migrant work' (see Castles *et al.*, 1984; Collins, 1991). The structural factors and discriminatory rules which led to initial low status caused enduring patterns of labour market segmentation. This applied particularly to migrant women, whose situation was affected both by patriarchal structures in the countries of origin and gender discrimination in the country of immigration. Their occupational status, wages and conditions were generally the worst of all groups in the labour market (Phizacklea, 1983, 1990).

Economic restructuring since the 1970s has brought significant changes. The pre-1973 entrants bore the brunt of restructuring, as low-skilled jobs in manufacturing declined. OECD research has found that once older unskilled foreign workers are laid off in recessions, they have little chance of being re-employed (OECD, 1995: 37). As unskilled manufacturing jobs disappear, many immigrants lack the language skills and basic education needed for retraining (Baker and Wooden, 1992). In Australia unemployment during the recessions of 1974–5, 1982–3 and 1990–2 was significantly higher for non-English-speaking background (NESB) immigrants than for other workers (Ackland and Williams, 1992). Workers from English-speaking-background (ESB) countries fare better. The 1996 Census revealed unemployment rates of 10.9 per cent for NESB workers, 6.4 per cent for ESB workers and 7.8 per cent for the population as a whole. The unemployment rate for foreigners in Germany in 1999 was 18.4 per cent, more than twice the 8.8 per cent for Germans (OECD, 2001: 171).

Many of the new migrants no longer fit the old stereotype of the low-skilled migrant worker. In Australia, many immigrants from countries such as Taiwan, Korea, China, Hong Kong and Malaysia find work in skilled, professional and managerial occupations. New migrants are needed to make up for skill deficits, and are seen as vital for linking the Australian economy to the fast-growing 'tiger economies' of the region. Germany, like other EU countries, experiences increasing international interchange of managers and experts.

At the same time, the past segmentation of migrant workers into specific forms of work means that they cannot easily be replaced by local labour, even at times of high unemployment. Newcomers who enter as dependants, asylum seekers or refugees often provide labour for insecure and poorly-paid casual jobs in the growing informal sector. Recent recognition of the decline in fertility in Western countries is changing attitudes towards immigration, with some economists and policy-makers arguing that low-skilled 'replacement migration' will be needed to maintain future demographic balances and economic growth (UNPD, 2000).

Children of migrants who have obtained education and vocational training in the immigration country often secure better jobs than their parents. Australian research indicates substantial intergenerational mobility. For example, the 1991 Census showed that 18.8 per cent of male Australians of Greek parentage had university degrees, compared with only 2.5 per cent of their fathers. Several European immigrant groups appeared to share this pattern, although usually to a lesser extent (Birrell and Khoo, 1995). However, second-generation immigrants with poor educational qualifications have little chance of steady employment. Up to the 1980s many foreign children were failing in German schools, while few employers were willing to give them training places (Castles *et al.*, 1984: Chapter 6). More recently school achievement rates have improved. However, a government commission found that there were still substantial deficits in both schooling and vocational training for children of immigrants (Süssmuth, 2001).

When the original immigrants arrived in the 1950s and 1960s, it was easy to find entry-level jobs in industry. Such jobs are now few and far between, so that many young ethnic minority members face a future of casual work and frequent joblessness. Some groups and individuals break out of this situation, but many are caught in a vicious circle: initial incorporation in low-skilled work and residential areas with poor educational opportunities is reinforced by processes of racialization. Young job-seekers may be rejected because the combination of poor educational credentials, ethnic appearance and living in certain neighbourhoods has become a stigma, denoting marginality and unreliability (Häussermann and Kazapov, 1996: 361).

One route out of factory work is self-employment: 'ethnic small business' has become significant in virtually all industrial countries (Waldinger *et al.,* 1990). In Australia, some migrant groups have higher rates of self-employment or business ownership than locally-born people. However, many small businesses fail, and the rate of self-exploitation (long working hours, poor conditions, use of family labour power, insecurity) is high. Many ethnic entrepreneurs are concentrated in 'ethnic niches' such as retail trade, catering, construction and transport (Waldinger *et al.,* 1990; Collins *et al.,* 1995). By 1999, 263 000 foreign residents of Germany were classified as self-employed, along with 23 000 family workers (OECD, 2001: 175).

Community formation

In Germany, many foreign workers were at first housed by employers in hostels or camps near the work site. In Australia, the Department of Immigration provided hostels for new arrivals, and they often sought work and longer-term housing in the area nearby. As the need for family accommodation grew, migrants had to enter the general housing market. Several factors put them at a disadvantage. Most had low incomes and few savings. Early arrivals lacked local knowledge and informal networks Some landlords refused to rent to migrants, while others made a business out of it, taking high rents for substandard and overcrowded accommodation. In some cases there was discrimination in the allocation of public housing, with rules that effectively excluded migrants, or put them at the end of long waiting lists (see Castles and Kosack, 1973: Chapter 7). Migrants therefore tended to become concentrated in the inner city or industrial areas where relatively low-cost housing was available. The quality of the accommodation and of the local social amenities (such as schools, health care facilities and recreational facilities) was often poor.

In Germany, such concentration persisted long after initial settlement. In Australia, where there is a strong tradition of owner-occupation, many migrants were able to improve their situation. By 1986, most Southern Europeans had bought their own homes by means of frugal living and working long hours, and their rate of owner-occupation was actually higher than for the Australian-born population (ABS, 1989). Many still remained in the original areas of settlement, though some had moved to outer suburbs. Newer groups, such as Indo-Chinese immigrants, were also following a similar trajectory.

There has been much debate in both countries about the formation of 'ethnic ghettos'. In fact, unlike the USA, there are very few areas with predominantly minority populations in either country. Rather we find class-based segregation, with migrants sharing certain areas with disadvantaged groups of the local population: low-income workers, the unemployed, social security recipients and pensioners. However, there are neighbourhoods where a specific ethnic group is large enough to have a decisive effect on its appearance, culture and social structure. The Turkish community of Kreuzberg in West Berlin is a well-known example. In Australia, a marked Italian atmosphere can be found in the Carlton area of Melbourne, or in the Leichhardt and Fairfield areas of Sydney. There are Chinatowns in the centre of Melbourne and Sydney, and Indo-Chinese neighbourhoods have developed in Richmond (Melbourne) and Cabramatta (Sydney).

Residential segregation of migrants has a double character: on the one hand, it can mean poor housing and social amenities, as well as relative isolation from the majority population; on the other hand, it offers the opportunity for community formation and the development of ethnic infrastructure and institutions. The most visible sign of this development is the establishment of shops, cafés, and agencies which cater for migrants' special needs. 'Ethnic professionals' – health practitioners, lawyers, accountants – also find opportunities in such areas. Small business owners and professionals form the core of ethnic middle classes, and take on leadership roles in associations. Welfare organizations cater for special needs of immigrants, sometimes compensating for gaps in existing social services. Social associations establish meeting places. Cultural associations aim to preserve homeland languages, folklore and tradition, and set up mother-tongue or religious classes. Political associations of all complexions struggle for influence within the community. Often their starting point is the politics of the country of origin, but with increasing length of stay their aims become more oriented to the country of immigration.

Religion plays a major part in community formation. Sometimes migrants can join existing structures: for instance, many Southern Europeans in Germany and Australia became attached to Catholic Churches. However, they sometimes found that religion was practised in ways not sensitive to their needs (Alcorso *et al.*, 1992: 110–12). Often priests or religious orders (such as the Scalabrinians from Italy) accompanied the migrants, giving churches in areas of migrant settlement a new character. Orthodox Christians from Greece, Yugoslavia and Eastern Europe had to establish their own churches and religious communities. In recent years, the most significant religious development has been connected with migrations of Muslims: Turks and North Africans to Germany; Lebanese, Turks and Malaysians to Australia. The establishment of mosques and religious associations has had a high priority. Bhuddist, Hindu and Bahá'í temples can also be found in what were formerly almost exclusively Christian countries.

Such developments can be found in all countries of immigration. They are at the nexus of the migratory process, where transitory migrant groups metamorphose into ethnic communities. Establishing community networks and institutions means an at least partially conscious decision to start 'placemaking' and building a new identity (Pascoe, 1992). Community formation is linked to awareness of long-term or permanent stay, to the birth and schooling of children in the country of immigration, to the role of women as 'cultural custodians' and above all to the coming of age of the second generation (Vasta, 1992). In Germany, growing awareness of the long-term nature of stay led to the formation of Turkish associations (Kastoryano, 1996). Though many had an Islamic character, they were increasingly aimed at obtaining social and political rights in Germany. Moreover, they were linked to a new collective identity which found expression in the demand for dual citizenship: that is, the recognition of being both Turkish and German.

The concept of the ethnic community plays a central part in debates on assimilation and multiculturalism. Community formation is not a mechanistic or predetermined process, and not all migrants form communities: for instance, one cannot speak of an English community in Australia, or an Austrian community in Germany. Community formation is not just concerned with cultural maintenance, but is also a strategy to cope with social disadvantage and to provide protection from racism. The relationships and institutions which make up the community are initially based on individual and group needs. However, as economic enterprises, cultural and social associations, and religious and political groups develop, a consciousness of the community emerges. This process is in no way homogeneous; rather it is based on struggles for power, prestige and control. The ethnic community can best be conceived as a changing, complex and contradictory network. It is most intense and easily identifiable at the neighbourhood level, but is in turn linked to wider networks at the national level and beyond. Ethnic communities may take on a transnational character, and provide the basis for communicative networks which unite people across borders and generations (see Chapter 2).

Legal frameworks and government policies

There are important parallels in the migratory process in Australia and Germany, but the laws and policies shaping the position of immigrants in society were very different until recently. The differences were linked to varying historical experiences of nation-state formation. The Australian model embodies a territorial concept of the nation building appropriate to a former settler colony: if immigration rules allow a person to become a permanent resident, then citizenship policy allows him or her to become a member of the political community and of the nation. By contrast, when the German Reich emerged as the first modern German state in 1871, nationality was defined not through territoritality, but through ethnicity as shown by language and culture. A person could only obtain German nationality by being born into the German community, so that 'blood' became a label for ethnicity. Until 2000, German citizenship was based on *ius sanguinis* (law of the blood, or nationality through descent). Millions of foreigners became part of society, but were excluded from the state and nation. This was the rationale behind the seemingly absurd slogan 'the FRG is not a country of immigration' (see Hoffmann, 1990).

In Australia, family reunion was accepted from the outset, and newcomers were encouraged to become Australian citizens. The initial five-year waiting period for naturalization was reduced to three and then two years. Today, two-thirds of immigrants who have been in Australia over two years are Australian citizens. Citizenship is based primarily on *ius soli* (law of the soil, or nationality based on birth in a territory), so that children born to legal immigrants in Australia are automatically citizens. Immigrants who become Australian citizens may retain their former nationality, so that dual citizenship is common.

For Germany there are two separate legal frameworks. The first applies to people of German ethnic descent such as post-1945 expellees and refugees and current *Aussiedler*. They have a right to citizenship, and are not considered foreigners. The second framework applies to foreigners. A long residence requirement, restrictive conditions and complicated procedures kept naturalizations low. Foreigners seeking naturalization had to renounce their previous nationality. By the mid-1980s over 3 million foreigners fulfilled the ten-year residence qualification, but only about 14 000 per year actually obtained citizenship (Funcke, 1991). Children born to foreign parents had no automatic right to German citizenship, and the great majority of immigrants and their descendants remained non-citizens.

In the period following German reunification in 1990, restrictive citizenship rules were identified as a major obstacle to social and political integration, especially for youths of immigrant background. Immigrants and anti-racist organizations campaigned for easier naturalization. A key demand was for the right to dual citizenship. Some immigrant groups, especially Turks, were not permitted by the government of origin to give up the previous nationality. In any case, renouncing the previous affiliation could involve both material loss (with regard to land ownership) and psychological difficulties.

The Social Democrat–Green Coalition that came to power in 1998 announced that it would reform naturalization law, and permit dual citizenship. This met with strong opposition from conservative groups. The 1999 citizenship law (which took effect in 2000) represents a major shift towards *ius soli*, but stops short of allowing dual citizenship. Foreign immigrants have a right to naturalization after eight years residence, providing they renounce their previous nationality, have not been convicted of a serious offence, are able to support themselves and their families, acquire basic proficiency in German and declare allegiance to the Basic Law. Children born to foreign parents in Germany acquire citizenship at birth if at least one parent has lived legally in Germany for at least eight years. They can hold dual citizenship until they reach maturity, but have to decide between their German and other citizenship by the age of 23. Despite the prohibition of dual citizenship, many exceptions are allowed – there were estimated to be 1.2 million dual citizens in the early 1990s (Çinar, 1994: 54), and the number has certainly grown since then. The number of naturalizations has increased sharply, reaching about 107 000 in 1998 – even before the new law came into force (Beauftragte der Bundesregierung, 2000: 33).

The fundamental differences in views on immigration and incorporation into society affected all aspects of public policy. The Australian model for managing diversity had two main stages. In the 1950s, the government introduced a policy of *assimilationism,* based on the doctrine that non-British immigrants could be culturally and socially absorbed, and become indistinguishable from the existing population. The central principle of assimilationism was the treatment of migrants as 'New Australians', who would live and work with Anglo-Australians and rapidly become citizens. There was no special educational provision for migrant children, who were to be brought up as Australians. Cultural pluralism and the formation of 'ethnic ghettos' were to be avoided at all costs.

By the 1960s, it became clear that assimilationism was not working, due to labour market segmentation, residential segregation and community formation. By the 1970s, political parties were also beginning to discover the political potential of the 'ethnic vote'. Assimilationism was replaced by *multiculturalism*: the idea that ethnic communities, which maintain the languages and cultures of the areas of origin, were legitimate and consistent with Australian citizenship, as long as certain principles (such as respect for basic institutions and democratic values) were adhered to. In addition, multiculturalism meant recognition of the need for special laws, institutions and social policies to overcome barriers to full participation of various ethnic groups in society (OMA, 1989).

In Germany there was no question of assimilation. The Federal Labour Office granted work permits and 'guestworkers' were controlled by a network of bureaucracies. The *Ausländerpolizei* (foreigners' police) issued residence permits, kept foreign workers under surveillance and deported those who offended against the rules. The personnel departments of the employers provided basic social services and managed company hostels. To deal with personal or family problems of foreign workers, the government provided funding to church and private welfare bodies (see Castles and Kosack, 1973). By the 1970s, foreign children were entering German schools in large numbers. The educational authorities introduced a 'dual strategy' designed both to integrate foreign children temporarily during their stay in Germany, and to prepare them for return to their country of origin. The result was a system of 'national classes', 'preparatory classes' and 'mother-tongue classes' which separated foreign from German students, and prevented many foreign children from achieving educational success (see Castles *et al.*, 1984: Chapter 6). Family reunion also meant that workers were leaving company accommodation and seeking housing in the inner cities. The result was a debate on a 'foreigners' policy' which was to continue up to the present.

After reunification in 1990, the myth of not being 'a country of immigration' became unsustainable (see Bade, 1994). After the murder of several Turkish immigrants in arson attacks in Mölln in 1992 and Solingen in 1993, there were large anti-racist demonstrations all over Germany. The demands for political and social integration of immigrants led to the reform of citizenship law mentioned above. However, the left remained ambivalent about multiculturalism. Some see it negatively as a model for cementing the identity of ethnic groups, and for maintaining cultures perceived as anti-modern and repressive, especially towards women, although others on the left support multiculturalism as a way of enriching German cultural life (Cohn-Bendit and Schmid, 1993). Conservatives now generally favour models designed to achieve economic and social integration, in the hope that this will lead to cultural assimilation in the long run.

In 2000, the German government appointed a commission to advise on immigration, chaired by a leading Christian Democratic Union (CDU) politician, Rita Süssmuth. The Commission's report declared that Germany had, in fact, been a country of immigration for many years, and that Germany needed immigrants for both demographic and economic reasons. The Commission proposed a planned immigration system to encourage recruitment of skilled personnel. It also advocated a comprehensive integration policy to overcome social disadvantages suffered by immigrants. The goal was to achieve full participation in society, while respecting cultural diversity. This policy required reforms and special measures in education, social welfare, the labour market, and other areas (Süssmuth, 2001). Together with the 1999 citizenship law, the Süssmuth Report can be seen as a major shift in German approaches to immigration and settlement. It comes close to the multicultural public policy concept typical of Australia in the past. However, it is not yet clear to what extent such measures will be implemented.

Interaction with the society of the receiving country

The attitudes and actions of the population and state of the country of immigration have crucial effects on the migratory process. In turn, community formation may modify or reinforce these effects. In both countries the control of migrant labour by the state, and its incorporation by employers, set the initial conditions for settlement processes. Discrimination in hiring and promotion, non-recognition of skills, and regulations explicitly designed to limit migrant workers' right to equal treatment in the labour market can be seen as forms of institutional racism. Local workers and their unions generally supported such discrimination, at least initially. Australian unions only agreed to recruitment of Eastern and Southern Europeans in the 1940s after being given official guarantees that they would not compete with local workers for jobs and housing (Collins, 1991: 22–3). German unions demanded that foreign workers should get the same work and conditions as Germans doing the same jobs, but also supported discriminatory 'guestworker' regulations, which ensured that most foreigners got inferior positions (Castles and Kosack, 1973: 129).

Attitudes of local workers towards migrants were part of a wider picture. Many Australians were highly suspicious of foreigners. In everyday terms this meant reluctance to rent housing to migrants or to have them as neighbours, hostility towards anyone speaking a foreign language in public, mistrust of visible foreign groups and resentment towards foreign children at school. However, the 1960s and the 1970s were years of growing acceptance of difference, due to the recognized contribution of immigration to economic growth and prosperity. This permitted such dramatic changes as the abolition of the White Australia Policy, the first large-scale Asian entries into the country and the introduction of multicultural policies.

In the 1980s, there were attacks by public figures and sections of the media against Asian immigration and multiculturalism. These corresponded with increases in racist violence against Aboriginal people and Asians (HREOC, 1991). However, the ALP government took a strong anti-racist line, and there was considerable public support for multiculturalism. After the 1996 Federal Election, however, anti-minority politics gained new prominence through the rise of the One Nation Party, and the Liberal-National government did little to combat racism or to support multiculturalism. Indeed, cuts in services for Aborigines and immigrants were a part of its policy, and the onslaught on multiculturalism seemed to reflect a desire to return to a more traditional monocultural identity (see Box 9.1).

In Germany, most social groups supported the 'guestworker' system at first, although there was informal discrimination, such as refusal to rent to foreigners, or exclusion from bars and dance halls. An escalation of racism came in the 1970s, when unemployment of Germans became a problem for the first time since the 'economic miracle'. The Turks had become the largest and most visible group, and fear of Islam remains a powerful historical image in central European cultures. Reunification in 1990 was accompanied by vast population movements and by growing economic and social uncertainty, particularly in the area of the former GDR. The result was widespread hostility towards immigrants and an upsurge of organized racist violence.

The visible emergence of ethnic communities influenced public attitudes, as well as state policies. For people who fear the competition of migrants or who feel threatened by difference, the existence of communities may confirm the idea that 'they are taking over'. Ethnic areas can become the target for organized racist attacks, and ethnic minorities can become the focus for extreme right mobilization. On the other hand, contact with new cultures and some of their more accessible symbols, such as food and entertainment, may break down prejudices. Where ethnic small business and community efforts rehabilitate inner-city neighbourhoods, good inter-group relationships may develop.

Complex links emerged between ethnic communities and the wider society. Local branches of political parties in ethnic community areas needed foreign members, and had to take account of their needs in order to attract votes. For instance, the ALP set up Greek branches in Melbourne in the 1970s. Unions needed to attract immigrant members; they hired immigrant organizers and published multilingual information material. Churches found that they had to overcome barriers, and work with mosques and Islamic associations if they were to maintain their traditional social role. Artists and cultural workers among minorities and the majority found they could enrich their creativity by learning from each other. Multifaceted new social networks developed in the ethnic community areas of the cities. These gave members of the majority population greater understanding of the social situation and culture of the minorities, and formed the basis for movements opposed to exploitation and racism.

Similar tendencies prevailed in the public sector. Both the German and the Australian state experimented with using ethnic associations as instruments for the delivery of social services, such as counselling, family welfare and youth work. Aims and methods differed so the relationship was often an uneasy one. Government agencies sometimes saw traditionalist ethnic organizations as effective instruments of social control of workers, young people or women. On the other hand, ethnic leaders might use their new role to preserve traditional authority, and slow down cultural and political change. The state could choose which ethnic leaders to work with and which to ignore, and reward desired behaviour through patronage and funding (Jakubowicz, 1989; Castles *et al.*, 1992c: 65–71). But cooption was a two-way process: state agencies tried to use ethnic community structures and associations, but had to make concessions in return. Cooption took place earlier in Australia than in Germany. In Australia, ethnic leaderships had an important power base and the 'ethnic vote' was thought to affect the outcome of elections (Castles *et al.*, 1992a: 131–3). In Germany, members of ethnic communities lacked political clout and were perceived as not really belonging to society.

The main expression of the Australian approach was to be found in the policy of multiculturalism, with its network of consultative bodies, special agencies and equal opportunities legislation which developed between 1973 and 1996. The current Liberal-National coalition government has strong reservations about multiculturalism, and has dismantled many of the special institutions and services. However, in 1999, the government launched *A New Agenda for Multicultural Australia* (DIMA, 1999), which largely endorsed the principles of the ALP's 1989 multicultural policy statement (OMA, 1989).

In Germany, a Federal Government Office for Foreigners has existed (in various guises) since the late 1970s. This agency had mainly provided public information, and appears to have had limited influence on policy. But there are trends towards recognizing ethnic associations and incorporating them into policy initiatives. In Berlin, Frankfurt and elsewhere 'commissions for foreigners' or 'offices for multicultural affairs' have been set up. These are seeking to build structures to work with ethnic community groups and to propagate legal and administrative reforms. A sort of 'de facto multiculturalism' has been emerging since the 1970s, especially in local educational and social work initiatives. This is not surprising, in view of the fact that foreigners make up a substantial proportion of the population in major cities, ranging from 30 per cent in

Frankfurt, to 24 per cent in Stuttgart and Munich, and 13 per cent in Berlin (Beauftragte der Bundesregierung, 2000). It is hard to see how social integration can be improved, without some form of policy designed to recognize cultural diversity.

Conclusions

The validity of a comparison between Australia and Germany can be questioned by arguing that Australia is a 'classical country of immigration', while Germany is a 'historical nation' and not a 'country of immigration'. This argument does not stand up to analysis. Both countries have experienced mass immigration. Germany has had over 20 million immigrants since 1945, one of the biggest population movements to any country ever. Australia – with immigration of 6 million people since 1945 – has had a very high inflow relative to its small population. Both countries initially recruited migrant workers in roughly the same areas at the same time. Whatever the intentions of policy-makers, both movements led to similar patterns of labour market segmentation, residential segregation and ethnic group formation. In both cases, racist attitudes and behaviour on the part of some sections of the receiving population have been problems for immigrants. Thus there are great similarities in the migratory process, despite the differences in policies and attitudes. These parallels are important because, if these apparently opposing examples show corresponding patterns, it should be possible also to find them for other countries, which are somewhere between these cases on the migration spectrum.

But the differences are also significant and require analysis. They go back to the different historical concepts of the nation and to the intentions of the post-war migration programmes. The Australian authorities wanted permanent settlement, and went to considerable lengths to persuade the public of the need for this. Chain migration and family reunion were therefore seen as legitimate, and the model for settlement was based on citizenship, full rights and assimilation. Assimilationism eventually failed in its declared goal of cultural homogenization, but it did provide the conditions for successful settlement and the later shift to multiculturalism.

The German government planned temporary labour recruitment without settlement, and passed this expectation on to the public. Official policies were unable to prevent settlement and community formation. But these policies (and the persistent failure to adapt them to changing conditions) did lead to marginalization and exclusion of immigrants. The results can be summed up by saying that the Australian model led to the formation of ethnic communities which are seen as an integral part of a changing nation, while the German model led to ethnic minorities, which are not seen as a legitimate part of a nation unwilling to accept a change in its identity. This is an oversimplification, for there are ethnic minorities in Australia too: Aboriginal people and some non-European groups. All the same, the distinction does capture an essential difference in the outcomes of the two models.

Herein lies the usefulness of the concept of the migratory process. It means looking at all the dimensions of migration and settlement, in relation to political, economic, and social and cultural practices and structures in the societies concerned. If the architects of the post-war European 'guestworker systems' had studied the migratory process in their own histories or elsewhere, they would not have held the naive belief that they could turn flows of migrant labour on and off as with a tap. They would have understood that movement of workers almost always leads to family reunion and permanent settlement. The very fears of permanent ethnic minorities held by some governments turned into self-fulfilling prophecies: by denying legitimacy to family reunion and settlement, governments ensured that these processes would take place under unfavourable circumstances, leading to deep divisions in society.

Since the mid-1990s, attitudes and policies in both Australia and Germany have changed dramatically. Australia has abandoned its traditional open attitude to refugees and family reunion. The government has taken a much narrower view of 'national interests' with regard to immigration, and has mobilized considerable public support for this change. Multicultural policies have been watered down, although not altogether abandoned. Germany has at last abandoned the ideology of not being a country of immigration, and seems to be moving towards a more open immigration policy and a more inclusive approach to integration. It would be misleading to see these changes in terms of convergence. Instead, they appear as specific political strategies, linked to complex domestic and international factors. Nonetheless, the changes reflect the breakdown of traditional national models of immigration and citizenship in the context of the globalization of migration.

Guide to further reading

Jupp (2001) is an up-to-date encyclopaedia of the migrant experience in Australia. The history of migration and ethnic group formation is discussed in Lever-Tracy and Quinlan (1988), Collins (1991) and Castles *et al.* (1992c). Reynolds (1987) gives a good account of the colonial treatment of indigenous people. Overviews of social and economic issues around migration are in Wooden *et al.* (1994) and Castles *et al.* (1998), while Vasta and Castles (1996) deals with racism. Davidson (1997) discusses citizenship and nationalism in Australia. Two important documents of changing government policies on multiculturalism are OMA (1989) and NMAC (1999). The recent changes in refugee policy are examined in Mares (2001) and Crock and Saul (2002).

For Germany see Castles and Kosack (1973), Castles *et al.* (1984) and Martin (1991). German readers will also find Hoffmann (1990), Leggewie (1990), Nirumand (1992), Cohn-Bendit and Schmid (1993) and Bade (1994) useful. The report of the Süssmuth Commission (2001) contains a great deal of information and analysis, and is available in German and English.

Chapter 10

New Ethnic Minorities and Society

Migration since 1945 has led to growing cultural diversity and the formation of new ethnic groups in many countries. Such groups are visible through the presence of different-looking people speaking their own languages, the development of ethnic neighbourhoods, the distinctive use of urban space, and the establishment of ethnic associations and institutions. In Chapter 9 we discussed the migratory process and its outcomes in two very different immigration countries. In this chapter we will examine a wider range of western societies. The topic is a broad one, and should entail detailed description of ethnic relations in each immigration country. That is not possible here for reasons of space. Instead brief summaries of the situation in selected countries – USA, Canada, the UK, France, the Netherlands, Sweden and Italy – will be presented as information boxes within the text. The comparative analysis will draw on these, as well as on the case studies of Australia and Germany in Chapter 9. Many immigration countries in Europe and the rest of the world are not discussed here, although there may well be parallels.

The aim of the chapter is to show similarities and differences in the migratory process, and to discuss why ethnic group formation and growing diversity have been relatively easily accepted in some countries, while in others the result has been marginalization and exclusion. We will then go on to examine the consequences for the ethnic groups concerned and for society in general. The argument is that the migratory process works in a similar way in all countries with respect to chain migration and settlement, labour market segmentation, residential segregation and ethnic group formation. Racism and discrimination are also to be found in all countries, although their intensity varies. The main differences are to be found in state policies on immigration, settlement, citizenship and cultural pluralism. These differences, in turn, are linked to different historical experiences of nation-state formation.

Figures on the ethnic minority population of the various countries will be discussed. This is not a precise term, and does not correspond with the statistical categories of national governments. For some countries we have to use figures on foreign residents, which may include children born to foreign parents in immigration countries which do not confer citizenship by right of birth in the territory. Foreign resident figures exclude immigrants who have been naturalized – constituting large groups in countries like France, Sweden and the Netherlands. Statistics for such countries show declining foreign populations, but that does not necessarily mean that ethnic minorities are shrinking. For other countries (USA, Canada, Australia), we use figures on people born overseas, many of whom may be citizens of the immigration country. Children born in the country are not included, although they may share their parents' ethnic minority status. These countries also often provide figures based on other criteria, such as self-assessment of ethnicity or race. The trend towards multiple data-sets is to be found elsewhere, too (e.g. UK, Sweden, and the Netherlands). In any case, foreign citizenship or birth overseas does not automatically indicate a certain social position, so use of other indicators and detailed analysis is always needed to identify ethnic minorities.

Immigration policies and minority formation

Three groups of countries may be distinguished. The so-called 'classical immigration' countries – the USA, Canada and Australia – have encouraged family reunion and permanent settlement and treated most legal immigrants as future citizens. Sweden, despite its very different historical background, has followed similar policies. The second group includes France, the Netherlands and the UK, where immigrants from former colonies were often citizens at the time of entry. Permanent immigration and family reunion have generally been permitted (though with some exceptions). Immigrants from other countries have had a less favourable situation, although settlement and naturalization have often been allowed. The third group consists of those countries which tried to cling to rigid 'guestworker' models, above all Germany, Austria and Switzerland. Such countries tried to prevent family reunion, were reluctant to grant secure residence status and had highly restrictive naturalization rules.

The distinctions between these three categories are neither absolute nor static. The openness of the USA, Canada and Australia only applied to certain groups: all three countries had exclusionary policies towards Asians until the 1960s. The USA tacitly permitted illegal farmworker migration from Mexico, and denied rights to such workers. France had very restrictive rules on family reunion in the 1970s. Germany and Switzerland could not completely deny the reality of settlement and therefore improved family reunion rules and residence status. For instance,

Box 10.1 Minorities in the USA

US society is a complex ethnic mosaic deriving from five centuries of immigration. The white population is a mixture of the White Anglo-Saxon Protestant (WASP) group, which achieved supremacy in the colonial period, and later immigrants, who came from all parts of Europe in one of the greatest migrations in history between 1850 and 1914. Assimilation of newcomers is part of the 'American creed', but in reality this process has always been racially selective.

Native American societies were devastated by white expansion westwards. The survivors, forced into reservations, still have a marginal social situation. Millions of African slaves were brought to America from the seventeenth to the nineteenth century to labour in the plantations of the South. Their African-American descendants were kept in a situation of segregation and powerlessness, even after the abolition of slavery in 1865. After 1914, many migrated to the growing industrial cities of the north and west. Since the changes to immigration law in 1965 (see Chapter 4), new settlers have come mainly from Latin America and Asia. The main Asian countries of origin are the Philippines, China, South Korea, Vietnam and India.

In 2000, 28.4 million people (10 per cent of the population) were foreign born. This compares with only 4.8 per cent in 1970, but is still lower than the 14.7 per cent foreign-born quota in 1910. In 1999, six states (California, New York, Florida, Texas, New Jersey and Illinois) received over two-thirds of all immigrants. The table below shows the classification by race according to the 2000 Census. Ethnic minorities now make up a quarter of the USA's population. The greatest divide in US society remains that between blacks (African-Americans) and whites. However, the number of Hispanics now exceeds that of blacks. Hispanics are the descendants of Mexicans absorbed into the USA through its south-western expansion, as well as recent immigrants from Latin American countries. Hispanics can be of any race, but are seen as a distinct group based on language and culture. The Asian population is also growing fast.

The movement of Europeans and African-Americans into low-skilled industrial jobs in the first half of the twentieth century led to labour market segmentation and residential segregation. In the long run, many 'white ethnics' achieved upward mobility, while African-Americans became increasingly ghettoized. Distinctions between blacks and whites in income, unemployment rates, social conditions and education are still extreme. Members of some recent immigrant groups, especially from Asia, have high educational and occupational levels, while many Latin Americans lack education and are concentrated in unskilled categories.

→

→

US population by race and Hispanic origin, 2000 Census

	Millions	*Per cent*
White	211.5	75.1
Black	34.7	12.3
American Indian, Eskimo or Aleut	2.5	0.9
Asian or Pacific Islander	10.6	3.7
Other race	15.4	5.5
Two or more races	6.8	2.4
Total population	281.4	100.0
Hispanic or Latino origin (of any race)	35.3	12.5

Source: US Census Bureau, DP-1 Profile of General Demographic Characteristics 2000.

Incorporation of immigrants into the 'American dream' has been largely left to market forces. Nonetheless, government has played a role by making it easy to obtain US citizenship, and through compulsory public schooling as a way of transmitting the English language and American values. Legislation and political action following the Civil Rights Movement of the 1950s and 1960s led to an enhanced role for a black middle class, and changes in stereotypes of blacks in the mass media. However, commitment to equal opportunities and anti-poverty measures declined during the Reagan–Bush era, leading to increased community tension. Racist violence remains a serious problem for minorities.

In the 1990s, illegal migration and the costs of welfare for immigrants became major political issues. There are estimated to be at least 9 million illegal residents, who are often seen as a new underclass. In 1994, Californian voters passed Proposition 187 which denied social services and education to illegal immigrants, although its implementation was largely blocked by the courts. In 1996, US Congress approved a law which drew a sharp line in welfare entitlements between US citizens and foreign residents. This encouraged many immigrants to apply for naturalization. Another 1996 law was designed to cut illegal entries. It authorized more border patrol agents, fences and security devices along the US–Mexico border.

Sources: Feagin (1989); Portes and Rumbaut (1996); IOM (2000b); OECD (2001).

by the early 1980s, three-quarters of foreign residents in Switzerland had establishment permits (*Niederlassungsbewilligungen*), conferring much greater protection from deportation.

One important change has been the erosion of the privileged status of migrants from former colonies in France, the Netherlands and the UK. Making colonized people into subjects of the Dutch or British crown, or citizens of France, was a way of legitimating colonialism. In the period of European labour shortage it also seemed a convenient way of bringing in low-skilled labour. But citizenship for colonized peoples became a liability when permanent settlement took place and labour demand declined. All three countries have removed citizenship from their former colonial subjects (with a few exceptions) and put these people on a par with foreigners.

There has been some convergence of policies: the former colonial countries have become more restrictive, while the former guestworker countries have become less so. But this has gone hand in hand with a new differentiation: the EC countries granted a privileged status to intra-Community migrants in 1968. The establishment of the EU in 1993 was designed to create a unified labour market, with all EU citizens having full rights to take up employment and to obtain work-related social benefits in any member country. At the same time, entry and residence have become more difficult for non-EU nationals, especially those from outside Europe.

Immigration policies have consequences for immigrants' status with regard to labour market rights, political participation and naturalization. If the original immigration policies were designed to keep migrants in the status of temporary mobile workers, then they make it likely that settlement will take place under discriminatory conditions. Moreover, official ideologies of temporary migration create expectations within the receiving population. If a temporary sojourn turns into settlement, and the governments concerned refuse to admit this, then it is the immigrants who are blamed for the resulting problems. Political discourses which portray immigration as threatening to the nation create problems for long-standing immigrants and even for their descendants born in the immigration country. Anyone who looks different becomes suspect.

One of the most important effects of immigration policies is on the consciousness of migrants themselves. In countries where permanent immigration is accepted and the settlers are granted secure residence status and most civil rights, a long-term perspective is possible. Where the myth of short-term sojourn is maintained, immigrants' perspectives are inevitably contradictory. Return to the country of origin may be difficult or impossible, but permanence in the immigration country is doubtful. Such immigrants settle and form ethnic groups, but they cannot plan a future as part of the wider society. The result is isolation, separatism and emphasis on difference. Thus discriminatory immigration policies cannot stop the completion of the migratory process, but they can be the first step towards the marginalization of the future settlers.

Labour market position

As Chapter 8 showed, labour market segmentation based on ethnicity and gender has developed in all immigration countries. This was intrinsic in the type of labour migration practised until the mid-1970s: obtaining cheaper labour or using increased labour supply to restrain wage growth was a major reason for recruiting immigrant workers. The situation has changed; new migrants are much more diverse in educational and occupational status. There is a trend towards polarization: highly-skilled personnel are encouraged to enter, either temporarily or permanently, and are seen as an important factor in skill upgrading and technology transfer. Low-skilled migrants are unwelcome as workers, but enter through family reunion, as refugees or illegally. Their contribution to low-skilled occupations, casual work, the informal sector and small business is of great economic importance, but is officially unrecognized.

Labour market segmentation is part of the migratory process. When people come from poor to rich countries, without local knowledge or networks, lacking proficiency in the language and unfamiliar with local ways of working, then their entry point into the labour market is likely to be at a low level. The question is whether there is a fair chance of later upward mobility. The answer often depends on state policies. Again it is possible to discern three groups. Some countries have active policies to improve the labour market position of immigrants and minorities through language courses, basic education, vocational training and anti-discrimination legislation. These countries include Australia, Canada, Sweden, the UK, France and the Netherlands. The USA has a special position: there are equal opportunities, affirmative action and anti-discrimination legislation, but little in the way of language, education and training measures. This fits in with the laissez-faire model of social policy and with cuts in government intervention in recent times.

The former 'guestworker' countries form a third category. Although there are education and training measures for foreign workers and foreign youth, there have also been restrictions on labour market rights. During the period of mass labour recruitment, work permits often bound foreign workers to specific occupations, jobs or locations. Swiss rules on frontier and seasonal workers still maintain such restrictions, as do the German rules for the temporary workers from Eastern Europe recruited in the 1990s. However, the overwhelming majority of workers in the two countries have now gained long-term residence permits, which give them virtual equality of labour market rights with nationals.

Box 10.2 Minorities in Canada

In 1996 the 5 million foreign-born residents made up 17.4 per cent of Canada's population of 28.5 million. Recent immigrants (those who arrived after 1980) are highly urbanized. In 1996, 70 per cent were living in Canada's three largest metropolitan areas, Toronto, Montreal and Vancouver, compared with only one-quarter of the total population.

The most notable feature is the decline in Europeans and the increase in immigrants from Asia, Latin America and the Caribbean since 1981 (see table below). The 1996 Census also asked respondents to state their 'ethnic origins'. (The complex categories used in the 1996 Census were not comparable with earlier ones.) The largest single category was 'multiple origins' (mainly British plus some other origin) with 10.2 million people (35.8 per cent of the population). Of the 'single origins' Canadian came first (18.7 per cent), followed by European (13.1 per cent), British Isles (11.5 per cent) and French (9.4 per cent). East and South-East Asian origins accounted for 4.5 per cent of the population, and people of Aboriginal origins made up 1.9 per cent.

Canadian history has been shaped by the struggle between the British and French. After 1945, separatist movements in French-speaking Quebec made language and culture into crucial areas of struggle. This led to devolution of power to the provinces and to a policy of bilingualism and two official languages. The most recent referendum on independence for Quebec was defeated in 1995, mainly because of fears by First Nation (Aboriginal) people and immigrants that they would be marginalized in a Francophone state. Conflicts on land rights and the social position of First Nation peoples play an important role in Canadian public life. The land claims of the Inuit people have been settled through the establishment of Nunavut, which give the Inuit control of one-fifth of Canada's land mass in the Arctic region. Other groups remain socially marginalized and are struggling for recognition of their claims.

In 1971, multiculturalism was proclaimed an official policy and a Minister of State for Multiculturalism was appointed. There were two main objectives: maintaining ethnic languages and cultures and combating racism. In 1982, equality rights and multiculturalism were enshrined in the Canadian Charter of Rights and Freedoms. The Employment Equity Act of 1986 required all federally regulated employers to assess the composition of their workforces, in order to correct disadvantages faced by women, 'visible minorities', native people and the disabled. The Multiculturalism Act of 1987 proclaimed multiculturalism as a central feature of Canadian citizenship and laid down principles for cultural pluralism.

→

→

Since the 1980s, public opinion on multiculturalism has become more negative. Community relations deteriorated through increased discrimination against Aboriginal people and racial assaults against blacks and Asians. The unwillingness of the authorities to respond to racist attacks has been a cause of politicization among 'visible minorities'. In 1993 the Federal Government took steps to reduce the emphasis on multiculturalism, most notably by merging the multicultural bureaucracy into an amorphous new department called Heritage Canada. The new emphasis was on living together in multi-ethnic cities. The government initiated the Metropolis Project – a major policy network including researchers and policy-makers from Canada and other immigration countries.

Canada: immigrant population by birthplace (Census figures in thousands)

Area of origin	*1981 total*	*1996 total*
Europe	**2568**	**2332**
of which:		
UK	879	656
Italy	385	332
Germany	155	182
Poland	149	193
America	**582**	**798**
of which:		
USA	302	245
Caribbean	173	279
South and Central America	107	274
Asia	**541**	**1563**
of which:		
India	109	236
China	52	231
Philippines	–	185
Africa	**102**	**229**
Oceania and other	**56**	**49**
Total	**3848**	**4971**

Sources: Stasiulis (1988); Breton *et al.* (1990); OECD (1992; 2001: Table B.1.4); Stasiulis and Jhappan (1995); CIC (2002); Statistics Canada (2002).

Residential segregation, community formation and the global city

Some degree of residential segregation is to be found in many immigration countries, though nowhere is it as extreme as in the USA, where in certain areas there is almost complete separation between blacks and whites, and sometimes Asians and Hispanics too. In the other countries there are city neighbourhoods where immigrant groups are highly concentrated, though they rarely form the majority of the population. The causes of residential segregation were discussed in Chapter 9. They are based partly on migrants' situation as newcomers, lacking social networks and local knowledge. Equally important is migrant workers' low social status and income. Another factor is discrimination by landlords: some refuse to rent to immigrants, while others make a business of charging high rents for poor accommodation.

Institutional practices may also encourage residential segregation. Many migrant workers were initially housed by employers or public authorities. There were migrant hostels and camps in Australia, barracks provided by employers in Germany and Switzerland, and hostels managed by the government Fonds d'Action Sociale (FAS, or Social Action Fund) in France. These generally provided better conditions than private rented accommodation, but led to control and isolation. Hostels also encouraged clustering: when workers left their initial accommodation they tended to seek housing in the vicinity.

In countries where racism is relatively weak, immigrants often move out of the inner-city areas to better suburbs as their economic position improves. However, where racism and social exclusion are strong, concentration persists or may even increase. In 2001, London contained 49 per cent of Britain's ethnic minority population. Twenty-eight per cent of London's population were members of ethnic minorities (ONS, 2001). Some groups are highly concentrated in disadvantaged areas, such as the housing estates inhabited mainly by Bangladeshis in Tower Hamlets (London). In Amsterdam, a quarter of the population belongs to ethnic minorities, such as Surinamese, Antilleans, Turks and Moroccans, and these are mainly concentrated in certain inner-city and peripheral locations. Such groups tend to have extremely high rates of unemployment: for instance, over half of all Turks in Amsterdam were jobless in 1991 (Cross, 1995).

Residential segregation is a contradictory phenomenon. In terms of the theory of ethnic minority formation set out in Chapter 2, it contains elements of both other-definition and self-definition. The relative weight of the two sets of factors varies from country to country and group to group.

Immigrants cluster together for economic and social reasons arising from the migratory process, and are often kept out of certain areas by racism. But they also frequently want to be together, in order to provide mutual support, to develop family and neighbourhood networks and to maintain their languages and cultures. Ethnic neighbourhoods allow the establishment of small businesses and agencies which cater for immigrants' needs, as well as the formation of associations of all kinds (see Chapter 8). Residential segregation is thus both a precondition for and a result of community formation.

Interestingly, the countries where community formation has taken place most easily have been those with open and flexible housing markets, based mainly on owner occupation, such as Australia, the UK and the USA. The continental European pattern of apartment blocks owned by private landlords has not been conducive to community formation, while large publicly-owned housing developments have frequently led to isolation and social problems, which are the breeding-ground for racism.

Immigration and ethnic minority formation are transforming post-industrial cities in contradictory ways. Sassen (1988) has shown how new forms of global organization of finance, production and distribution lead to 'global cities'. These attract influxes of immigrants, both for highly-specialized activities and for low-skilled service jobs which service the luxurious lifestyles of the elites. In turn, this leads to a spatial restructuring of the city, in which interacting factors of socio-economic status and ethnic background lead to rapidly changing forms of differentiation between neighbourhoods.

Many immigrants are forced by powerful social and economic factors into isolated and disadvantaged urban areas, which they share with other marginalized social groups (Dubet and Lapeyronnie, 1992). Some local people perceive residential segregation as a deliberate and threatening attempt to form 'ethnic enclaves' or 'ghettos'. Areas of concentration of specific immigrant groups are often the focus of conflicts with other disadvantaged sections of the population. The extreme right has mobilized around fears of ghettos in Western European countries since the 1970s. Ethnic minority neighbourhoods can also be the site of confrontations with the state and its agencies of social control, particularly the police (see Chapter 11). One official reaction has been dispersal policies, designed to reduce ethnic concentrations.

Ethnic clustering and community formation may be seen as necessary products of migration to the global cities. They may lead to conflicts, but they can also lead to renewal and enrichment of urban life and culture. Specific ethnic groups can never be completely isolated or self-sufficient in modern cities. Cultural and political interaction is negotiated around complex processes of inclusion and exclusion, and of cultural transference.

Box 10.3 Minorities in the United Kingdom

No table is given for the UK, due to the difficulty in aggregating the foreign population and the ethnic minority population. In 2000, there were 2.3 million foreign residents in the UK (4 per cent of the total population). Just over one-third were EU citizens. The largest groups were the Irish (404 000), Indians (153 000), US citizens (114 000), Italians (95 000), Pakistanis (94 000) West Africans (85 000), French (85 000) and Australians (75 000). The ethnic minority population, most of whom are British-born people of non-European ancestry (originating from the Commonwealth), totalled 4 million (7.1 per cent of the total population). Nearly half of them have their origins in the Indian subcontinent (India, Pakistan and Bangladesh), 13 per cent in the Caribbean and 11 per cent in Africa.

Commonwealth immigrants who came before 1971 were British subjects, who enjoyed all rights (including electoral rights) once admitted. This situation was ended by the 1971 Immigration Act and the 1981 British Nationality Act, which put Commonwealth immigrants who arrived thereafter on a par with foreigners. Irish immigrants enjoy virtually all rights, including the right to vote. It is relatively easy for foreigners to obtain citizenship after five years of legal residence in the UK.

In the 1950s and 1960s, immigrants from the Commonwealth became concentrated in the least desirable jobs. Since then, the various immigrant groups and their descendants born in the UK have had diverse experiences. Pakistanis and Bangladeshis remain concentrated in semi-skilled manual work, and have very high unemployment rates. Women from these groups have low participation rates in paid work. Indians are well represented in professional, managerial and other non-manual occupations. Indians also have high rates of self-employment. Caribbean women have done reasonably well in gaining access to non-manual work, while Caribbean men have not been so successful.

Race Relations Acts were passed in 1965, 1968 and 1976, outlawing discrimination in public places, employment and housing. A Commission for Racial Equality (CRE) was set up in 1976 to enforce these laws and promote good community relations. However, the National Front and other racist groups grew rapidly in the 1970s and recruited members of violent youth

→

Much of the energy and innovative capacity within the cities lies in the cultural syncretism of the multi-ethnic populations, as Davis (1990) has shown in the case of Los Angeles. This syncretism can be seen as a creative linking and development of aspects of different cultures, in a process of community formation, which always has political dimensions (Gilroy, 1987). Just as there can be no return to mono-ethnic populations (always a myth in any case), so there is no way back to static or homogeneous cultures. The global city with its multicultural population is a powerful laboratory for change.

→

sub-cultures, such as skinheads. Racist violence became a major problem for Asians and Afro-Caribbeans. Black youth discontent exploded into riots in inner-city areas in 1980–1 and again in 1985–6 and 1991. Such problems led to government measures to combat youth unemployment, improve education, rehabilitate urban areas and change police practices. However, these measures had limited success in the climate of growing social inequality during the Thatcher years.

A strong contrast persists between the formal equality enjoyed by ethnic minorities, and their everyday experience of unemployment, incarceration, inequality in education and social exclusion. Black people continue to experience discrimination by the police and other public service organizations. The 1999 Stephen Lawrence Inquiry (set up to analyse the poor police response after the murder of a young black man by a white gang) revealed the strength of institutional racism. The Race Relations (Amendment) Act 2000 gave the CRE stronger enforcement powers and increased funding. All public bodies are now obliged to introduce race equality schemes and eliminate discrimination. In 2002, riots broke out involving youth of Asian origin in the northern cities of Oldham, Burnley and Bradford. The extreme-right British National Party sought to exploit local conflicts, and had some electoral success amongst white voters who felt disadvantaged by special programmes for ethnic minorities.

Since the late 1990s, the main immigration issue has been the number of asylum seekers, with applications increasing from 28 000 in 1993 to 93 600 in 2000. The media and right-wing politicians have made this an issue of national sovereignty. Successive governments have introduced three new laws on immigration and asylum since 1993, with a fourth currently under discussion. Each law has tightened up entry rules, and introduced deterrent measures such as detention, restrictions on welfare, and limitation of the right to work. Despite these policies, asylum seekers continue to come, and a large proportion are likely to remain permanently. It is not clear how the integration strategies developed for the earlier Commonwealth immigrants will be applied, and how effective they will be for these diverse groups.

Sources: Benyon (1986); Layton-Henry (1986; 2003); Home Office (1989); OECD (1992, 2001); Solomos (1993); Solomos and Back (1995); ONS (2001).

Social policy

As migrants moved into the inner cities and industrial towns, they were blamed for rising housing costs, declining housing quality and deteriorating social amenities. In response, a whole set of social policies developed. Sometimes policies designed to reduce ethnic concentrations and ease social tensions achieved the opposite.

Nowhere were the problems more severe than in France. After 1968, measures were taken to eliminate *bidonvilles* (shanty-towns) and make

Box 10.4 Minorities in France

The 3.3 million foreign residents made up 5.6 per cent of France's total population in 1999 (see Table 10.1) – a decline from 3.6 million (6.3 per cent of total population) in 1990. This does not indicate a reduction in the population of immigrant origin. Well over a million former immigrants have become naturalized. In addition there were up to half a million French citizens of African, Caribbean and Pacific Island origin from overseas departments and territories. France's population of immigrant origin has changed from one of mainly Southern European origins in the 1970s to one of mainly North and West African background.

Despite the dominant 'republican model' which offers naturalization and equal social rights to immigrants, there is considerable differentiation between groups. EU citizens enjoy all basic rights, except the right to vote. Immigrants from non-EU European countries (such as Poland and former Yugoslavia) lack many rights, and often have an irregular legal situation. People of non-European birth or parentage (whether citizens or not) constitute the ethnic minorities. These include Algerians, Tunisians and Moroccans, young Franco-Algerians, black Africans, Turks and settlers from the overseas departments and territories. They may have formal rights as French citizens, but they still suffer socio-economic exclusion and racism.

There is marked residential concentration in inner-city areas and in the public housing estates on the periphery of the cities. The work situation of ethnic minorities is marked by low status, insecure jobs and high unemployment rates, especially for youth. Racist discrimination and violence, especially against North Africans, have been a problem for many years. However, some young people of immigrant background have secured upward mobility through education. A new middle class of ethnic professionals and business people is emerging, popularly known as the *beurgeoisie* (which derives from the slang term *beurs*, which is used for people of Arab origins).

In the 1970s, policies towards immigration (especially family reunion) became increasingly restrictive. Police raids, identity checks and deportations of immigrants convicted of even minor offences were common. In the 1980s, the Socialist government improved residence rights, granted an amnesty to illegals and allowed greater political participation. Growing racism and serious social problems in areas of immigrant concentration led to a series of special programmes to improve housing and education and combat youth unemployment.

→

public housing more accessible to immigrants. The concept of the *seuil de tolérance* (threshold of tolerance) was introduced, according to which the immigrant presence should be limited to a maximum of 10 or 15 per cent of residents in a housing estate or 25 per cent of students in a class (Verbunt, 1985: 147–55; MacMaster, 1991: 14–28). The implication was that immigrant concentrations presented a problem, and that dispersal was the precondition for assimilation. Subsidies to public housing societies

→

The position of ethnic minorities in French society has become highly politicized. Immigrants have taken an active role in major strikes, and demanded civil, political and cultural rights. Second-generation North African immigrants and Muslim organizations are emerging political forces. Youth discontent with unemployment and police practices led to riots in Lyons, Paris and other cities in the 1980s and 1990s. More recently, campaigns by the *beurs* have asserted the need for a new type of 'citizenship by participation', based on residence rather than nationality or descent. This means demanding a form of pluralism quite alien to the French assimilationist model.

In the 1990s, the centre–right government became increasingly restrictive towards minorities. This was partly due to the increasing influence of the extreme-right Front National (FN), which regularly got around 15 per cent of the votes in national elections, and which gained control of the local authorities of several cities. The 1993 Loi Pasqua (named after the Minister of the Interior) tightened up immigration and nationality rules. Conditions for entry and family reunion became stricter, while deportation was facilitated. Citizenship rules for children of immigrants were changed, undermining the republican principle of *ius soli* (citizenship through birth in France). Fears about Islamic fundamentalism turned into near panic when violence in Algeria spilled over into bomb attacks on the Paris Métro in 1995. There were mass deportations of *sans papiers* (people without documents) – immigrants who had lost their legal status through the Loi Pasqua.

The Socialist government of 1997–2002 reasserted the republican model through laws which reinstated *ius soli* for descendants of immigrants and liberalized rules on entry and residence. The aim was to create a secure situation for legal immigrants, while combating illegal entry. However, claims that society was threatened by immigrant criminality and that French identity was being undermined by African and other non-European minorities remained powerful rallying points for the extreme right. Le Pen, leader of the FN, shocked Europe by gaining one-fifth of the votes in the first round of the 2002 presidential election, defeating the Socialist candidate, Jospin. The new Centre–Right government pledged to cut immigration and strengthen law and order.

Sources: Verbunt (1985); Wihtol de Wenden (1988, 1995), Wihtol de Wenden and Levau (2001); Noiriel (1988); Costa-Lascoux (1989); Lapeyronnie *et al.* (1990); Weil (1991b); OECD (2001); Hollifield (2003).

(*habitations à loyer modéré,* or HLMs) were coupled to quotas for immigrants. The HLMs used the subsidies to build new estates, where mainly French families were housed. In order to minimize conflicts with the French, immigrant families were concentrated in specific estates. The HLMs could claim that they had adhered to the quotas – on an average of all their dwellings – while in fact creating new ghettos (Weil, 1991b: 249–58).

By the 1980s, the central social policy issue was therefore the situation of ethnic minorities in the inner-city areas and in the great public housing estates surrounding the cities. These were rapidly turning into areas of persistent unemployment, social problems and ethnic conflicts. Social policies focused on urban youth, and the Socialist government developed a range of programmes to improve housing and social conditions, to boost educational outcomes and to combat youth unemployment, especially for youth of North African background.

Patrick Weil (who was to head the Socialist government's review of immigration and nationality policy in 1997) found that the social policy measures of the 1980s had failed: they were designed to achieve integration into French society, but in fact they 'linked all the problems of these towns and neighbourhoods to immigration'. Thus social policy encouraged concentration of minorities, which slowed integration, encouraged the formation of ethnic communities, and strengthened group religious and cultural affiliations (Weil, 1991b: 176-9). Weil's interpretation fits into the assimilationist French republican tradition (see below). In fact, the extent to which the state should provide special social policies to facilitate immigrant integration is controversial in most immigration countries.

Special policies for immigrants can reinforce tendencies to segregation. For instance, the 'dual strategy' pursued in German education (see Chapter 9) led to special classes for foreign children, causing social isolation and poor educational performance. Housing policies in the UK are intended to be non-discriminatory, yet they have sometimes led to the emergence of 'black' and 'white' housing estates. In the Netherlands, critics of the Minorities Policy of the 1980s argued that culturally-specific social measures actually increased socio-economic marginalization. Sweden's special public housing schemes for immigrants have led to a high degree of ethnic concentration and separation from the Swedish population. On the other hand, multicultural social policies are based on the idea that immigrants do need special services that address their special needs with regard to education, language learning and housing.

Again it is possible to suggest a rough classification of social policy responses. Australia, Canada, Sweden and the Netherlands have pursued active social policies for immigrants and minorities. The basic assumption has been that such policies do not lead to separatism but, on the contrary, form the precondition for successful integration. This is because the situation of immigrants and ethnic minorities is seen as the result both of cultural and social difference, and of barriers to participation based on institutional and informal discrimination.

A second group of countries reject special social policies for immigrants. US authorities regard special policies for immigrants as unnecessary government intervention. Nonetheless, equal opportunities, anti-discrimination and affirmative action measures have benefited immigrants, and

Table 10.1 ***Foreign resident population in France, the Netherlands, Sweden and Switzerland (thousands)***

Country of origin	*France 1999*	*Netherlands 1999*	*Sweden 1999*	*Switzerland 1999*
Italy	202	18	–	328
Portugal	554	9	–	135
Spain	162	17	–	86
UK	–	40	12	20
Finland	–	–	99	–
Poland	–	–	16	–
Former Yugoslavia	–	16	57	332
Algeria	478	–	–	–
Morocco	504	120	–	–
Tunisia	154	1	–	–
Turkey	208	101	16	80
Others	1002	330	287	388
Total	**3263**	**652**	**487**	**1369**
of which: EU	1196	196	–	800

Notes: A blank indicates that a certain country of origin is not among the main sources of immigrants for a given receiving country. Smaller groups are aggregated under 'others'. French figures are from the Census, the others are from population registers.
Sources: OECD (2001: Table B.1.5).

special social and educational measures are to be found at the local level. French governments have rejected special social policies on the principle that immigrants should become citizens, and that any special treatment would hinder that. Yet despite this there have been some special programmes, as already described. The UK developed a range of social policies in response to the urban crisis, racist violence and youth riots, despite the ideological rejection of such measures by the Conservative government from 1979 to 1997.

The third group of countries is, again, the former 'guestworker' recruiters. Germany has pursued contradictory and changeable polices concerning the access of immigrants to the highly-developed welfare system. In the early years, the government delegated the provision of special social services to charitable organizations linked to the churches and the labour movement. Although foreign workers were guaranteed equal rights to work-related health and pension benefits, they were excluded from some welfare rights. Claims for social security payments on the grounds of long-term unemployment or disability could lead to

Box 10.5 Minorities in the Netherlands

In 1999, there were 652 000 foreign residents in the Netherlands: 4.1 per cent of the total Dutch population of 15.6 million (see Table 10.1). The foreign population has declined from a 1993 peak of 780 000. However, many immigrants have become citizens. The foreign-born population was 1.6 million in 1999 (10 per cent of total population), of whom 58 per cent were Dutch citizens. The top five countries of origin of the foreign-born were Surinam (185 000), Turkey (178 000), Indonesia (168 000), Morocco (153 000) and Germany (124 000).

The ethnic minority population became overwhelmingly concentrated in urban areas, especially the four biggest cities (Amsterdam, Rotterdam, The Hague and Utrecht), where they often live in distinct neighbourhoods. Mediterranean, Surinamese and Antillean workers mainly got unskilled jobs in manufacturing and the services. Immigrants bore the brunt of economic restructuring in the 1980s, with unemployment rates of 20–40 per cent for some groups. By 1999, unemployment had declined for all groups, but was much higher for workers of immigrant origin (9 per cent) than for non-immigrants (3 per cent). Minority groups with high unemployment rates included Moroccans (18 per cent) and Turks (13 per cent).

The revised Constitution of 1983 introduced municipal voting rights for resident non-citizens. The 1983 Minorities Policy was based on multicultural principles, declaring the need for social policies to integrate minorities as groups rather than as individuals. Preservation of cultural identity was a key aim. The Minorities Policy covered Mediterranean workers and their families, people of Surinamese and Antillean origins, Moluccans, refugees (but not asylum seekers), gypsies and caravan dwellers. These groups were estimated to add up to 876 385 people in 1990. However, the Minorities Policy was

→

deportation. In the meantime, settlement and family reunion have made it necessary for family and youth services, education, health and aged care to take account of the needs of immigrants. Anti-discrimination legislation or affirmative action programmes have little place in either Germany or Switzerland: many laws specifically provide for preferential treatment of nationals over foreigners. Yet the racist violence of the early 1990s caused German authorities to seek ways of overcoming social exclusion of immigrants, which was seen as a major factor encouraging attacks by the extreme right.

Racism and minorities

Three categories may be distinguished in immigration countries. First, some settlers have merged into the general population and do not

→

criticized on the grounds that it did little to overcome unemployment, poor educational performance and social disadvantage.

In 1994, a new Integration Policy was introduced, covering persons of Turkish, Moroccan, Surinamese and Antillean descent, as well as refugees. This policy aims at reducing social and economic deprivation, and has two elements: a 'newcomers'' or 'reception policy', and an 'integration policy'. The newcomers' policy consists of Dutch language courses, social orientation and vocational training, plus individual case management to secure entry into further education or the labour market. Immigrants who fail to participate may be deprived of social security benefits. Integration policy is concerned with improving the educational and labour market position of minority youth, and ameliorating the safety and the living conditions of neighbourhoods. This approach was confirmed and reinforced by the Law on Civic Integration for Newcomers of 1998.

Citizenship is fairly easy to obtain, with a five-year qualification period. The Netherlands has laws which prohibit racial defamation, incitement to racial hatred, discrimination and violence, and discrimination at work or in public places. Organizations which call for racial discrimination can be forbidden. Nonetheless, racism and racist violence are still problems in the Netherlands. The rise of the politician Pim Fortuyn and his anti-immigration party in 2001 took many observers by surprise. After the assassination of Fortuyn in 2002, the Pim Fortuyn List (PFL) became the second strongest party, and part of a coalition government where it held the immigration portfolio. However, the PFL fell apart due to internal squabbling by the end of 2002. In the January 2003 elections, the PFL lost most of its seats, and the Christian Democrats and Social Democrats re-emerged as the main parties.

Sources: Entzinger (1985, 2002); Muus (1991, 1995); OECD (2001: 210–15).

constitute separate ethnic groups. These are generally people who are culturally and socio-economically similar to the majority of the receiving population: for instance, British settlers in Australia, French in Switzerland and Austrians in Germany.

Second, some settlers form ethnic communities: they tend to live in certain neighbourhoods and to maintain their original languages and cultures although they are not excluded from citizenship, political participation and opportunities for economic and social mobility. The ethnic community may have developed partly due to initial discrimination, but the principal reasons for its persistence are cultural and psychological. Examples are Italians in Australia, Canada or the USA; the Irish in the UK; and people of Southern European background in France or the Netherlands. Such communities are likely to decline in salience over time, as later generations intermarry with other groups, and move out of initial areas of concentration.

Third, some settlers form ethnic minorities. Like the ethnic communities they tend to live in certain neighbourhoods and to maintain their languages and cultures of origin. But, in addition, they usually share a disadvantaged socio-economic position and are partially excluded from the wider society by such factors as weak legal status, refusal of citizenship, denial of political and social rights, ethnic or racial discrimination, racist violence and harassment. Examples are Asian immigrants in Australia, Canada or the USA; Hispanics in the USA; Afro-Caribbeans and Asians in the UK; North Africans and Turks in most Western European countries; and asylum seekers of non-European background just about everywhere.

All the countries examined have all three categories, but our concern here is with the second and third categories. It is important to examine why some immigrants take on the character of ethnic communities, while others become ethnic minorities. A further important question is why more immigrants take on minority status in some countries than in others. Two groups of factors appear relevant: those connected with the characteristics of the settlers themselves, and those connected with the social structures, cultural practices and ideologies of the receiving societies.

Looking at the settlers, it is inescapable that phenotypical difference (skin colour, racial appearance) is the main marker for minority status. A survey carried out on behalf of the Commission of the European Community in all member countries in 1989 indicated widespread acceptance of fellow Europeans. However, there were strong feelings of distance and hostility towards non-Europeans, particularly Arabs, Africans and Asians. Overall one European in three believed that there were too many people of another nationality or race in his or her country, with such feelings being most marked in Belgium and Germany (Commission of the European Communities, 1989). This emphasis on phenotypical difference applies even more to non-immigrant minorities, such as aboriginal peoples in the USA, Canada and Australia, or African-Americans in the USA. They, together with non-European immigrants, make up the most marginalized groups in all the countries.

There are four possible explanations for this: phenotypical difference may coincide with recent arrival, with cultural distance, with socio-economic position, or, finally, it may serve as a target for racism.

The first explanation is partly correct: in many cases, black, Asian or Hispanic settlers are among the more recently arrived groups. Historical studies reveal examples of racism and discrimination against white immigrants quite as virulent as against non-whites today (see Chapter 3 above). Recent arrival may make a group appear more threatening, and new groups tend to compete more with local low-income groups for jobs and housing. But recent arrival cannot explain why aboriginal populations are victims of exclusionary practices, nor why African-Americans and other long-standing minorities are discriminated against. Neither can it explain why racism against white immigrant groups tends to disappear in time, while that against non-whites continues over generations.

What about cultural distance? Its significance depends partly on how culture is defined. Some non-European settlers come from rural areas with pre-industrial cultures, and may find it hard to adapt to industrial or post-industrial cultures. But many Asian settlers in North America and Australia are of urban background and highly educated. This does not protect them from racism and discrimination. If culture is defined in terms of language, religion and values, then some non-European migrants are perceived as different by the receiving populations. This applies particularly to Muslims. Fear of Islam has a tradition going back to the medieval crusades. In recent years, fears of fundamentalism and loss of modernity and secularity have played a major role. But it could be argued that such fears are based on racist ideologies rather than social realities. The strengthening of Muslim affiliations is often a protective reaction of discriminated groups, so that fundamentalism is something of a self-fulfilling prophecy.

As for the third explanation, phenotypical difference does frequently coincide with socio-economic status. Some immigrants from less-developed countries lack the education and vocational training necessary for upward mobility in industrial economies. But even highly-skilled immigrants may encounter discrimination. Many immigrants discover that they can only enter the labour market at the bottom, and that it is subsequently hard to move up the ladder. Thus low socio-economic status is as much a result of processes of marginalization as it is a cause of minority status.

We may therefore conclude that the most significant explanation of minority formation lies in practices of exclusion by the majority populations and the states of the immigration countries. We refer to these practices as racism and to their results as the racialization of minorities (see Chapter 2). Traditions and cultures of racism are strong in all European countries and former European settler colonies. The increased salience of racism and racist violence since the late 1970s is linked to growing insecurity for many people resulting from rapid economic and social change.

Racist violence

German reunification in 1990 was followed by an outburst of racist violence. Neo-Nazi groups attacked refugee hostels and foreigners on the streets, sometimes to the applause of bystanders. At first the violence was worst in the area of the former GDR, but in 1992 and 1993 several Turkish immigrants were killed in arson attacks in West Germany. Altogether 2600 racially-motivated acts of violence were officially recorded in 1992, leading

Box 10.6 Minorities in Sweden

Until 1945 Sweden was a fairly homogeneous country, with only a small aboriginal minority: the Sami or Lapps (about 10 000 people today). After 1945, labour migration was encouraged. Foreign worker recruitment was stopped in 1972, but family reunion and refugee entries continued. In 1999, the 487 000 foreign residents made up 5.5 per cent of Sweden's total population of 8.9 million (see Table 10.1). In fact, 918 000 persons had been born abroad, but 40 per cent of them had been in Sweden for over 20 years, and 582 000 had acquired Swedish citizenship. Including children born in Sweden to at least one immigrant parent, the population of immigrant origin is 1.8 million, or about 20 per cent of the population. A third of the foreign population come from Nordic countries (Finland, Denmark and Norway), while the rest are of very diverse origins, including former Yugoslavia, Iraq, Iran, Turkey, Somalia and Chile.

Immigrant workers are overrepresented in manufacturing, and in lower-skilled services occupations. They are underrepresented in agriculture, health care and social work, administrative and clerical work, and commerce. In 1993, unemployment in Sweden reached a historical peak of 8 per cent. The rate for foreigners was 21 per cent, while for non-Europeans it reached 37 per cent. Immigrants have mainly settled in the cities, and people of the same nationality cluster in certain neighbourhoods, allowing linguistic and cultural persistence. In some city neighbourhoods, non-Europeans make up 75 per cent of the population – among the highest rates of ethnic concentration in Europe.

The increase in asylum-seeker entry in the late 1980s led to anti-immigrant campaigns by extreme-right groups. In 1988, a referendum in the small town of Sjöbo decided to keep refugees out. This was followed by an increase in racist violence, including arson and bomb attacks on refugee centres. From 1989, the government introduced a series of measures to restrict the entry of asylum seekers. In 1992 the inflow – particularly from former Yugoslavia – peaked at 84 000, but had declined to 11 200 by 1992. Today, violent extremist groups have few members and supporters, but public opinion has become much more sceptical about immigration and asylum.

→

to 17 deaths (Baringhorst, 1995: 225). But such incidents are neither new nor confined to Germany: racist harassment and attacks have become major issues in all the countries of immigration.

In Britain, racist violence led by the National Front and the British Movement grew from the 1970s. Research by the Home Office (1989) showed the extent to which violence had become a major constraint on the lives of ethnic minorities. Since 1986, all police forces have collected data on racist incidents. The Crime and Disorder Act 1998 introduce new categories of 'racially aggravated offences'. The British Crime Survey found that the number of incidents considered to be racially motivated by the victim was 280 000 in 1999 (a decline of 27 per cent from 1995). The

→

In 1975, Parliament set out an immigrant policy with three basic objectives: *equality,* which refers to giving immigrants the same living standards as Swedes; *freedom of choice,* which means giving members of ethnic minorities a choice between retaining their own cultural identities or assuming Swedish cultural identity; and *partnership,* which implies that minority groups and Swedes benefit from working together. Since 1975, foreign residents have had the right to vote and stand for election in local and regional elections. It was planned to extend such rights to national elections, but it proved impossible to get the parliamentary majority required for a change in the Constitution. Anti-discrimination laws were introduced in 1986 and 1994. The waiting period for naturalization is two years for Scandinavians and five years for everybody else, while children born to foreign resident parents can obtain Swedish citizenship upon application. A new citizenship law was enacted in 2001. It recognized dual citizenship and made naturalization procedures easier for some groups. Naturalization rates are high.

Immigrants enjoy the benefits of Sweden's highly-developed welfare state, as well as a number of special services including language courses for adult immigrants. Children of immigrants can receive pre-school and school instruction in their own language, within the normal curriculum. Other measures include translator and interpreter services, information services, grants to immigrant organizations and special consultative bodies. However, such policies do not seem to have been very successful in achieving greater equality and participation in Swedish society for minorities. In 1998, a new integration policy was introduced, and a National Integration Office was set up. The new policy put less emphasis on ethnic identity and multiculturalism, and more on integration and equal opportunities. In view of the high degree of ethnic concentration in urban areas, a 'metropolitan policy' was introduced, to reduce segregation and to help achieve equal living conditions in cities.

Sources: Hammar (1985b); Ålund and Schierup (1991); Larsson (1991); OECD (1995, 2001); Westin (2000).

number of racially motivated offences reported to the police in 1999 was 47 814, although few ever led to a prosecution. Offences included harassment, wounding, common assault and criminal damage (Home Office, 2000).

In France, there was a series of murders of North Africans in 1973. In the 1980s and 1990s, the Front National was able to mobilize resentments caused by unemployment and urban decline, and to crystallize them around the issues of immigration and cultural difference. The USA has a long history of white violence against African-Americans. Despite the anti-racist laws secured by the Civil Rights Movement, the Ku Klux Klan remains a significant force. Asians, Arabs and other minorities are also frequent targets (ADL, 1988). Police violence against minorities is common. The Los Angeles riots of May 1992 were provoked by police brutality towards a black motorist, which went unpunished by the courts.

Even countries which pride themselves on their tolerance, like Canada, Sweden and the Netherlands, report a growing incidence of racist attacks. In Sweden 2622 offences of a racist, xenophobic or Nazi nature were reported to the police in 1998 (Westin, 2000). In the mid-1980s, the European Parliament's Committee of Inquiry into Fascism and Racism in Europe found that 'immigrant communities ... are daily subject to displays of distrust and hostility, to continuous discrimination ... and in many cases, to racial violence, including murder' (European Parliament, 1985). A decade later, such aggressions had become so commonplace that they barely made the headlines. According to a study of racist violence in Europe: 'By the early 1990s, many groups of people have had to face racist violence and harassment as a threatening part of everyday life' (Björgo and Witte, 1993: 1). These groups included immigrants and asylum seekers, but also long-standing minorities such as Jews and gypsies.

The escalation of racist violence has not gone unchallenged. Anti-racist movements have developed in most countries quite early in the settlement process, often based on coalitions between minority organizations, trade unions, left-wing parties, churches and welfare organizations. Anti-racist organizations have helped to bring about equal opportunities and anti-discrimination legislation, as well as policies and agencies designed to curb violence. However, the anti-racist movement was not effective in preventing the increase in racist violence in the early 1990s, though it may have prevented matters getting even worse. Similarly, political campaigns against asylum seekers at the beginning of the twenty-first century have been countered by non-governmental organizations. However, as long as politicians are eager to make electoral capital out of such issues, racism will continue to be a problem.

The political implications of anti-immigrant campaigns will be discussed in Chapter 11. Racist campaigns, harassment and violence are important factors in the process of ethnic minority formation. By isolating minorities and forcing them into defensive strategies, racism may lead to self-organization and separatism, and even encourage religious fundamentalism. Conversely, anti-racist action may help overcome the isolation of minorities, and facilitate their social and political incorporation into mainstream society.

Minorities and citizenship

Why do some countries turn most of their settlers into ethnic minorities, while others marginalize only far more limited groups? The answer does not lie primarily in the characteristics of the migrants, but rather in the histories, ideologies and structures of the societies concerned. Varying models of the nation-state lead to different concepts of citizenship (see Chapter 2). Some countries of immigration make it very difficult for immigrants to become citizens, others grant citizenship but only at the price of cultural assimilation, while a third group makes it possible for immigrants to become citizens while maintaining distinct cultural identities.

Citizenship is more than just formal status, as demonstrated by possession of a passport. It is important to consider the contents of citizenship, in terms of civil, political and social rights. Moreover possession of citizenship is not an either/or question. With increasing length of residence, immigrants sometime acquire forms of 'quasi-citizenship', which confer some, but not all, rights of citizenship. One important form of quasi-citizenship is the citizenship of the EU, introduced through the 1991 Maastricht Treaty. The rules for becoming a citizen in various countries are complex and have undergone considerable change in recent years (see Çinar, 1994; Guimezanes, 1995; Aleinikoff and Klusmeyer, 2000, 2001).

Laws on citizenship or nationality derive from two competing principles: *ius sanguinis* (literally: law of the blood) which is based on descent from a national of the country concerned, and *ius soli* (law of the soil) which is based on birth in the territory of the country. *Ius sanguinis* is often linked to an ethnic or folk model of the nation-state (typical of Germany and Austria), while *ius soli* generally relates to a nation-state built through incorporation of diverse groups on a single territory (such as the UK), or through immigration (USA, Canada, Australia, Latin American countries). In practice, all modern states have citizenship rules based on a combination of *ius sanguinis* and *ius soli*, although one or the other may be predominant.

Naturalization of immigrants

Table 10.2 shows acquisitions of citizenship in various countries. This includes naturalizations and some other procedures, such as declarations by people born to foreign parents in the immigration county. There is a clear trend towards increasing naturalization of foreign residents: numbers and rates were higher in all countries (except the UK) in 1999 compared with 1988. The absolute number of naturalizations was high in Australia, Canada and the USA, but it has not been possible to calculate acquisition rates in these countries due to lack of data on the foreign resident population. In several European countries, notably France, Sweden and the Netherlands, naturalization rates have increased sharply. This represents changes in naturalization rules and conscious efforts to encourage

Box 10.7 Minorities in Italy

Italy, like other Southern European countries, has made a rapid transition from a country of emigration to one of immigration. Due to poor economic conditions, 7 million Italians emigrated in the 30 years after 1945. Today important Italian communities remain in the USA, Argentina, Brazil, Australia, Germany, Switzerland, and elsewhere. Since the 1970s, rapid economic growth and declining fertility have reversed former patterns, raising the question whether a former emigration country is better able to deal with new immigrant populations than other receiving countries.

Italy's legally-resident foreign population rose from just 423 000 in 1985 to 1 million in 1995, and continued to grow to 1.7 million in 2000. In addition there are estimated to be up to 300 000 undocumented foreign residents. Foreigners make up roughly 3.5 per cent of Italy's total population of 58 million. The table below shows only a slight increase in residents from other EU countries, while the populations from Eastern and Central Europe, Africa and Asia have risen sharply. The three largest communities are Moroccans, Albanians and Filipinos.

The majority of immigrants are admitted as workers, although the numbers of family reunions and asylum seekers are also increasing. Immigrants provide a source of low-skilled labour, which plays an important role in sustaining agriculture and industry, at a time when few young Italians are available for such work. Undocumented workers are concentrated in the 'underground economy', which is significant in Italy. However, the proportion of legal immigrants has increased, partly due to four regularization programmes since the 1980s. A range of indicators show that Italy is becoming a permanent settlement county: increased family reunion, a higher female share (46 per cent of residence permit holders in 1999), longer duration of residence, more births to foreign women (4 per cent of all births in 1999) and increasing numbers of children entering Italian schools. This development is controversial, and opinion polls show that many Italians are very concerned about immigration.

The right-wing Northern League and National Alliance campaign against immigration as a threat to law and order, and there has been considerable violence, especially against non-Europeans. The Left calls for integration and even multiculturalism. Trade unions, left-wing parties and advocacy groups

→

immigrants to become citizens. However, naturalization rates remain fairly low in Germany, Japan and Switzerland. Germany has improved, and is likely to so further due to the citizenship law of 1999 (see Chapter 9).

Dual citizenship has become important for immigrants, because it seems an appropriate way of managing the multiple identities which arise from globalization. The principle of singular citizenship is in fact being eroded everywhere by mixed marriages. In many countries, the nationality of a child born abroad used only to be transmitted through the father. Such rules were changed to achieve equality of the sexes in the 1970s and 1980s.

→

Italy: legal foreign residents by region and country of origin

	1994 (thousands)	*1999 (thousands)*	*Percentage increase*
Europe	**239**	**458**	**79**
European Union	120	143	19
East/Central Europe	103	270	162
Albania	25	94	279
Yugoslavia	33	52	37
Africa	**211**	**366**	**73**
Morocco	79	148	88
Tunisia	36	51	42
Egypt	19	31	61
Senegal	19	32	62
Asia	**106**	**209**	**97**
Philippines	26	59	126
China	16	41	161

Note: These figures give lower total foreign resident figures than other sources, but still provide a good indication of trends in distribution by origins.

Source: Istat (2001); ISMU-Cariplo (2001); quoted here from Calavita (2003).

support migrant rights, while employers' associations campaign for increased immigration. The Centre-Left's 1998 Immigration Act liberalized entry rules and contained a range of integration measures, though little has been done to implement these. However, at the municipal level, left-wing local authorities have introduced policies to recognize diversity and to improve services for minorities, although there is often a large gap between rhetoric and implementation. In 2002 the Centre-Right Berlusconi government passed a new immigration law, which repealed many of the 1998 measures. It emphasized recruitment of seasonal workers and introduced tough measures against illegal immigration.

Sources: King (2000); Mingione and Qassoli (2000); OECD (2001: 192–7); Però (2001); Calavita (2003).

Once mothers obtain the same right to transmit their nationality as fathers, bi-national marriages automatically lead to dual citizenship.

Austria, Denmark, Finland, Germany, Japan, Luxembourg and Norway legally require renunciation of the former nationality upon naturalization, while other OECD countries, including Australia, Canada, France, the UK and the USA, permit dual citizenship (Guimezanes, 1995: 165). Sweden generally accepted dual citizenship in practice and has now changed its law to permit it formally. The Netherlands introduced the right to dual citizenship in 1991, but withdrew it again in 1997, following claims that

Table 10.2 *Acquisition of nationality in selected countries, 1988 and 1999*

Country	*Acquisition of nationality 1988*	*Acquisition rate 1988*	*Acquisition of nationality 1999*	*Acquisition rate 1999*
Australia	81 218	n.a.	76 474	n.a.
Belgium	8 366	10	24 273	27
Canada	58 810	n.a.	158 753	n.a.
France	46 351	13	145 435	45
Germany	16 660	4	107 000	15
Italy	7 442	12	11 291	9
Japan	5 767	6	16 120	11
Netherlands	9 110	14	52 090	94
Sweden	17 966	43	37 777	76
Switzerland	11 356	11	20 363	15
UK	64 600	35	54 902	25
USA	242 063	n.a.	463 060	n.a.

n.a. = not available.

Notes: The acquisition of nationality rate is defined here as the number of acquisitions of nationality per thousand foreign residents. The figure for France includes children born to foreign parents in France who declared their intention to become French. The German figure is for 1998 and excludes naturalization based on legal entitlement, which applies mainly to 'ethnic Germans' from Eastern Europe. The US figure is for 1998. The comparison has only indicative value, as definitions and procedures vary from country to country.

Sources: OECD (1997, 2001: Tables A.1.6 and B.1.6). Germany: Beauftragte der Bundesregierung (2000: 33).

immigrants were using Dutch citizenship in an instrumental way to facilitate travel within the EU, rather than because of commitment to the Netherlands (Entzinger, 2002). The increasing acceptance of dual citizenship represents a major shift since 1963, when most European countries signed the Strasbourg Convention on the Reduction of Cases of Multiple Nationality.

Legal requirements for naturalization (such as 'good character', regular employment, language proficiency, evidence of integration) seem quite similar in various countries, but actual practices are very different. Switzerland, Austria and (until recently) Germany impose long waiting periods and complex bureaucratic practices, and make it clear to applicants that naturalization is an act of grace by the state. At the other end of the spectrum, classical immigration countries encourage newcomers to become citizens. The Australian Government declared a Year of

Citizenship in 1989, while US Vice-President Gore launched a Citizenship USA drive in 1995. Over 1 million persons were naturalized in 1996. In Texas, 10 000 new citizens were sworn in at once through mass ceremonies at the Dallas Cowboys' football stadium and 6000 at the Houston convention centre (*Guardian Weekly*, 22 September 1996). The act of becoming American (or Australian or Canadian) is seen as an occasion for celebration and part of the national myth. In contrast, to be Austrian or Swiss means to have been born so, and naturalization is still an exception.

Status of the second generation

The transmission of citizenship to the second generation (the children of the original immigrants) and subsequent generations is the key issue for the future. National variations parallel those found with regard to naturalization. In principle, *ius soli* countries confer citizenship on all children born in their territory. *Ius sanguinis* countries confer citizenship only on children of existing citizens. However, most countries actually apply models based on a mixture of the two principles. Increasingly, entitlement to citizenship grows out of long-term residence in the country: the *ius domicili*.

Ius soli is applied most consistently in Australia, Canada, New Zealand, the USA, the UK and Ireland. A child born to immigrant parents in the USA becomes a US citizen, even if the parents are visitors or illegal residents. In Australia, Canada and the UK, the child obtains citizenship if at least one of the parents is a citizen or a legal permanent resident. Such countries only use the *ius sanguinis* principle to confer citizenship on children born to their citizens while abroad (Çinar, 1994: 58–60; Guimezanes, 1995: 159).

A combination of *ius soli* and *ius domicili* has emerged in France, Italy, Belgium and the Netherlands. Children born to foreign parents in the territory obtain citizenship, providing they have been resident for a certain period, and fulfil other conditions. Since 2000, Germany has had similar arrangements – a significant move away its past *ius sanguinis* model (see Chapter 9). France, Belgium and the Netherlands also apply the so-called 'double *ius soli*'. Children born to foreign parents, at least one of whom was also born in the country, acquire citizenship at birth. This means that members of the 'third generation' automatically become citizens, unless they specifically renounce this right upon reaching the age of majority (Çinar, 1994: 61).

Ius sanguinis is still the dominant legal principle in some European countries as well as in Japan (Guimezanes, 1995: 159). However, in Western Europe, only Austria and Switzerland still apply the principle strictly: children born in these countries to foreign parents have no entitlement to citizenship, even if they have lived there all their lives

(Çinar, 1994: 68–9). Other *ius sanguinis* countries have taken cautious steps towards *ius domicili*. This means giving an *option* of facilitated naturalization to young people of immigrant origin. Foreigners who have been resident in Sweden for five years before reaching the age of 16, and who have lived there since that age, can become citizens by declaration between the ages of 21 and 23. Similar rules also exist in Belgium (Çinar, 1994: 61; Guimezanes, 1995: 178).

The distinction between *ius sanguinis* and *ius soli* countries remains significant, though less so than ten years ago. In the former, children who have been born and grow up in the country may be denied not only security of residence, but also a clear national identity. They are formally citizens of a country they may have never seen, and can even be deported there in certain circumstances. In countries with *ius soli* the second generation still generally have multiple cultural identities, but they have a secure legal basis on which to make decisions about their life perspectives. Dual citizenship seems the best solution, as it would avoid decisions which can be extremely difficult for many individuals.

Linguistic and cultural rights

Maintenance of language and culture is seen as a need and a right by most settler groups. Many of the associations set up in the process of ethnic community formation are concerned with language and culture: they teach the mother tongue to the second generation, organize festivals and carry out rituals. Language and culture not only serve as means of communication, but take on a symbolic meaning which is central to ethnic group cohesion. In most cases, language maintenance applies in the first two to three generations, after which there is a rapid decline. The significance of cultural symbols and rituals may last much longer.

Many members of the majority see cultural difference as a threat to a supposed cultural homogeneity and to national identity. Migrant languages and cultures become symbols of otherness and markers for discrimination. Giving them up is seen as essential for success and integration in the country of immigration. Failure to do so is regarded as indicative of a desire for separatism. Hostility to different languages and cultures is rationalized with the assertion that the official language is essential for economic success, and that migrant cultures are inadequate for a modern secular society. The alternative view is that migrant communities need their own languages and cultures to develop identity and self-esteem. Cultural maintenance helps create a secure basis which assists group integration into the wider society, while bilingualism brings benefits in learning and intellectual development.

Policies and attitudes on cultural and linguistic maintenance vary considerably. Some countries have histories of multilingualism. Canada's policy of bilingualism is based on two 'official languages', English and French. Multicultural policies have led to limited recognition of – and support for – immigrant languages, but they have hardly penetrated into mainstream contexts, such as broadcasting. Switzerland has a multilingual policy for its founding languages, but does not recognize immigrant languages. Australia and Sweden both accept the principle of linguistic and cultural maintenance, and have multicultural education policies. They provide language services (interpreting, translating, mother-tongue classes) and support for ethnic community cultural organizations, although in both countries funding has been cut in recent years. In Australia, multicultural radio and television are funded by the government.

In the USA, language has become a contentious issue. The tradition of monolingualism is being eroded by the growth of the Hispanic community: in major cities like Los Angeles and Miami, the number of Spanish speakers is overtaking that of English speakers. This led to a backlash in the late 1980s, in the form of 'the US English movement' which called for a constitutional amendment to declare English the official language. Several states passed referenda to introduce this measure, but it proved extremely hard to implement, and public agencies and private companies continued to provide multilingual information material and services. Monolingualism is also the basic principle in France, the UK, Germany and the Netherlands. Nonetheless, all these countries have been forced to introduce language services to take account of migrant needs in communicating with courts, bureaucracies and health services. The multilingual character of inner-city school classes has also led to special measures for integration of immigrant children, and to a gradual shift towards multicultural education policies.

Minorities and nation

On the basis of the preceding comparisons, it is possible roughly to divide immigration countries into three categories (see Castles, 1995).

The differential exclusionary model

Differential exclusion is to be found in countries in which the dominant definition of the nation is that of a community of birth and descent (referred to as the 'folk or ethnic model' in Chapter 2). The dominant group is unwilling to accept immigrants and their children as members of the nation. This unwillingness is expressed through exclusionary immigration policies (especially limitation of family reunion and refusal to grant secure residence status), restrictive naturalization rules and the ideology of not being countries of immigration. Differential exclusion means that immigrants are incorporated into certain areas of society (above all the labour market) but denied access to others (such as welfare systems, citizenship and political participation). Immigrants become ethnic minorities, which are part of civil society (as workers, consumers and parents) but are excluded from full participation in economic, social, cultural and political relations. Since ethnic minorities are usually socio-economically disadvantaged, there is a strong and continuing link between class and ethnic background. The differential exclusion model applies to former 'guestworker' recruiting countries in Western Europe, such as Germany, Switzerland and Austria.

The assimilationist model

Assimilation may be defined as the policy of incorporating migrants into society through a one-sided process of adaptation: immigrants are expected to give up their distinctive linguistic, cultural or social characteristics and become indistinguishable from the majority population. The role of the state is to create conditions favourable to this process, through insistence on use of the dominant language and attendance at normal schools for migrant children. In most cases, assimilation has been abandoned over time, and replaced with more flexible 'integration policies'. This happened in the 1960s in Australia, Canada and the UK as it became clear that immigrants were forming ethnic communities. Integration strategies stress that adaptation is a gradual process in which group cohesion plays an important part. Nonetheless, the final goal is still absorption into the dominant culture, so that integration policies are often simply a slower and gentler form of assimilation.

Essentially the assimilationist model permits people who have become members of civil society to join the nation and the state at the price of cultural assimilation. The assimilationist model has been used in all highly-developed immigration countries. In some countries there has been an evolution, starting with differential exclusion, progressing to assimilationism, moving on to gradual integration, and finally to multicultural models (Australia is a case in point). Today, of all the highly-developed immigration countries, France comes closest to the assimilationist model. Some European immigration countries are ambivalent: policies of assimilation in some areas (such as labour market or social policy) may coexist with multiculturalism in other sectors (for instance, education and cultural policy).

In some cases, assimilationism seems to derive from a merging of what we referred to in Chapter 2 as the 'imperial model' and the 'republican model'. France, the UK and the Netherlands all have aspects of both models. All three were imperial powers, which turned colonial subjects into citizens. France introduced the notion of citizenship as a political community after the 1789 Revolution, yet its policies towards colonized peoples maintained elements of the 'imperial model'. The UK and the Netherlands moved away from the 'imperial model' as their empires crumbled after 1945. Introduction of a more modern form of citizenship, based on membership of the political community, often meant depriving former colonial subjects of citizenship. The ambiguity of the situation of minorities reflects the contradictory and transitional nature of these post-imperial states.

A closer look at the republican model in France shows some of its contradictions. The essence of the model is to be found in the first report of the Haut Conseil à l'Intégration (High Council for Integration, or HCI):

> French conceptions of integration should obey a logic of equality and not a logic of minorities. The principles of identity and equality which go back to the Revolution and the Declaration of Rights of Man and of Citizens impregnate our conception, thus founded on the equality of individuals before the law, whatever their origin, race, religion ... to the exclusion of an institutional recognition of minorities. (HCI, 1991: 10, quoted here from Lloyd, 1991: 65)

The central idea is that immigrants can (and should) become integrated into the political community as French citizens, and that this will bring about cultural integration. Exponents of the model see France as temporarily multi-ethnic, but not as permanently multicultural. According to Weil (1991b), citizenship is essentially a political relationship, most simply expressed by the statement: '*Celui qui vote est français et citoyen*' ('A person who votes is French and a citizen'). Any granting of rights (such as local voting rights) to non-citizens means watering down this principle, and could lead to new identifications, not only by migrants but also by French people, on the basis of 'origins, blood, race or culture'. In this view, rights for minorities lead directly to racism. Despite the emphasis on political integration, the implication of cultural homogenization is very strong. Weil goes on to lament that the 'great Republican institutions' have become too weak to produce national identity (Weil, 1991b: 300–2; see also Schnapper, 1994).

The multicultural model

The final category is multiculturalism (or pluralism), which implies that immigrants should be granted equal rights in all spheres of society, without being expected to give up their diversity, although usually with an expectation of conformity to certain key values. In a multicultural country, membership of civil society, initiated through permission to immigrate, should lead to full participation in the state and the nation. There are two main variants. In the laissez-faire approach typical of the USA, cultural difference and the existence of ethnic communities are accepted, but it is not seen as the role of the state to ensure social justice or to support the maintenance of ethnic cultures. The second variant is multiculturalism as a government policy, as in Canada, Australia and Sweden – as well as in certain policy areas in other countries. Here, multiculturalism implies both the willingness of the majority group to accept cultural difference, and state action to secure equal rights for minorities.

For the classical countries of immigration, multiculturalism appears as the best way of rapidly incorporating large groups of culturally diverse immigrants into society. Moreover, the imperative of making immigrants into citizens reinforces the pressure for a multicultural policy: if immigrants become voters, ethnic groups can gain political clout. Sweden, by contrast, appears as an anomaly: it is a society which was unusually homogeneous until recently. Yet it has had large-scale settlement, and adopted multicultural policies very similar to those of Australia and Canada. The reason seems to lie in the strongly state interventionist model of Swedish social democracy, which has used the same approaches to integrating immigrants into civil society and the state as were used earlier to integrate the working class and reduce class conflict. Since the late 1980s, the Swedish model has faced increasing difficulties: economic constraints have made it harder to finance generous social policies, while public support for admission of refugees has declined. The new integration policy introduced in 1998 seems to represent a move away from multiculturalism, in favour of greater emphasis on social and economic integration – a trend very similar to the new integration policy introduced in the Netherlands in 1994 (see Box 10.5).

Multiculturalism has always been controversial: even in Australia and Canada, there has been a continual debate, with significant groups calling for a return to assimilationist policies. To many members of the dominant ethnic group, multiculturalism appears as a threat to their culture and identity. Others criticize multiculturalism for leading to a superficial acceptance of cultural difference, without bringing about real institutional change (see Vasta, 1996). Indeed, the trend of the 1990s was away from multiculturalism. In Canada, the emphasis has shifted to 'heritage and citizenship'. In Australia, the Centre-Right government elected in 1996 severely cut multicultural programmes and services, although it has reasserted the basic principles of multiculturalism (see Chapter 9). However, it is too early to say that multiculturalism is being abandoned: growing ethnic diversity generates pressure for measures to recognize cultural rights and to prevent social exclusion, whatever the label given to such policies. Debates on such issues play an important role in virtually all immigration countries, as the difficulties of exclusionary and assimilationist models become apparent.

Conclusion

Our comparison of various immigration countries can be summed up by saying that ethnic group formation takes place everywhere, but under conditions which vary considerably. This leads to different outcomes: in some countries ethnic groups become marginalized and excluded minorities, whereas in others they take the form of ethnic communities which are accepted as part of a pluralist society.

The reality in each country is much more complex and contradictory than our brief account can show. Nonetheless, these distinct experiences provide some useful conclusions. The first is that temporary migrant labour recruitment is likely to lead to permanent settlement of at least a proportion of the migrants. The second is that the character of these future ethnic groups will, to a large measure, be determined by what the state does in the early stages of migration. Policies which deny the reality of immigration lead to social marginalization, minority formation and racism. Third, the ethnic groups arising from immigration need their own associations and social networks, as well as their own languages and cultures. Policies which deny legitimacy to these needs lead to isolation and separatism. Fourth, the best way to prevent marginalization and social conflicts is to grant permanent immigrants full rights in all social spheres. This means making citizenship easily available, even if this leads to dual citizenship.

This last point has far-reaching consequences: removing the link between citizenship and ethnic origin means changing the defining principle of the nation-state. This applies particularly to nations based on the ethnic model. The principles of citizenship also need redefining in countries with post-imperialist or republican models, such as France, the UK and the Netherlands. The classical countries of immigration – Australia, Canada and the USA – have already moved towards citizenship based on territoriality and are capable of incorporating newcomers of varying origins.

Migration both to developed countries and to NICs is likely to continue in the years ahead, so the presence of ethnic communities will be an inescapable part of society. The most direct impact will be felt in 'global cities' such as Los Angeles, Paris, Berlin, Sydney, Singapore and Hong Kong. The formation of ethnic groups and the spatial restructuring of the city are powerful forces for change, which may give rise to conflicts and violence, but can also be a great source of energy and innovation.

Globalization is leading to multiple identities and transnational belonging. Exclusionary models of immigrant rights and nationhood are questionable, because they lead to divided societies. Similarly, assimilationist models are not likely to succeed, because they fail to take account of the cultural and social situation of settlers. The multicultural model is a combination of a set of social policies to respond to the needs of settlers, and a statement about the openness of the nation to cultural diversity. Multicultural citizenship appears to be the most viable solution to the problem of defining membership of a nation-state in an increasingly mobile world.

Guide to further reading

Feagin (1989) gives a good overview of ethnic relations, with special emphasis on the USA. Breton *et al.* (1990) present a thorough empirical study on ethnic identity and class in Canada, while Reitz (1998) discusses comparative dimensions for Canada, the USA and Australia. Koopmans and Statham (2000) is an excellent collection on the politics of immigrant incorporation in Europe. Portes and Rumbaut (1996) is a good study on the USA, while Carnoy (1994) analyses race relations. Favell (1998) compares French and British approaches. King *et al.* (2000) provides studies of Southern European immigration countries. Stasiulis and Yuval-Davis (1995) gives an interesting comparison of ethnic relations in settler societies. Björgo and Witte (1993) and Hargreaves and Leaman (1995) look at racism and responses to it in Europe. Solomos (1993) and Solomos and Back (1995) are good on the UK. Ålund and Schierup (1991) provide a sceptical account of Swedish multiculturalism. Davis's (1990) book on Los Angeles is a fascinating case study of the 'post-modern' city and the role of minorities within it. Comparative urban sociological studies are to be found in Cross and Keith (1993) and Mingione (1996). The Metropolis Website (www.international.metropolis.net) is a valuable source of information on integration policies and urban issues in many countries.

Chapter 11

Migrants and Politics

As international migration reshapes societies, it inevitably and often profoundly affects political life. Yet, paradoxically, international migration is frequently viewed as a socio-economic phenomenon largely devoid of political significance. This viewpoint is part and parcel of the temporary worker idea examined earlier. Relatively few people foresaw that the decision to recruit foreign labour in the wake of the Second World War would one day affect the political landscape of Western Europe. But immigration did lead to a significantly altered political environment: one that now includes Islamic fundamentalist movements composed mainly of immigrants and their offspring, as well as extreme-right, anti-immigrant parties.

The most lasting significance of international migration may well be its effects upon politics. However, much depends on how immigrants are treated by governments, and on the origins, timing, nature and context of a particular migratory flow. It makes a difference whether migrants were legally admitted and permitted to naturalize or whether their entry (legal or illegal) was seen as merely temporary but they then stayed on permanently. On the one hand, immigrants can quickly become citizens without a discernible political effect, save for the addition of more potential voters. On the other hand, international migration may lead to an accretion of politically disenfranchised persons whose political marginality is compounded by various socio-economic problems.

The universe of possible political effects of international migration is vast and characteristically intertwines the political systems of two states: the homeland and the receiving society. The political significance of international migration can be active or passive. Immigrants can become political actors in their own right or manifest apoliticism, which itself can be important to maintenance of the status quo. On the other hand, immigrants often become the object of politics: allies for some and foes for others. Chapter 10 has already dealt with one key political issue: the extent to which immigrants and their descendants can become citizens with full rights of political participation. This chapter cannot hope to do justice to all the other facets of immigration-related politics. Only a few themes can be considered. The emphasis is on emergent forces that have rendered politics within and between states more complex and volatile.

Homelands and expatriates

Most states have significant populations of citizens or subjects living abroad. For many, if not most, expatriates, the country of origin and its politics remains the foremost concern. Likewise, governments of migrant-sending societies often nurture a relationship with citizens or subjects abroad. Such policies can be driven by economic concerns such as facilitating the sending of remittances. They can also be driven by national security concerns in cases where political opposition forces become active in expatriate populations and are perceived as a threat by a homeland government.

Perhaps the best example of a sending state projecting a form of governance to its citizens abroad was Algeria from 1962 to 1990. Algeria achieved independence from France only after an eight-year-long conflict that cost 1 million lives. During the war of independence, the major Algerian revolutionary party, the National Liberation Front, had created a substantial organizational membership and infrastructure in metropolitan France. After the cessation of hostilities, the National Liberation Front organization in France was transformed into the Amicale des Algériens en France (AAE).

The head of the AAE was usually also a high-ranking official of the National Liberation Front and of the Algerian government. The AAE enjoyed a quasi-diplomatic status in France. It represented the interests of Algerian emigrants in Algerian policy-making circles as well as vis-à-vis the French government. Throughout the 1960s and 1970s, the AAE virtually monopolized the representation of Algerians in France, although it was opposed by rival groups, like the outlawed Movement of Arab Workers, a revolutionary communist organization with ties to radical Palestinian factions, which played a key role in organizing protests against attacks on Algerians and other North Africans in 1973 (Miller, 1981: 89–104). The AAE opposed the French government's decision in 1981 to grant aliens the right to form associations (Weil, 1991b: 99–114). Prior to 1981, associations of foreigners required government authorization in order to operate, which condemned anti-Algerian regime parties to clandestinity. The 1981 reform undercut the virtual AAE monopoly, and open opposition to the Algerian regime soon flourished.

The Algerian government was particularly concerned by the ability of Muslim fundamentalist groups, such as the Islamic Salvation Front, to operate openly in France, which they could not do in Algeria. This concern was shared with other non-Islamic governments in predominantly Muslim societies such as Turkey and Tunisia. Political dissidence expressed on French soil presaged the fundamentalist victory in the December 1991 elections, although many Algerians who voted for the Islamic Salvation Front were not so much voting for an Islamic republic as protesting against

National Liberation Front rule. The influence of French policies with regard to aliens' associations illustrates how international migration binds together the politics of two societies.

More recently, Mexico, under the leadership of President Fox, has sought to bolster the Mexican government's relationship with the large Mexican-background population in the USA, estimated to number over 8 million, about half of whom are legally resident in the USA. Even prior to Fox's arrival in power, activist Mexican consular officials played an important role in leading opposition to Proposition 187 in California, which aimed to deny government services to non-US citizens, such as schooling for illegally-resident children. President Fox's visit to the USA shortly before 11 September 2001 had all the trappings of an election campaign and featured impassioned calls for legalization and increased legal admissions of Mexican workers (see Box 1.2).

One factor influencing homeland government efforts to improve conditions for expatriates obtains only in cases where expatriates can vote in homeland elections. The modalities for expatriate voting vary considerably. Some homelands, like Turkey, Italy and Mexico, require emigrants to return home in order to vote. Other states, like Algeria and Israel, permit consular voting. Still others permit absentee voting, as in the USA. Indeed, absentee balloting by Floridians abroad played a key role in the contested outcome of the 2000 US presidential election. Electoral campaigning increasingly reflects the weight of voters abroad. Ecuadorian and Dominican Republic presidential candidates campaign for votes in New York City, just as Italian and Portuguese parties campaign for votes in Paris.

Nevertheless, the potential for emigrants to influence electoral outcomes at home does not necessarily translate into effective representation of their interests by homeland governments. Overall, the track record of diplomatic representation of migrants' interests is deficient, in part because homelands are often reluctant to criticize treatment of emigrants for fear of offending the host government and jeopardizing the homeland flow of remittances. The asymmetrical power of homelands and immigrant-receiving states was clearly demonstrated when one Western European state after the other unilaterally curbed further foreign worker recruitment in the 1970s.

Immigrants as political actors in Europe

The inadequacy of diplomatic representation of foreign residents was one reason for the emergence of distinctive channels of alien political participation and representation in Europe. Indeed there is reason to believe that nascent immigrant participation in Western European politics contributed to the decisions to curb foreign worker recruitment. By the early 1970s, supposedly politically-quiescent aliens had become involved in a number of significant industrial strikes and protest movements. In some instances, extreme leftist groups succeeded in mobilizing foreigners. Largely foreign worker strikes in French and German car plants (see Chapter 8) demonstrated the disruptive potential of foreign labour and constrained trade unions to do more to represent foreign workers. Generally speaking, foreigners were poorly integrated into unions in 1970. By 1980, significant strides had been made towards integration. This was reflected in growing unionization rates among foreign workers and the election of foreign workers to works councils and union leadership positions.

Immigrants also sought participation and representation in local government. In several countries, advisory councils were instituted to give immigrants a voice in local government. Experiences with these advisory councils varied and some were discontinued. Some people contested them as efforts to coopt aliens, while others saw them as illegitimate interference by aliens in the politics of the host society. In certain countries, aliens were accorded a right to vote in local and regional elections. Sweden was the pacesetter in this regard, but alien participation in Swedish local and regional elections declined over time. The Netherlands was the second country to accord qualified aliens voting rights. However, the results of alien voting there have also been somewhat disappointing (Rath, 1988: 25–35). Proposals to grant local voting rights to legally-resident aliens became important domestic political and constitutional issues, particularly in France and Germany. By 2001, Belgium had the most extensive network of local government consultative structures for resident aliens. Luxembourg and Switzerland were also noteworthy for the variety and extent of consultation of foreign populations at the local level (Oriol, 2001: 20).

By the 1980s, the stakes involved in the granting of voting rights were quite high in many Western democracies. Aliens were often spatially concentrated in major cities and certain neighbourhoods. Enfranchising them would dramatically affect political outcomes in many local elections. Supporters of the granting of municipal voting rights generally regarded it as a way to foster integration and as a counterweight to the growing influence of parties like the FN in France. However, many immigrants were already politically enfranchised, particularly in the UK. This did not prevent the eruption of riots involving immigrants and their British-born children in the mid-1980s. The granting of local voting rights was thus not in itself a panacea for the severe problems facing immigrants in Western Europe.

Since the 1970s, immigrants have increasingly articulated political concerns, participated in politics and sought representation. Immigrant protest movements became part of the tapestry of Western European politics and frequently affected policies. Persistent hunger strikes by undocumented immigrants and their supporters, for example, brought pressure to bear on French and Dutch authorities to liberalize rules regarding legalization (see Chapter 5). There was great variation in patterns of alien political participation and representation from country to country, with some countries, like Sweden, succeeding in institutionalizing much of it.

Mobilization of immigrants and ethnic minorities outside the normal channels of political representation is often linked to experience of exclusion, either through racist violence or institutional discrimination. For instance, members of ethnic minorities often feel that the police are more concerned with social control than with protecting them from racist violence. In the UK, Afro-Caribbean and Asian youths have organized self-protection groups against racist attacks. Sub-cultures of resistance developed around reggae music and rastafarianism for the West Indians, and Islam and other religions for the Asians (Gilroy, 1987). The reaction by government and the media was to see ethnic minority youth as a problem of public order: a 'social time bomb' on the verge of explosion. There was widespread panic about the alleged high rates of street crime ('mugging') by black youth and a tendency to see black people as an 'enemy within' who threatened British society (CCCS, 1982).

Black youth discontent exploded into uprisings in many inner-city areas in 1981 and 1985 (Sivanandan, 1982; Benyon, 1986). Later, there were new disturbances, in which 'joyriders' stole cars and publicly raced them to destruction in inner-city streets, to the acclaim of crowds of onlookers. After the riots, the initial official response was to insist that the central issue was one of crime and to lament the breakdown of parental control (Solomos, 1988). Newspapers blamed the problems on 'crazed Left-wing extremists' and 'streetfighting experts trained in Moscow and Libya'. The disturbances were generally labelled as 'black youth riots', but, in fact, there was a high degree of white youth involvement (Benyon, 1986).

The riots were caused by a number of interrelated factors. Deteriorating community relations and lack of political leadership against racism were major causes of alienation of black youth. There were concentrations of disadvantaged people (both white and black) in inner-city areas, marked by high unemployment, poor housing, environmental decay, high crime rates, drug abuse and racist attacks. As Benyon (1986: 268) points out, the areas where riots took place were politically disadvantaged: they lacked the institutions, opportunities and resources for putting pressure on those with political power. Finally these areas had suffered repressive forms of policing, experienced by young people as racism and deliberate harassment. The riots may be seen as defensive movements of minority youth, connected with protection of their communities as well as assertion of identity and culture (Gilroy, 1987; Gilroy and Lawrence, 1988). Similar urban uprisings have taken place in France, where a second generation has emerged of young people of mainly Arab descent who feel French, but find themselves excluded by discrimination and racism. The spectre of insecurity became a major campaign issue for mainstream French parties of the right and the left, not only the FN.

Fear of further unrest has led to a multifaceted response by governments. Some of the French and British measures were mentioned in Chapter 10. However, it is doubtful whether such policies can effectively combat the powerful economic, social and political forces which marginalize ethnic minority youth. Moreover, the capacity and willingness of Western European governments to carry out social policy measures in favour of immigrants has in many cases declined, resulting in situations of severe and persistent social exclusion.

After the reinforcement of the powers attributed to the European Union in the 1990s, the EU became a more important factor affecting migrants and politics in Europe. Citizens of the EU residing in other member states were empowered to vote in local and European elections in their countries of residence. In the 2001 French municipal elections, statistics compiled for municipalities of more than 3500 people indicated that 204 non-French EU citizens were elected as municipal concillors, less than 1 per cent of all elected. EU citizens comprised about 2 per cent of the total population (*La lettre de la citoyenneté*, 2001: 1). Many pro-immigrant activists hoped that the strengthening of federal-level governance would facilitate enfranchisement of third country nationals and extend to them freedom of movement. But proposals such as these were blocked. Nevertheless, European institutions became an important focus of migrant political activism in the 1990s.

New issues and new political forces: Islam in Western Europe

By 1970, Islam was the second religion of France. By 1990, it was the second religion of the French. There were 15 million Muslims in Europe in 2002, including over 4 million in France (Hunter, 2002: xiii; Leveau *et al.*, 2002: 140–1). Nevertheless, as late as 1970 Islam was largely invisible in France. According to Kepel and Leveau (1987), the affirmation of Islam since then was an essential part of the settlement of foreign workers, the progression of the migratory chain. It was manifested primarily through construction of mosques and prayer-rooms and through formation of Islamic associations. In turn, the reaction to Islamic affirmation, manifested above all through the emergence of the FN, has undermined governmental integration policies. By 1990, immigration had become one of the key political issues in France and the politicization of immigration issues at times appeared to threaten the stability of French democratic institutions. The paradox was that this was largely unforeseen.

This politicization became apparent around 1970 (Wihtol de Wenden, 1988: 209–19), when extreme-right student groups began to demonstrate against *immigration sauvage*, or illegal migration. Counter-demonstrations took place and violence erupted. By 1972, the extremist groups principally involved in this violence – the Trotskyist Communist League and the neo-fascist New Order – were banned. Leftist groups continued to mobilize immigrants in various struggles such as the long rent strike in the SONACOTRA housing for foreign workers (Miller, 1978). Elements of the extreme right, however, began to mobilize on anti-immigration themes. Françoise Gaspard (1990) has related how the FN began to campaign in local elections in the Dreux area near Paris. The FN gradually increased its share of the vote before scoring a dramatic breakthrough in 1983, when it obtained 16.7 per cent of the vote. In 1989, an FN candidate in a by-election in Dreux won a seat in the Chamber of Deputies, with 61.3 per cent of the vote. In the space of 11 years, the number of FN voters increased from 307 to 4716 (Gaspard, 1990: 205). By 1997, the FN dominated municipal governments in four southern cities, including Toulon, and was supported by about 15 per cent of the national electorate. Nearly 4 million French citizens voted for FN candidates in the first round of the 1997 legislative elections. Hence, the second place finish of FN candidate Le Pen in the first round of the presidential elections of 2002 did not reflect a major, sudden increase in support for the FN.

The French reaction to Islam was irrational but grounded in concrete immigration-related problems. The irrational dimension stemmed from the trauma of the Algerian war and the association of Islam and terrorism. In 1982, following a series of crippling strikes in major car plants in the Paris region which principally involved North African workers (see Chapter 8), Prime Minister Mauroy insinuated that Iran was trying to destabilize French politics by backing Islamic fundamentalist groups (*Le Monde*, 1 February 1983). While no evidence of an Iranian involvement was produced, it was clear that Islamic groups were heavily involved, with the French Communist Party and its trade union affiliate, the CGT, desperately trying to regain control of the strike movement (Miller, 1986: 361–82). Islam was seen by many as incompatible with democracy because it made no distinction between religious faith and state. France's Muslims were portrayed as heavily influenced by Islamic fundamentalism when, in fact, only a small minority of them considered themselves fundamentalists and these were divided into multiple and often competing organizations (Kepel and Leveau, 1987).

The integration problems affecting France's Islamic minority were perhaps more central to the politicization of immigration issues. Gaspard (1990) recounts how tensions over housing exacerbated French-immigrant relations in Dreux. Immigrants, particularly those of North African origin, were living in disproportionate numbers in inadequate housing. As settlement and family reunification proceeded, more and more immigrants applied for subsidized governmental housing, causing severe friction when their numbers grew, while the number of non-immigrant residents diminished. Before long, entire buildings came to be viewed as immigrants' quarters. The physical isolation of many immigrants in substandard housing, along with the educational problems faced by schools with a disproportionately high number of immigrant children, contributed to a malaise on which the FN fed. By the 1980s, primarily North African Muslim-origin youths, most of whom were French citizens, became involved in urban unrest that was deeply unsettling to the French.

The French Socialist Party sought to galvanize support by appealing to North African-origin voters. The pro-immigrant SOS-Racisme organization was largely an initiative of the Socialist Party. The party clearly appealed to voters as a bulwark against the FN. Yet support for the Socialists had plunged dramatically by 1992 and they were booed at pro-immigrant rallies. The socialist message of integration of resident aliens and curbing illegal immigration was viewed as pro-immigrant by the extreme right and as anti-immigrant by the extreme left.

Illustrative of the broader issue was the question of the *foulards* or Islamic headscarves worn by some young girls to school in the late 1980s. In a country where the tradition of the separation of religion and state is deeply rooted and politically salient, the wearing of headscarves appeared to many people as incompatible with the very principles of the French Republic, which prohibited the wearing of religious articles. On the other side of the debate was the claim that the choice of wearing a headscarf was an individual's prerogative, a private matter of no consequence to public authorities. In the end, French authorities ruled in favour of the girls, but not before the question had become a *cause célèbre*. Should school cafeterias serve *halal* food (that is, food prepared in accordance with Islamic ritual prescriptions)? Should Muslims be granted representation in French politics, as Catholics are through governmental consultations with the bishops, and Jews are through the consistory? Should factories honour Islamic holidays in addition to Catholic feast days? As Islam was affirmed, a host of long latent issues came to the fore, with major consequences for the French political system.

Paradoxically, the French government had encouraged the creation of mosques and prayer-rooms back in the 1960s and 1970s. Islam was seen by some businesses and public authorities as a means of social control. Prayer-rooms were constructed in factories in the hope that North African workers would be less likely to join left-wing trade unions. Moreover, supporting Islam was part of a policy of cultural maintenance, designed to encourage the eventual repatriation of migrant workers and their families. Saudi Arabia and Libya financed the construction of mosques across Western Europe, hoping to influence the emerging Islam of Western

Europe (Kepel and Leveau, 1987). Despite the separation of faith and state in France, many local governments supported the construction of mosques as part of integration policy. The building of mosques was often violently opposed, and several were bombed. Other Western European governments also fostered Islam through policies which brought Islamic teachers to Western Europe. These policies usually stemmed from provisions of bilateral labour agreements which granted homeland governments a role in educating migrant children. Many Koran schools in Germany were controlled by Islamic fundamentalists. Such institutionalization of Islam in Western Europe has probably progressed the furthest in Belgium.

Most of Western Europe's Muslims saw their religion as a private matter. The Rushdie affair made Islamic identity more of a political problem than, say, Catholicism or Protestantism. Salman Rushdie is an Indian-born Muslim citizen of the UK, who scandalized many Muslims with his book *The Satanic Verses* and was condemned to death by Iran's Ayatollah Khomeini. Much-publicized anti-Rushdie demonstrations by Muslims in England, France and Belgium confirmed the incompatibility of Islam with Western institutions in the eyes of some people.

The crisis over the Iraqi occupation of Kuwait in 1990 also prompted fear of Islamic subversion. Muslims in France were more supportive of Iraq than was French society as a whole (Perotti and Thepaut, 1991: 76–9). But many Muslims opposed the Iraqi invasion, and the Gulf War did not produce the terrorism and mass unrest in Western Europe that some had predicted. Nonetheless it was clear that French governmental support for the war effort alienated it from France's Islamic community. Tensions between Muslims and non-Muslims reached new heights. Similarly in Australia, incidents of abuse and harassment against Muslims, particularly women, increased during the Gulf crisis, prompting government action to improve community relations.

The renewal of Israeli-Palestinian violence in 2001 had major repercussions for France. Over 120 acts of violence and vandalism against French Jewish targets occurred in one month. Youths of North African Muslim background were thought to have perpetuated most of the attacks (Perotti and Thepaut, 2001).

While the vast majority of Muslim immigrants eschewed fundamentalism, Western Europe certainly was affected by the upsurge in religious fervour that swept the Muslim world. Fundamentalism often had the greatest appeal among groups suffering from various forms of social exclusion. Across Western Europe, Muslims are affected by disproportionately high unemployment rates. In areas where unemployment is compounded by educational and housing problems, and these underlying socio-economic tensions are overlaid with highly politicized religious identity issues, the ingredients for socio-political explosions are strong. All it takes is an incident, usually a violent encounter between a Muslim youth and the police, for violence to erupt. This was the pattern behind urban unrest in France in the 1980s and 1990s. The profound problems facing Western Europe's Muslims are one reason why integration will remain the top immigration policy priority for Western European governments and for the EU for the foreseeable future (Commission of the European Communities, 1990).

In the aftermath of 11 September 2001, there were hundreds of arrests of Muslims living in North America or Western Europe for suspected involvement with Al-Qaida and its confederates. And the role played by certain mosques in radicalizing worshippers became better understood and documented. Some observers portrayed the West as a fertile ground for the spread of Islamic fundamentalism. However, such views did not withstand close scrutiny. The Al-Qaida network was present in many Western democracies, including Canada and the USA, but it did not have a significant base of support. Most Muslims in the West condemned Al-Qaida and its terrorism.

Immigrants as objects of politics: the growth of anti-immigrant extremism

The French were not alone in finding it difficult to come to grips with the emergent Islamic reality in their midst. Belgium became the scene of urban unrest in 1991, when youths who were largely of Moroccan origin clashed with police following a rumour that an anti-immigrant political party, the Flemish Vlaams Blok, was going to organize a political rally in an area heavily populated with immigrants (*The Bulletin*, 1991: 20). Partly as a result of this violence, support for the Vlaams Blok increased sharply in the 1991 Belgian elections.

Similarly, in the 1991 Austrian municipal and regional elections, the anti-immigrant Freedom Party scored an important breakthrough by increasing its share of the vote to almost one-quarter. Eventually, the Freedom Party achieved a rough parity with the Austrian Socialist Party and the People's Party and formed a government with the latter. This precipitated a crisis in EU-Austria relations, as other EU member states regarded the Freedom Party's position on immigration as unacceptable. In reality, Freedom Party preferences for migration were not that different from those of the EU mainstream. Nevertheless, the Freedom Party's leader Jörg Haider resigned as chairman in 2000 as a result of the imbroglio. In Italy, a backlash against immigration figured importantly in the political convulsions of the 1990s. The Northern Leagues, Forza Italia and the neo-fascist National Alliance expressed frustration over immigration to varying degrees. Meanwhile, the politically influential Catholic clergy and the Pope himself voiced support for humanitarian initiatives such as legalization. Many Italian voters supported right-wing parties and protested against the deeply-embedded corruption of the Christian Democrats and the Socialists. Protest voting against a discredited *partitocrazia* was far more prevalent than anti-immigrant voting. But the second Berlusconi government announced a crackdown on illegal immigration by 2002.

Certainly support for anti-immigrant parties involved an element of protest voting. While 15 per cent of the electorate voted for the FN in France and one-third of all voters sympathized with FN positions on immigration (Weil, 1991a: 82), it also was clear that the FN was picking up part of the protest vote traditionally received by the French Communist Party. The FN did particularly well in areas with concentrations of *Pieds-Noirs*, Europeans repatriated from Algeria in 1962 and their offspring. FN opposition to the European institutions was also a major point of attraction to some of its electorate (Marcus, 1995).

In 1997, it was estimated that the FN candidates would be strong contenders in 200 of France's 577 legislative districts. However, in most French elections, if one candidate does not win a seat outright with more than 50 per cent of the vote, there is a second run-off election. After the first round of the 1997 elections, 133 FN candidates became eligible for the second round by garnering more than 12.5 per cent of the vote. Candidates from parties which have fashioned national electoral alliances – generally there is one for left-leaning and one for right-leaning parties – are favoured in second-round voting. The FN had encountered difficulty forging such electoral pacts outside the Marseilles area and therefore stood little chance of winning seats in the National Assembly. It won only one seat in the 1997 election and none in 2002. Its candidates received under 400 000 votes in the second round of the 2002 legislative elections, less than 2 per cent of votes cast (de Montvalon, 2002). The major exception to this pattern occurred in the 1986 elections when the FN won 35 seats in the National Assembly. That election was conducted according to modified proportional representation rules, unlike other national legislative elections during the Fifth Republic (Tiersky, 1994: 111–12).

By 2000, anti-immigrant political movements had developed virtually across Europe, even in formerly communist states like the former Czechoslovakia and Hungary. Many of these movements had historical precedents. Part of the hardcore support for the FN, for example, came from quarters traditionally identified with the anti-republican right. These political forces had been discredited by the Second World War and their programmes and policies were generally viewed as illegitimate until the anti-immigrant reaction of the 1980s and 1990s. Immigration issues have served as an entrée for extreme right-wing parties into mainstream politics across Europe, even in Scandinavia.

It would be a mistake to dismiss the upsurge in voting for anti-immigrant parties as simply an expression of racism and intolerance. As pointed out in Chapter 2, support for extreme-right groups is often the result of bewilderment in the face of rapid economic and social change. The erosion in organizational and ideological strength of labour organizations due to changes in occupational structures is also important. Extreme-right parties also attract support as a result of public dissatisfaction with certain policies, such as those concerning asylum seekers and illegal immigration. Other extremist parties have fared less well. The National Front in the UK, for example, appeared to be gaining strength in the mid-1970s before the Conservative Party, under the leadership of Margaret Thatcher, pre-empted it by adopting key parts of its programme (Layton-Henry and Rich, 1986: 74–5). The UK's two party system and its 'first past the post' electoral law make it very difficult for any new party to win seats in the House of Commons.

Some scholars have suggested that the emergence of right-wing parties has had anti-immigrant effects across the political spectrum (Messina, 1989). It has been argued, for example, that French socialist stands on immigration shifted to the right as support for the FN increased. However, it is difficult to reconcile the Socialist Party's celebration of its anti-racism with such a thesis. As Patrick Weil observed, 'immigration can appear as the ideal arena for differentiating politics of the Left from the Right. The Socialist Party, first and foremost François Mitterrand, found in immigration and antiracism a privileged domain for political intervention' (Weil, 1991a: 95).

Migrants and ethnic voting blocs

The most extreme example of international migration transforming politics is the case of Palestine and Israel. From 1920 to 1939, British-authorized immigration increased the Jewish share of the population in the British mandate of Palestine from roughly 10 per cent to one-third, despite fierce Palestinian Arab opposition including strikes, riots and revolts. The goal of the mainstream Zionists was the creation of a Jewish homeland, which did not necessarily connote the creation of a Jewish state. A minority Zionist current – the so-called Revisionists, led by Vladimir Jabotinsky – proclaimed the creation of a Jewish state on all the territory of the Palestine Mandate, including the East Bank of the Jordan, as the goal of Zionism (Laqueur, 1972). During the Holocaust and the Second World War, the Zionist movement was radicalized and the creation of a Jewish state in Palestine became the paramount goal of the movement. Palestinians and other Arabs had long opposed the Zionist project because they feared it would displace the Palestinian Arabs or reduce them from majority to minority status. Fighting broke out in 1947 and the worst fears of the Arabs were realized.

The politics of the Jewish state of Israel, created in 1948, remain heavily influenced by immigration. As a result of the inflow of Oriental or Sephardic Jews primarily from largely Muslim societies during the 1950s and 1960s, the Sephardic-origin Jewish population surpassed that of European-origin Ashkenazi Jews in the mid-1970s. This demographic shift benefited the modern-day followers of Jabotinsky and the Revisionists: the Likud bloc led by Menachem Begin, who was elected prime minister in 1977 with the support of Sephardic-origin Jews. In 1990, a new wave of Soviet Jewish immigration began, again affecting the balance between Ashkenazi and Sephardic Jews as well as the Arabs. Nearly 1 million Soviet Jews arrived in Israel between 1990 and 2002.

Now comprising about 15 per cent of the Israeli electorate, the Soviet Jewish vote decisively affected the outcomes of general elections beginning in 1992. By the 1996 election, an immigrant party led by the former Soviet dissident Natan Sharansky won seven seats in the Knesset and joined the coalition government dominated by the Likud. In the 2001 Israeli election there were several predominately Soviet Jewish parties competing for votes, with Sharansky's again receiving the most. A number of Soviet Jewish political leaders called for mass expulsion of Israeli Arabs and Palestinians from the West Bank and Gaza. In 2002, the government of Jordan sought reassurance that an attack on Iraq would not lead to mass deportation of Palestinians. Polls revealed growing support for 'transfer', the Israeli euphemism for ethnic cleansing of Palestinian Arabs.

The Israeli case illustrates in the extreme the potential impact of an immigrant voting bloc upon electoral outcomes. Immigrants generally are not such an important factor as in Israel and immigrants do not necessarily vote in ethnic blocs. Yet immigration clearly is affecting electoral politics across Western democracies as growing numbers of aliens naturalize, and as immigrant-origin populations are mobilized to vote. In the 1996 referendum over the future of Quebec and the Canadian Federation, Quebec's immigrant voters overwhelmingly voted against the referendum and for maintenance of the status quo. They decisively affected the outcome, prompting angry anti-immigration remarks by Quebecois leaders. In the close 2002 German elections, the 350 000 Germans of Turkish background emerged as a potentially decisive voting bloc whose backing may have enabled the Social Democratic–Green coalition to scrape through to victory. Although only 1 per cent of the electorate in 2002, the Turkish-German voting bloc is expected to double in size by 2006. Naturalized German citizens from Eastern Europe, on the other hand, strongly favour conservative parties (Wûst, 2002).

The growing mass of immigrant voters has made many political parties and their leaders more sensitive to multicultural concerns and issues. In some instances, immigration policy debates have been influenced by electoral calculations. In general, political parties on the left side of the political spectrum appear to take the lead in appealing to immigrant voters and are rewarded for their efforts. Conservative parties often benefit electorally from anti-immigrant backlash. And a number of conservative parties have begun to compete in earnest for the immigrant-origin electorate, particularly in the UK and the USA. Following the 1996 US elections, some Repubicans felt that President Clinton and the Democrats had outmanoeuvred the Republicans by encouraging a naturalization campaign while several Republican presidential candidates embraced anti-immigrant positions. Subsequently, George W. Bush ran a campaign in 2000 that courted Hispanic voters and electoral concerns drove his immigration initiative towards Mexico in 2001 (see Box 1.2). The endorsement of multiculturalism by the conservative coalition which was in power in Australia from 1975 to 1982 also seemed to be connected with concern about the 'ethnic vote' (Castles *et al.*, 1992a). In several countries, new ethnic voting blocs have come into existence. This is seen as normal in some democracies, but as a problem in others.

Many naturalized immigrants do not register to vote nor exercise their voting rights. DeSipio suggests that claims that naturalized Americans participated more than other Americans in the 1920s and 1930s are not empirically grounded. His review of studies of naturalized American participation in the 1996 election led him to conclude that the naturalized participate less than other Americans, even with controls for socio-economic differences in place. Naturalized US citizens are more likely than the population as a whole to have socio-demographic characteristics associated with political marginalization. Moreover, he argues that the process of immigrant political adaptation to the USA is not a group process but a highly individual one. Participation of immigrants is shaped by class and education factors that shape participation in US politics in general (DeSipio, 2001).

Abstention rates are very high among Asian-American voters despite voter registration and participation drives aimed specifically at them. In 1992, only 350 000 of the estimated 2.9 million Asian-Americans living in California were registered to vote (Choo, 1992). The 33rd US Congressional district in California, downtown Los Angeles, was thought to be one of the districts with the highest number of non-citizens: 225 116 out of the 384 158 adults resident in 1992. Only 13 per cent of the adult population voted in the 1992 election. Twenty-eight per cent did not exercise their voting rights and 59 per cent were ineligible as non-citizens. This compared to a national average of 52 per cent of adults voting, 44 per cent not voting and 4 per cent ineligible to vote due to non-citizenship (*Washington Post*, 22 May, 1994). Abstention rates are also high for French citizens of immigrant background. In the first round of the 1997 legislative elections 32.1 per cent of the electorate abstained and 3.3 per cent cast blank ballots. Abstention rates were especially high in the Saint-Denis district which is heavily populated by immigrants.

The immigration trends of the last several decades have significantly affected electoral politics in many Western democracies. The ability of legal immigrants to naturalize, and eventually to vote, constitutes a major concern for any democracy. That immigrant political participation is viewed as legitimate and as an anticipated outcome demarcates the USA, Australian and Canadian experiences from those of many Western European nations. Political exclusion is inherent in the concept of temporary worker policy. That is one reason why such policies are commonplace in authoritarian and undemocratic settings, such as in the Arab monarchies of the Gulf. Post-Second World War Western European guestworker policies created a conundrum when unplanned, unforeseen mass settlement occurred. The guestworkers and their families could not be excluded from Western European democracies without grievous damage to the fabric of democracy.

The UK constitutes an exception to the Western European pattern in that most post-1945 immigrants – those from the Commonwealth up to 1971, and the Irish – entered with citizenship and voting rights. However, as previously suggested, this seemed to have little effect upon the socio-economic position of immigrants which was quite similar to that of guestworkers on the continent. Immigration became an object of political debate as early as 1958 when there were race riots in Notting Hill in London and other areas. In 1962, British immigration law was tightened up, setting a precedent for even more restrictive measures in the future. In the late 1960s, right-wing politicians such as Enoch Powell warned of looming racial conflict, of 'rivers of blood' on the horizon. Immigration became increasingly politicized in the mid-1970s. The National Front, a descendant of Sir Oswald Mosley's British Union of Fascists of the 1930s, played a key role in provoking immigration-related violence. British neo-fascists and leftists battled over immigration in the 1970s in much the same way that French neo-fascists and left-wing, pro-immigrant groups had confronted one another several years earlier (Reed, 1977).

The frequently violent clashes, which then were regarded as uncharacteristic of normally civil British politics, combined with the mounting numbers of immigrants to make immigration a key issue in the 1979 general election. Margaret Thatcher adroitly capitalized on the immigration backlash to deflate support for the National Front and to score a victory over the Labour Party, which was supported by most immigrant voters. A 1975 pamphlet published by the Community Relations Commission underscored the growing electoral significance of immigrant ethnic minorities, and even the Conservative Party began efforts to recruit ethnic minority members as a result (Layton-Henry, 1981).

In subsequent general elections and in local elections, black and Asian Briton participation became more conspicuous. In 1987, four black Britons were elected to Parliament and three of them joined a black member of the House of Lords to form a black parliamentary caucus styled upon the then 24-member US Congressional black caucus. Hundreds of black and Asian Britons were elected to positions in local government. In the industrial city of Birmingham, for example, the first Muslim municipal councillor (a Labourite, like most immigrant-origin elected officials) was elected in 1982. In 1983, two more Muslim Labourites were elected. By January 1987, there were 14 municipal councillors of immigrant origin, including six Muslims. Within the Birmingham Labour Party, in the Sparkbrook area, 600 of the 800 local party members were Muslims (Joly, 1988: 177-8). But, Studlar and Layton-Henry argue, this growing black and Asian participation and representation in British politics generally did not result in greater attention being paid to immigrant issues and grievances (Studlar and Layton-Henry, 1990: 288).

Part of the difficulty faced by immigrants in getting their concerns onto the Labour Party agenda stems from the necessity of defining group issues in terms of class. However, formation of an alternative immigrants' party is not a viable option. Hence, even in a Western European country where most immigrants are enfranchised, their participation and representation remains problematic. Immigrant-origin voters can significantly affect electoral outcomes in 30–60 of the UK's 650 parliamentary constituencies. These are located in cities.

In Australia, there has been considerable debate on the impact of post-war immigration on politics. Most observers argue that the effects have been very limited (Jupp *et al.*, 1989: 51). McAllister states that post-war immigration 'has not resulted in any discernible change in the overall pattern of voting behaviour. Despite large-scale immigration, social class, not birthplace, has remained the basis for divisions between political parties' (McAllister, 1988: 919). A study of the role of Italo-Australians in political life found that they did not have a high profile, and argued that this was typical for immigrants in Australia (Castles *et al.*, 1992a). Italians have not established their own political parties or trade unions, and neither have they gained significant representation in Parliament. On the other hand, there are a large number of local councillors and some mayors of Italian origin. Yet even at this level of government Italo-Australians are underrepresented in comparison with their share in the population. Despite this, the 'ethnic vote' appears to be an 'issue' which influences the behaviour of Australian political leaders. Not only do they make efforts to approach ethnic associations to mobilize electoral support, they also make concessions to ethnic needs and interests in their policies.

The explanation for this combination of a low political profile with fairly successful interest articulation seems to lie in the relative openness of the Australian political system for immigrant groups, at least at a superficial level. As a reaction to mass immigration since 1947, governments and trade unions combined to guarantee orderly industrial relations, which would not threaten the conditions of local workers. The policy of assimilation from the 1940s to the early 1970s provided civil rights and citizenship, which laid the groundwork for political integration. Multicultural policies after 1972 accepted the legitimacy of representation of special interests through ethnic community associations. Their leaderships were granted a recognized, if limited, political role in government consultative bodies. Thus immigrant groups obtained some political influence in ethnic affairs and welfare policies. Yet they were still far from gaining significant influence in central political and economic decision-making processes. The question is whether this situation will change as the mainly poorly-educated migrant generation leaves the centre stage and is replaced by a new self-confident second generation, which has passed through the Australian educational system (Castles *et al.*, 1992b: 125–39).

Migration and security

The duality of international migrants as political actors and as targets of politics is perhaps most vivid when broaching the increasingly important topic of migration and security, and particularly political terrorism. Involvement of a minority of immigrants in political violence contributed to international migration's new salience on post-Cold War security agendas. Skewed and insufficiently informed discussion of immigration and terrorism was only to be expected in the post-Cold War period and new measures and laws against terrorism eroded the legal status of aliens in several Western democracies, most notably Germany and the USA. Three cases can be summarized to elucidate why migration and security concerns have become so important.

It has long been known that Turks of Kurdish background were overrepresented in the ranks of Turkish emigrants. There were nearly 1 million Kurds in Europe in 2000 (Boulanger, 2000: 20). However, during the mass recruitment period, this seemed to be of little political consequence. In the 1980s Kurdish aspirations for independence or autonomy from Turkey galvanized (see Box 6.2). The PKK, or Kurdish Workers Party, emerged as an important force and began an armed insurrection against the Turkish Republic.

It is estimated that one-quarter to one-third of the over 2 million Turkish citizens resident in Germany are of Kurdish origin. Perhaps 50 000 of these sympathize with the PKK and up to 12 000 are active members of the party or its front organizations (Boulanger, 2000: 23). As argued above, it is not unusual for migrants to be actively engaged in homeland-oriented political parties. What is unsettling to German and Turkish authorities is that the PKK transformed Germany into a second front in its struggle and frequently struck at Turkish consulates, airlines and businesses in Germany and elsewhere in Western Europe. Moreover, Turkish repression of the PKK-led insurgency, which has taken tens of thousands of lives, has seriously complicated its diplomatic ties with EU member states. Turkish counter-insurgency measures have included mysterious death squads and the uprooting and forced relocation of millions of Kurdish civilians. This backdrop renders the PKK activities on German soil and German and Turkish counter-measures highly emotive and significant. Indeed, the PKK became one of the primary German national security concerns by the mid-1990s, particularly after the PKK leader Öcalan threatened to send Hamas-style suicide bombers against German targets in retaliation for German cooperation with Turkey in its war against the PKK.

Despite the decision to outlaw the PKK and its front organizations, the PKK possesses an extensive organizational infrastructure in Germany and nearby European states. PKK tactics have featured protest marches and hunger strikes. German authorities have routinely banned street demonstrations on Kurdish and Turkish issues. Nonetheless, many have been staged and they frequently result in violent clashes. Formerly, participation in such events was legally only a minor offence and resident aliens apprehended at them did not become subject to deportation. In 1996, the German government sought to strengthen its ban on PKK street protests by making participation in banned events a major offence. Several Kurdish protesters were subsequently apprehended and recommended for deportation, at a time when hunger strikes in Turkish prisons had cost the lives of numerous prisoners and torture and ill-treatment of prisoners appeared to be commonplace. Hence, deportation of Kurdish activists raised important legal and human rights issues which polarized German public opinion.

The arrest of Abdallah Öcalan by Turkish authorities in 1999 sparked a massive wave of Kurdish protests in Europe and as far away as Australia. Three Kurds were killed after trying to enter the Israeli consulate in Berlin and scores of protesters were injured. During his subsequent trial, Öcalan called upon his followers to abandon armed struggle. This resulted in a large decrease in Kurdish militant activities on German soil but the unresolved Kurdish question remained and with it the potential for renewed conflict.

Such concerns undoubtedly contributed to Germany's opposition in 2002 to a US-led attack upon Iraq. German–US divergence over Iraq damaged overall diplomatic relations between long-time allies. This dramatic turn of events had much to do with differing German and US perspectives on the migration and security nexus in the Middle East. With an estimated 3.3 million Muslims in a total population of 83 million in Germany, only about 1 per cent of Muslims were estimated to be political extremists (Johnson, 2002).

Across the Rhine, French authorities grappled with the spillover of the fundamentalist insurgency in Algeria to French soil. An offshoot of the Islamic Salvation Front, the Armed Islamic Group, pursued an insurgency against the Algerian government, which is dominated by the military. As seen in Chapters 1 and 6, tens of thousands have died in a merciless war of terrorism and counter-terrorism in Algeria. France has given military and economic support to the Algerian government which became the pretext for the extension of Armed Islamic Group operations to French soil. A network of militants waged a bombing campaign, principally in the Paris region in 1995, before being dismantled. In late 1996, the Armed Islamic Group was thought to have been behind another fatal bombing, although no group took responsibility for the attack.

French authorities undertook numerous steps to prevent bombings and to capture the bombers. Persons of North African appearance were routinely subjected to identity checks. These checks were accepted by most French citizens and resident aliens of North African background as a necessary inconvenience. Indeed, information supplied by such individuals greatly aided in the dismantling of the terrorist group, several of whom were killed in shoot-outs with French police. From time to time, French police have rounded up scores of suspected sympathizers with the Armed Islamic Group.

Several guerrillas involved in attacks against hotels in Morocco, designed to disrupt the economically-important tourism industry so despised by some Islamic militants, were French citizens of North African background. They had been recruited into a fundamentalist network in the Parisian suburbs and their involvement was deeply disturbing to the French population, including most of the Islamic community. All available evidence suggests that mobilization of North African-origin citizens and permanent residents of France into violence-prone Islamicist organizations is extremely rare, but this particular incident stoked concern over the potential for that to happen more frequently.

Such fears appeared warranted in the aftermath of 11 September 2001. Scores of Armed Islamic Group and Al-Qaida-linked individuals, mainly of North African background, were detained for involvement in various plots, including one to attack the US embassy in Paris. Several of those arrested were French citizens of North African background including Zacarias Moussaoui, who was accused of plotting with the perpetrators of thc 11 September attacks. At least one French citizen of North African background died during the Allied military campaign against the Taliban and Al-Qaida in Afghanistan. Algerians and other individuals of North African Muslim background with links to the Armed Islamic Group figured prominently in the hundreds of arrests in the transatlantic area. The anti-Western resentment of some of those arrested was linked to perceived injustices endured by migrants and their families and their exclusion.

The USA has also received large inflows of Islamic immigrants. The total Muslim population of the USA in 2002 was estimated at 3 million, of whom 2 million were immigrants (Pipes and Durán, 2002: 1). Most Muslim immigrants have arrived legally, but many others have come as asylum seekers or illegally. Islam is one of the fastest growing religions in the USA. The Iranian-origin population alone, for instance, was estimated to number over 1 million by one Iranian-American organization (*The New York Times*, 29 June 1995). Major influxes of Palestinians have occurred since 1967 which were connected to the turmoil and difficult living conditions faced by Palestinians in the Middle East.

Middle East-origin immigrants were principally involved in the 1993 bombing of the World Trade Centre. That attack led to the adoption of a new counter-terrorism law in 1996 with significant implications for the rights of immigrants. The new law empowered the government to expedite detention and removal of aliens without customary judicial review. Fund-raising activities by terrorist organizations and their fronts were also targeted. A major target of the 1996 law was Hamas, a Palestinian Islamist movement which had emerged during the Intifada of the 1980s. Israeli authorities complained that the USA had become a major base of operations and fund-raising for Hamas.

In late 1999, the capture of Ahmed Ressam and the discovery of a plot to bomb Los Angeles airport indicated that the Armed Islamic Group from Algeria had extended its range of activities to North America, putting the USA onto a state of alert. A 2002 study of 48 Muslim radicals arrested since 1993, including the perpetrators of the 11 September attacks, indicated that one-third of them were tourists in the USA, another third were lawful permanent resident aliens, a quarter were illegal residents and three were asylum seekers (Camarota, 2002).

In the wake of those attacks, hundreds of immigrants, mainly of Middle Eastern background, were detained by US authorities. The legality of many of the arrests was challenged. In some instances, charges were dropped and detainees released. But critics charged that there was widespread violation of the civil rights of detainees. Middle East-background applicants for visas were subjected to additional scrutiny and Middle East students were systematically questioned by federal authorities.

These three cases illustrate why migration increasingly is viewed as germane to national security policies in the transatlantic area. However, as Myron Weiner in particular has demonstrated, concern over the security implications of international migration has been a significant factor around the world (Weiner, 1993). A common thread in transatlantic regional responses to immigrant violence has been increased repression and erosion of the legal status of resident aliens. Too often there has been conflation of immigration with terrorism when, in fact, very few immigrants have been involved in it.

Organized, politically-motivated violence against immigrants and foreign-born populations is distressingly commonplace. This is why Turkey, for instance, declares the security of its citizens abroad to be a top foreign policy objective. The vulnerability of migrants to political violence appears to have grown in the post-Cold War period. Yet discussions of political terrorism and counter-terrorism strategy pay little heed to anti-immigrant violence.

The risk of overreaction to violence by a handful of immigrants appears quite high. Without minimizing the importance of the integration barriers faced by Islamic-origin immigrants in Western democracies, the overall pattern is one of integration and incorporation. The results of several surveys of France's Muslim population, citizen and non-citizen, confirm this (Tribalat, 1995). The most important survey found that persons of Algerian Muslim background are quite secular in orientation. The Moroccan-origin population is considerably more religiously oriented. However, like Catholics, Protestants and Jews in France, only a minority practise their faith regularly. There is considerable dating and socializing with non-Muslims, and intermarriage is not uncommon. The surveys confirm the importance of unemployment and educational problems faced by France's Muslims, but the unmistakable overall thrust of the findings is that integration is occurring.

Integration, of course, provides security for immigrants and the host population. It is the decisive long-term factor affecting migration and security. The risk taken in measures like the 1996 counterterrorism legislation and those adopted after 11 September 2001 is that they will adversely affect integration without appreciably enhancing deterrence of terrorism. As argued elsewhere in this book, policies that exclude immigrants by denying them equal rights risk generating conflict over the long run. Hence, the inescapable conclusion is that democracies have a vital security interest in immigrant rights.

The growing, if belated, interest in migration and security involves a paradox. Many states have a profoundly important security interest in effective implementation of their immigration laws and regulations. Yet, all too often, insufficient appropriations are made or too few personnel engaged to enforce laws and regulations effectively. Stated or proclaimed policies are often undermined by funding or staffing decisions or undercut by political pressure to protect certain constituents, clients or (especially in the USA) contributors to election campaigns. One result is a credibility gap which potentially endangers legally-admitted aliens who have an extraordinary stake in well-managed policies which ensure that immigration and settlement are consented to and are legitimate.

Thus far, there is little evidence that racism and violence against immigrants have provoked violent counter-measures by immigrants. However, the potential for racism and anti-immigrant violence begetting terrorism in response should not be underestimated, once again underscoring everyone's stake in vigilance against violent opponents of immigrants.

Conclusion

International migration has played a major role in fostering multicultural politics. Migration can dramatically affect electorates, as witnessed in the Israeli case, and immigrants can influence politics through non-electoral means as well. Immigrants have fostered transnational politics linking homeland and host-society political systems in fundamental ways. Migrants and minorities are both subjects and objects of politics. An anti-immigrant backlash has strengthened the appeal of right-wing parties in Western Europe. One way in which migration has fundamentally altered the Western European political landscape is through the constitution of increasingly vocal Islamic organizations, which present a dilemma for democratic political systems: refusal to accept their role would violate democratic principles, yet many people see their aims and methods as intrinsically anti-democratic. International migration has fostered new constituencies, new parties and new issues. Many of Western Europe's newer political parties, such as the FN in France, feature anti-immigrant themes. Violence against immigrants is also a factor in ethnic minority formation and political mobilization.

In the USA, Canada and Australia immigrant political participation and representation is less of a problem, partly because of the preponderance of family-based legal immigration. However, disenfranchisement of legally-resident aliens and illegally-resident aliens in major US cities increasingly troubles authorities. Much of New York's population cannot vote, either because they are not naturalized or because they are illegally resident. Virtually everywhere, international migration renders politics more complex. Ethnic mobilization and the ethnic vote are becoming important issues in many countries. Another new issue may be seen in the politics of naturalization. One or two decades ago, virtually no one knew naturalization law or considered it important. The changing nature of international migration and its politicization has changed that. Most democracies now face a long-term problem stemming from growing populations of resident aliens who are unable or unwilling to naturalize; the status of illegal immigrants is particularly problematic (Rubio-Marin, 2000).

Immigrant politics are in a continual state of flux, because of the rapid changes in migratory flows as well as the broader transformations in political patterns which are taking place in many Western societies. As migratory movements mature – moving through the stages of immigration, settlement and minority formation – the character of political mobilization and participation changes. There is a shift from concern with homeland politics to mobilization around the interests of ethnic groups in the immigration country. If political participation is denied through refusal of citizenship and failure to provide channels of representation, immigrant politics is likely to take on militant forms. This applies particularly to the children of immigrants born in the countries of immigration. If they are excluded from political life through non-citizenship, social marginalization or racism, they are likely to present a major challenge to existing political structures in the future.

Guide to further reading

Miller (1981) provides one of the first comparative studies on the political role of migrant workers. Layton-Henry (1990) looks at the political rights of migrant workers in Western Europe. The *International Migration Review* Special Issue 19:3 (1985) gives information on the political situation in several countries. The comparative studies on citizenship by Brubaker (1989) and Hammar (1990) are of great value. Solomos and Wrench (1993) examine racism and migration in several European counties, while Layton-Henry and Rich (1986) focus on Britain. Shain (1989) looks at the consequences of refugee movements for the nation-state.

More recent important books include Ireland (1994) on immigrant politics in France and Switzerland. Feldblum (1999), Geddes and Favell (1999) and Joppke (1999) examine dimensions of migrant politics, citizenship matters and implications of migration for political identities and for the state. Gerstle and Mollenkopf (2001) provide contemporary and historical reflections on immigrant political incorporation in the USA while Rubio-Marin (2000) explores the challenge posed by migration, and particularly illegal migration, to democratic institutions in the USA and Germany. Koslowski (2000) offers seminal insight into the implications of international population movements for the study of international relations, a theme also explored in Miller (1998). Cohen and Layton-Henry (1997) provide an invaluable collection of key contributions to the study of the politics of migration in the *International Library of Studies of Migration* series.

Chapter 12

Conclusion: Migration in the Post-Cold War Era

This book has argued that international migration is a constant, not an aberration, in human history. Population movements have always accompanied demographic growth, technological change, political conflict and warfare. Over the last five centuries, mass migrations have played a major role in colonialism, industrialization, the emergence of nation-states and the development of the capitalist world market. However, international migration has never been as pervasive, or as socio-economically and politically significant, as it is today. Never before have politicians accorded such priority to migration concerns. Never before has international migration seemed so pertinent to national security and so connected to conflict and disorder on a global scale. The violence of 11 September 2001, which US Secretary of State Powell claimed marked the beginning of a new era in international relations, bore mute testimony to these central trends.

The hallmark of the age of migration is the global character of international migration: the way it affects more and more countries and regions, and its linkages with complex processes affecting the entire world. This book has endeavoured to elucidate the principal causes, processes and effects of international migration. Contemporary patterns, as discussed in Chapters 4, 6 and 7 are rooted in historical relationships and shaped by a multitude of political, demographic, socio-economic, geographical and cultural factors. These flows result in greater ethnic diversity within countries and deepening transnational linkages between states and societies. International migrations are greatly affected by governmental policies and may, in fact, be precipitated by decisions to recruit foreign workers or to admit refugees.

Yet, international migrations may also possess a relative autonomy and be impervious to governmental policies. As we have seen, official policies often fail to achieve their objectives, and may even bring about the opposite of what is intended. People as well as governments shape international migration. Decisions made by individuals, families and communities – often with imperfect information and constrained options – play a vital role in determining migration and settlement. The social networks which arise through the migratory process help shape long-term outcomes. The agents and brokers who make up the burgeoning 'migration industry' may have their own interests and aims. Despite the growth in migratory movements, and the strength of the factors which cause them, resistance to migration is also of growing importance. Large sections of the populations of receiving countries may oppose immigration. Governments sometimes react by adopting strategies of denial, hoping that the problems will go away if they are ignored. In other instances, mass deportations and repatriations have been carried out. Governments vary greatly in their capacities to regulate international migration and in the credibility of their efforts to regulate unauthorized migration.

In Chapter 2 we provided some theoretical perspectives on the reasons why international migrations take place and discussed how they often lead to permanent settlement and the formation of distinct ethnic groups in the receiving societies. We suggested that the migratory process needs to be understood in its totality as a complex system of social interactions with a wide range of institutional structures and informal networks in both sending and receiving countries, and at the international level. In a democratic setting, legal admission of migrants will almost always result in some settlement, even when migrants are admitted temporarily.

The comparison of two very different immigration countries – Australia and Germany – in Chapter 9 showed how the migratory process takes on its own dynamics, sometimes leading to consequences unforeseen and unwanted by policy-makers. Acceptance of the seeming inevitability of permanent settlement and formation of ethnic groups is the necessary starting point for any meaningful consideration of desirable public policies. The key to adaptive policy-making in this realm (as in others) is an understanding of the causes and dynamics of international migration. Policies based on misunderstanding or mere wishful thinking are virtually condemned to fail. Hence, if governments decide to admit foreign workers, they should from the outset make provision for the legal settlement of that proportion of the entrants that is almost sure to remain permanently: a consideration that needs to be taken to heart by the governments of countries as diverse as Japan, Malaysia, Italy and Greece at present.

Today the governments and peoples of immigration countries have to face up to some very serious dilemmas. The answers they choose will help shape the future of their societies, as well as their relations with the poorer countries of the South. Central issues include:

- regulating legal immigration and integrating settlers;
- policies to cope with illegal migration;
- finding 'durable solutions' to emigration pressure through improved international relations;
- the role of ethnic diversity in social and cultural change, and the consequences for the nation-state.

Legal migration and integration

Virtually all democratic states, and many not so democratic states as well, have growing foreign populations. As shown in Chapters 4, 6 and 7, the presence of these immigrants is generally due to conscious labour recruitment or immigration policies, or to the existence of various linkages between sending and receiving countries. In some cases (notably the USA, Canada and Australia) policies of large-scale immigration still exist. Invariably they are selective: economic migrants, family members and refugees are admitted according to certain quotas which are politically determined.

*There is considerable evidence that planned and controlled entries are conducive to acceptable social conditions for migrants as well as to relative social peace between migrants and local people. Countries with immigration quota systems generally decide on them through political processes which permit public discussion and the balancing of the interests of different social groups. Participation in decision-making increases the acceptability of immigration programmes. At the same time this approach facilitates the introduction of measures to prevent discrimination and exploitation of immigrants, and to provide social services to support successful settlement. There is therefore a strong case for advocating that all countries which actually continue to have immigration should move towards planned immigration policies.

As Chapter 10 showed, governmental obligations towards immigrant populations are shaped by the nature of the political system in the host society, as well as the mode of entry of the newcomers. Governments possess an internationally recognized right to regulate entry of aliens, a right that may be voluntarily limited through governmental signature of bilateral or multilateral agreements (for example, in the case of refugees). Clearly it makes a difference whether an alien has arrived on a territory through legal means of entry or not. In principle, the proper course for action with regard to legally-admitted foreign residents in a democracy is straightforward. They should be rapidly afforded equality of socio-economic rights and a large measure of political freedom, for their status would otherwise diminish the quality of democratic life in the society. However, this principle is frequently ignored in practice. As Chapters 5, 8 and 11 showed, unauthorized immigration and employment makes immigrants especially vulnerable to exploitation. The perceived illegitimacy of their presence can foster conflict and anti-immigrant violence.

Guestworker policy-style restrictions on the employment and residential mobility of legally-admitted aliens appear anachronistic and, in the long run, administratively unfeasible. They are difficult to reconcile with prevailing market principles, to say nothing of democratic norms. The same goes for restrictions on political rights. Freedom of speech, association and assembly should be unquestionable. Under normal circumstances of international cooperation, there is no reason to restrict recent immigrants' ability to participate in their homelands' political system. The only restriction on the rights of legally-admitted aliens which seems compatible with democratic principles is the reservation of the right to vote and to stand for public office to citizens. This is only justifiable if resident aliens are given the real opportunity of naturalization, without daunting procedures or high fees. But, even then, some foreign residents are likely to decide not to become citizens for various reasons. A democratic system needs to secure their political participation, too. This can mean setting up special representative bodies for resident non-citizens, or extending voting rights to non-citizens who fulfil certain criteria of length of stay (as in Sweden and the Netherlands).

The global character of international migration results in the intermingling and cohabitation of people from increasingly different physical and cultural settings. The severity of integration problems is highly variable. In some instances, public authorities may regard it as unnecessary to devise policies to facilitate integration. In most cases, however, specific and selective measures are necessary in order to forestall the development of socio-economic, cultural or political cleavages which can be conflictual. Here the multicultural models developed in Australia, Sweden and Canada deserve careful scrutiny. Their common thread is an acceptance of the cultural diversity and social changes wrought by immigration. Immigrants are not forced to conform to a dominant cultural or linguistic model but instead can maintain their native languages and cultural life if they choose to do so. The diversity produced by immigration is seen as an enrichment rather than as a threat to the predominant culture.

A multicultural approach enhances democratic life in that it allows for choice. It can mean a redefinition of citizenship to include cultural rights, along with the widely accepted civil, political and social rights. With the passage of time, it is expected that most immigrants and their offspring will reconcile their cultural heritage with the prevailing culture, and the latter will be somewhat altered, and most likely richer, for that. Conflict will have been minimized. Nonetheless, multicultural models have their contradictions and are often the subject of heated debate and renegotiation. For instance, multiculturalism may proclaim the right to use of an immigrant's mother tongue, but those who do not learn the dominant language can find themselves disadvantaged in the labour market. Maintenance of some cultural norms may be a form of discriminatory social control, particularly for women and youth. The dividing line between real participation of ethnic leaderships in decision-making and their cooperation through the patronage of government agencies can be very thin.

Many people argue that an explicitly multicultural approach to immigrant integration may not be appropriate for all societies. This applies particularly in states like the USA and France, with established traditions of immigrant incorporation and assimilation. In these countries it is believed that political integration through citizenship provides the essential precondition for social and cultural integration. Special cultural or social policies for immigrants are thought to perpetuate distinctions and lead to formation of ghettos. Thus the basic principle of the US model is a policy of benign (but not unthinking) neglect in the public sphere and a reliance on the integrative potential of the private sphere: families, neighbourhoods, ethnic solidarity organizations and so on. Yet the formation of racial or ethnic cleavages in the USA and the growth of ethnic conflict in France present a challenge to such approaches at the present time.

However, whether the choice is made for benign neglect or for more explicitly multicultural policies, certain preconditions must be met if marginalization and isolation of minorities are to be avoided. The state needs to take measures to ensure that there is no long-term link between ethnic origin and socio-economic disadvantage. This requires legal measures to combat discrimination, social policies to alleviate existing disadvantage and educational measures to ensure equal opportunities and to provide the chance of upward mobility. The state also has the task of eliminating racism, combating racial violence and, above all, of dealing with organized racist groups. As pointed out in Chapter 10, racist discrimination and violence are major factors leading to formation of ethnic minorities in all the countries of immigration examined. There is clearly a great need for action in this area.

Regulating 'unwanted' immigration

Prospects are slim for significantly increased legal immigration flows to Western democracies over the short to medium term. In Germany and similar countries some observers have suggested a need for increased immigration to compensate for low birth rates and an ageing population: foreign workers might provide the labour for age-care and other services as well as the construction industry. But immigration cannot effectively counteract the demographic ageing of Western societies unless it is substantially increased. Political constraints will not permit this. There will be some room for highly-skilled labour, family reunification and refugees, but not for a resumption of massive recruitment of foreign labour for low-level jobs. Most industrial democracies will have to struggle to provide adequate employment for existing populations of low-skilled citizen and resident alien workers. The generally adverse labour market situation will make any recruitment of foreign labour politically controversial.

One of the most pressing challenges for many countries today is therefore to find ways of coping with 'unwanted' migratory flows.

'Unwanted immigration' is a somewhat vague blanket term, which embraces:

- illegal border-crossers;
- legal entrants who overstay their entry visas or who work without permission;
- family members of migrant workers, prevented from entering legally by restrictions on family reunion;
- asylum seekers not regarded as genuine refugees.

Most such migrants come from poor countries and seek employment, but generally lack work qualifications. They compete with disadvantaged local people for unskilled jobs, and for housing and social amenities. Many regions throughout the world have had an enormous increase in such 'unwanted immigration' in the last 30 years or so. Of course, the migration is not always as 'unwanted' as is made out, employers often benefit from cheap workers who lack rights, and some governments tacitly permit such movements, but 'unwanted immigration' is often seen as being at the root of public fears of mass influxes. It is therefore a catalyst for racism and is at the centre of extreme-right agitation.

Stopping 'unwanted immigration' is increasingly regarded by governments as essential for safeguarding social peace. In Western Europe, the result has been a series of agreements designed to secure international cooperation in stopping illegal entries, and to speed up the processing of applications for asylum (see Chapters 4, 5 and 6). In the USA, Canada and Australia, measures have also been taken to improve border control and to speed up refugee determination. Several African and Asian countries have carried out quite draconian measures, such as mass expulsions of foreign workers (for example, Nigeria, Libya, Malaysia, USA), building fences and walls along borders (South Africa, Israel, Malaysia), severe punishments for illegal entrants (corporal punishment in Singapore; imprisonment or a bar on future admission in many countries) and sanctions against employers (South Africa, Japan and other countries). In addition, non-official punishments such as beatings by police are routinely meted out in Italy, some Arab countries and elsewhere. As seen in Chapter 5, the effectiveness of these measures is hard to assess; however, unauthorized migration clearly remains a concern almost everywhere.

The difficulty in achieving effective control is not hard to understand. Barriers to mobility contradict the powerful forces of globalization which are leading towards greater economic and cultural interchange. In an increasingly international economy, it is difficult to open borders for movements of information, commodities and capital and yet close them to people. Global circulation of investment and know-how always means movements of people too. Moreover, flows of highly-skilled personnel tend to encourage flows of less-skilled workers too. The instruments of border surveillance cannot be sufficiently fine-tuned to let through all those whose presence is wanted, but to stop all those who are not.

The matter is further complicated by a number of factors: the eagerness of employers to hire foreign workers (whether documented or not) for menial jobs, when nationals are unwilling to take such positions; the difficulty of adjudicating asylum claims and of distinguishing economically-motivated migrants from those deserving of refugee status; and the inadequacies or insufficiencies of immigration law. The weakening of organized labour and declining trade union membership in many Western democracies has also tended to increase unauthorized foreign employment. Similarly, policies aimed at reducing labour market rigidities and enhancing competitiveness may result in expanded employer hiring of unauthorized foreign workers. Social welfare policies also have unintended consequences, making employment of unauthorized alien workers more propitious.

Thus, despite the claimed desire of governments to stop illegal migration, many of the causes are to be found in the political and social structures of the immigration countries, and their relations with less-developed areas. This has led to the call for 'durable solutions' to address the root causes of mass migration. But such measures are not likely to bring a quick reduction in 'unwanted immigration'. In the current political climate there is no doubt that receiving countries will continue to regulate migration and attempt to curb illegal immigration. This will require investing more personnel and budgetary resources into enforcement of employer sanctions and adjudication of asylum claims than in the past. Enforcement of immigration laws will probably be accorded higher priority in the future, if only because of growing apprehension over the possible political consequences of continuing illegal migration and implications for security. How successful such measures can be remains to be seen.

Durable solutions and international relations

Clearly, international migration is not the solution to the North–South gap. Migration will not resolve North Africa's unemployment problem, or appreciably reduce the income and wage gap between the USA and Mexico, or make a significant impact on rural poverty in India. The only realistic long-term hope for reduction of international migration is broad-based, sustainable development in the less-developed countries, enabling economic growth to keep pace with growth in the population and labour force. Growing realization in the highly-developed countries that border control alone cannot stop 'unwanted immigration' has led to a discussion on 'durable solutions', to achieve a long-term reduction of migration pressures. Such measures are wide-ranging and closely linked to the debate on development strategies for the countries of the South. They include trade policy, development assistance, regional integration and international relations.

Reform of trade policies could help encourage economic growth in the less-developed countries. The most important issue is the level of prices for primary commodities as compared with industrial products. This is linked to constraints on world trade through tariffs and subsidies. The conclusion of the GATT Uruguay Round and the formation of the WTO in 1994–5 may help to improve the trade perspectives of less-developed countries, though it is hard to be certain as the USA, Japan and the EU insisted on safeguards for their own primary producers. Reform of the European Common Agricultural Policy could bring important benefits for less-developed countries. But trade policies generally operate within tight political constraints: few politicians are willing to confront their own farmers, workers or industrialists, particularly in times of economic recession. Reforms favourable to the economies of the less-developed countries will only come gradually, if at all.

Development assistance is a second strategy which might help to reduce 'unwanted' migration over the long term. Some states have good records in this respect, but international assistance generally has not been at a level sufficient to make a real impression on problems of underdevelopment. Indeed, the balance of nearly five decades of development policies is not a positive one. Although some countries have managed to achieve substantial growth, in general the gap between the poor and rich countries has grown. Income distribution within the countries of the South has also become more inequitable, increasing the gulf between the wealthy elites and the impoverished masses. The problems of rapid demographic growth, economic stagnation, ecological degradation, weak states and abuse of human rights still affect many countries of Africa, Asia and Latin America. Moreover, control of world finance by bodies such as the International Monetary Fund and the World Bank has led to credit policies which have increased the dependency and instability of many countries of the South.

Regional integration – the creation of free-trade areas and regional political communities – is sometimes seen as a way of diminishing 'unwanted' migration by reducing trade barriers and spurring economic growth, as well as by legalizing international movement of labour. But successful regional integration usually takes place between states which share political and cultural values and which resemble one another economically. Consequently, as seen in Chapter 5, the world's most successful regional integration unit, the EU, has witnessed low labour mobility between member-states.

Perhaps the shock of 11 September 2001 and UN Secretary General Annan's subsequent call for a global response to it through lessening of global disparities will provide the motivation to achieve real change through international cooperation. This would mean restricting the international weapons trade, altering the terms of trade between the North and the South, and changing world financial systems so that they encourage a real transfer of resources from the rich to the poor countries, rather than the other way round, as at present. It would also mean basing development assistance programmes on criteria of human rights, environmental protection, ecological sustainability and social equity.

Yet, however successful such policies might be – and sadly they seem fairly utopian in the light of the current world disorder – they will not bring about substantial reductions of international migration in the short term. As shown in Chapters 3, 6 and 7, the initial effect of development and integration into the world market is to increase migration from less-developed countries. This is because the early stages of development lead to rural–urban migration, and to acquisition by many people of the financial and cultural resources needed for international migration. The 'migration transition' – through which emigration ceases, and is eventually replaced by immigration – requires specific demographic and economic conditions, which may take generations to develop. Neither restrictive measures nor development strategies can totally curb international migration, at least in the short term, because there are such powerful forces stimulating population movement. These include the increasing pervasiveness of a global culture and the growth of cross-border movements of ideas, capital, commodities and people. The world community will have to learn to live with mass population movements for the foreseeable future.

Ethnic diversity, social change and the nation-state

The age of migration has already changed the world and many of its societies. Most highly-developed countries and many less-developed ones have become far more culturally diverse than they were even a generation ago. A large proportion, indeed the majority, of nation-states must face up to the reality of societal pluralism. In fact, few modern nations have ever been ethnically homogeneous. However, the nationalism of the last two centuries strove to create myths of homogeneity. In its extreme forms, nationalism even tried to bring about such homogeneity through expulsion of minorities and genocide. The appalling spectacle of 'ethnic cleansing' in the ruins of former Yugoslavia shows that such tendencies still exist. But the reality for most countries today is that they have to contend with a new type of pluralism, and that – even if migration were to stop tomorrow – this will affect their societies for generations.

One reason why immigration and the emergence of new ethnic groups have had such an impact in most highly-developed countries is that these trends have coincided with the crisis of modernity and the transition to post-industrial societies. The labour migration of the pre-1973 period appeared at the time to be reinforcing the economic dominance of the old industrial nations. Today we can interpret it as part of a process of capital accumulation which preceded a seminal change in the world economy. Growing international mobility of capital, the electronic revolution, the decline of old industrial areas and the rise of new ones are all factors which have led to rapid economic change in Western Europe, North America and Australia. The erosion of the old blue-collar working class and the increased polarization of the labour force have led to a social crisis in which immigrants find themselves doubly at risk: many of them suffer unemployment and social marginalization, yet at the same time they are often portrayed as the cause of the problems. That is why the emergence of the 'two-thirds society', in which the top strata are affluent while the bottom third is socially excluded, is everywhere accompanied by ghettoization of the disadvantaged and the rise of racism.

Nowhere is this more evident than in the global cities of the early twenty-first century. Los Angeles, Toronto, Paris, London, Berlin and Sydney – to name just a few – are crucibles of social change, political conflict and cultural innovation. They are marked by enormous gulfs: between the corporate elite and the informal sector workers who service them, between rich, well-guarded suburbs and decaying and crime-ridden inner cities, between citizens of democratic states and illegal non-citizens, between dominant cultures and minority cultures. The gulf may be summed up as that between inclusion and exclusion. The included are those who fit into the self-image of a prosperous, technologically-innovative and democratic society. The excluded are the shadow side: those who are needed to do the menial jobs in industry and the services, but who do not fit into the ideology of the model.

Both groups include nationals and immigrants, though the immigrants are more likely to belong to the excluded. But the groups are more closely bound together than they might like to think: the corporate elite need the illegal immigrants, the prosperous suburbanites need the slum-dwellers they find so threatening. It is out of this contradictory and multi-layered character of the post-modern city that its enormous energy, its cultural dynamism and its innovative capability emerge. But these coexist with potential for social breakdown, conflict, repression and violence. It is here that the complex social and cultural interaction between different ethnic groups may in future give birth to new forms of society.

The new ethnic diversity affects societies in many ways. Among the most important are issues of political participation, cultural pluralism and national identity. As Chapter 11 showed, immigration and formation of ethnic groups have already had major effects on politics in most developed countries. These effects are potentially destabilizing. The only resolution appears to lie in broadening political participation to embrace immigrant groups, which in turn may mean rethinking the form and content of citizenship, and decoupling it from ideas of ethnic homogeneity or cultural assimilation.

This leads on to the issue of cultural pluralism. Processes of marginalization and isolation of ethnic groups have gone so far in many countries that culture has become a marker for exclusion on the part of some sections of the majority population, and a mechanism of resistance by the minorities. Even if serious attempts were made to end all forms of discrimination and racism, cultural and linguistic difference will persist for generations, especially if new immigration takes place. That means that majority populations will have to learn to live with cultural pluralism, even if it means modifying their own expectations of acceptable standards of behaviour and social conformity.

This move towards cultural pluralism corresponds with the emergence of a global culture, which is fed by travel, mass media and commodification of cultural symbols, as well as by migration. This global culture is anything but homogeneous, but the universe of variations which it permits has a new meaning compared with traditional ethnic cultures: difference need no longer be a marker for strangeness and separation, but rather an opportunity for informed choice among a myriad of possibilities. The new global culture is therefore passionately syncretistic, permitting endless combinations of elements with diverse origins and meanings. The major obstacle to the spread of the global culture is that it coincides with a political, economic and social crisis in many regions. Where change is fast and threatening, narrow traditional cultures seem to offer a measure of defence. Hence the resurgence of exclusionary nationalism in areas like the former Soviet Union and Yugoslavia, which had been cut off so long from global influences that change, when it came, was experienced as a cataclysm. And hence the resurgence of racism in highly-developed societies among those groups who find themselves the main victims of economic and social restructuring.

Clearly, trends towards political inclusion of minorities and cultural pluralism can threaten national identity, especially in countries in which it has been constructed in exclusionary forms. If ideas of belonging to a nation have been based on myths of ethnic purity or of cultural superiority, then they really are threatened by the growth of ethnic diversity. Whether the community of the nation has been based on belonging to a *Volk* (as in Germany) or on a unitary culture (as in France), ethnic diversity inevitably requires major political and psychological adjustments. The shift is far smaller for countries that have seen themselves as nations of immigrants, for their political structures and models of citizenship are geared to incorporating newcomers. However, these countries, too, have historical traditions of racial exclusion and cultural homogenization which still need to be worked through. Assimilation of immigrants, as epitomized in the American Dream, seems less viable in view of continuing population movements and strong trends towards cultural and linguistic maintenance by ethnic communities.

That means that all countries of immigration are going to have to re-examine their understanding of what it means to belong to their societies. Monocultural and assimilationist models of national identity may no longer be adequate for the new situation. Immigrants may be able to make a special contribution to the development of new forms of identity. It is part of the migrant condition to develop multiple identities, which are linked to the cultures both of the homeland and of the country of origin. Such personal identities possess complex new transcultural elements.

Immigrants are not unique in this; multiple identities are becoming an almost general characteristic of the people of late modern societies. But it is above all migrants who are compelled by their situation to have multi-layered socio-cultural identities, which are constantly in a state of transition and renegotiation. Moreover, migrants frequently develop a consciousness of their transcultural position, which is reflected not only in their artistic and cultural work, but also in social and political action. Despite current conflicts about the effects of ethnic diversity on national cultures and identity, immigration does offer perspectives for change. The hope must be that new principles of identity can emerge, which will be neither exclusionary nor discriminatory, and which will provide the basis for better intergroup cooperation. Moreover, multilingual capabilities and intercultural understanding are beginning to be seen as important economic assets in the context of international trade and investment.

Inevitably transcultural identities will affect our fundamental political structures. The democratic nation-state is a fairly young political form, which came into being with the American and French revolutions and achieved global dominance in the nineteenth century. It is characterized by principles defining the relationship between people and government which are mediated through the institution of citizenship. The nation-state was an innovative and progressive force at its birth, because it was inclusive and defined the citizens as free political subjects, linked together through democratic structures. But the later nationalism of the nineteenth and twentieth centuries turned citizenship on its head by equating it with membership of a dominant ethnic group, defined on biological or cultural lines. In many cases the nation-state became an instrument of exclusion and repression.

National states, for better or worse, are likely to endure. But global economic and cultural integration and the establishment of regional agreements on economic and political cooperation are undermining the exclusiveness of national loyalties. The age of migration could be marked by the erosion of nationalism and the weakening of divisions between peoples. Admittedly there are countervailing tendencies, such as racism, or the resurgence of nationalism in certain areas. Coming transformations are likely to be uneven, and setbacks are possible, especially in the event of economic or political crises. But the inescapable central trends are the increasing ethnic and cultural diversity of most countries, the emergence of transnational networks which link the societies of emigration and immigration countries and the growth of cultural interchange. The globalization of migration provides grounds for optimism, because it does give some hope of increased unity in dealing with the pressing problems which beset our small planet.

Bibliography

Abella, M. I. (1995) 'Asian migrant and contract workers in the Middle East', in R. Cohen (ed.), *The Cambridge Survey of World Migration* (Cambridge: Cambridge University Press).

Abella, M. I. (2002) *Complexity and Diversity of Asian Migration* (Geneva: Unpublished manuscript).

ABS (1989) *Overseas Born Australians: A Statistical Profile* (Canberra: Australian Government Publishing Service).

Ackland, R. and Williams, L. (1992) *Immigrants and the Australian Labour Market: The Experience of Three Recessions* (Canberra: Australian Government Publishing Service).

Adepoju, A. (1988) 'Links between internal and international migration: the African situation', in C. Stahl (ed.), *International Migration Today*, Vol. 2 (UNESCO/University of Western Australia).

Adepoju, A. (2001) 'Regional integration, continuity and changing patterns of intra-regional migration in Sub-Saharan Africa', in M. A. B. Siddique (ed.), *International Migration into the 21st Century* (Cheltenham/Northampton, MA: Edward Elgar).

ADL (Anti-Defamation League) (1988) *Hate Groups in America* (New York: ADL of B'nai B'rith).

Alcorso, C., Popoli, C. and Rando, G. (1992) 'Community networks and institutions', in S. Castles, C. Alcorso, G. Rando and E. Vasta (eds), *Australia's Italians: Culture and Community in a Changing Society* (Sydney: Allen & Unwin).

Aleinikoff, T. A. and Klusmeyer, D. (eds) (2000) *From Migrant to Citizens: Membership in a Changing World* (Washington, DC: Carnegie Endowment for International Peace).

Aleinikoff, T. A. and Klusmeyer, D. (eds) (2001) *Citizenship Today: Global Perspectives and Practices* (Washington, DC: Carnegie Endowment for International Peace).

Alexandre, G. (2001) *La Question Migratoire entre la Republique Dominicaine et Haiti* (Dominican Republic: IOM).

Ålund, A. and Schierup, C.-U. (1991) *Paradoxes of Multiculturalism* (Aldershot: Avebury).

Amjad, R. (1996) 'Philippines and Indonesia: on the way to a migration transition', *Asian and Pacific Migration Journal*, 5: 2–3.

Anderson, B. (1983) *Imagined Communities* (London: Verso).

Andreas, P. (2001) 'The transformation of migrant smuggling across the US-Mexico border', in D. Kyle and R. Koslowski (eds) *Global Human Smuggling* (Baltimore, MD: The Johns Hopkins Press).

Anthias, F. and Yuval-Davis, N. (1989) 'Introduction', in N. Yuval-Davis and F. Anthias (eds), *Woman–Nation–State* (London: Macmillan).

Appenzeller, G. *et al.* (2001) 'Kardinal Sterzinsky in Gespraech: Die Union fragt nur, was tut uns Deutschen gut' (*Tagesspiegel*, 19 May 2001).

Appleyard, R. T. (ed.) (1988) *International Migration Today: Trends and Prospects* (Paris: UNESCO).

Appleyard, R. T. (1991) *International Migration: Challenge for the Nineties* (Geneva: IOM).

Appleyard, R. T. (ed.) (1998a) *Emigration Dynamics in Developing Countries*, Vol. I: *Sub-Saharan Africa* (Aldershot: Ashgate).

Appleyard, R. T. (ed.) (1998b) *Emigration Dynamics in Developing Countries*, Vol. II: *South Asia* (Aldershot: Ashgate).

Appleyard, R. T. (ed.) (1999a) *Emigration Dynamics in Developing Countries*, Vol. III: *Mexico, Central America and the Carribean* (Aldershot: Ashgate).

Appleyard, R. T. (ed.) (1999b) *Emigration Dynamics in Developing Countries*, Vol. IV: *The Arab Region* (Aldershot: Ashgate).

Archdeacon, T. (1983) *Becoming American: An Ethnic History* (New York: The Free Press).

Arnold, F., Minocha, U. and Fawcett, J. T. (1987) 'The changing face of Asian immigration to the United States', in J. T. Fawcett and B. V. Cariño (eds), *Pacific Bridges: The New Immigration from Asia and the Pacific Islands* (New York: Center for Migration Studies).

Aronson, G. (1990) *Israel, Palestinians and the Intifada: Creating Facts on the West Bank* (Washington, DC: Institute for Palestine Studies).

Bade, K. J. (ed.) (1994) *Das Manifest der 60: Deutschland und die Einwanderung* (Munich: Beck).

Baganha, M. (ed.) (1997) *Immigration in Southern Europe* (Oeiras: Celta Editora).

Baganha, M. and Reyneri, E. (2001) 'New migrants in South European countries: An assessment', in R. Leveau, C. Wihtol de Wenden and K. Mohsen-Finan (eds), *Nouvelles citoyennetés: refugiés et sans-papiers dans l'espace européen* (Paris: Institut français des relations internationales).

Baker, M. and Wooden, M. (1992) *Immigrant Workers in the Communication Industry* (BIR, Canberra: Australian Government Publishing Service).

Balibar, E. (1991) 'Racism and nationalism', in E. Balibar and I. Wallerstein (eds) *Race, Nation, Class: Ambiguous Identities* (London: Verso).

Balibar, E. and Wallerstein, I. (eds) (1991) *Race, Nation, Class: Ambiguous Identities* (London: Verso).

Baringhorst, S. (1995) 'Symbolic highlights or political enlightenment? Strategies for fighting racism in Germany', in A. Hargreaves and J. Leaman (eds), *Racism, Ethnicity and Politics in Contemporary Europe* (Aldershot: Edward Elgar).

Barlán, J. (1988) *A System Approach for Understanding International Population Movement: The Role of Policies and Migrant Community in the Southern Cone* (IUSSP Seminar, Genting Highlands, Malaysia) September 1988.

Bartram, D. (1999) *Foreign Labor and Political Economy in Israel and Japan* (Madison: Dissertation, Department of Sociology, University of Wisconsin).

Basch, L., Glick-Schiller, N. and Blanc, C. S. (1994) *Nations Unbound: Transnational Projects, Post-Colonial Predicaments and Deterritorialized Nation-States* (New York: Gordon and Breach).

Battistella, G. and Assis, M. M. B. (1998) *The Impact of the Crisis on Migration in Asia* (Quezon City, Philippines: Scalabrini Migration Center).

Bauböck, R. (1991) 'Migration and citizenship', *New Community*, 18: 1.

Bauböck, R. (1994) 'Changing the Boundaries of Citizenship: the inclusion of immigrants in democratic polities', in R. Bauböck (ed.), *From Aliens to Citizens* (Aldershot: Avebury).

Bauböck, R. and Rundell, J. (eds) (1998) *Blurred Boundaries: Migration, Ethnicity, Citizenship* (Aldershot: Ashgate).

Bauman, Z. (1998) *Globalization: the Human Consequences* (Cambridge: Polity).

BBC (http://news.bbc.co.uk/2/hi/asia-pacific/2163440.stm), 1 August 2002.

Beauftragte der Bundesregierung für Ausländerfragen (2000) *Daten und Fakten zur Ausländersituation* (Berlin: Beauftragte der Bundesregierung für Ausländerfragen).

Bedzir, B. (2001) 'Migration from Ukraine to Central and Eastern Europe', in C. Wallace and D. Stola (eds), *Patterns of Migration in Central Europe* (Basingstoke: Palgrave).

Belguendouz, A. (2001) 'La Dimension Migratoire Maroco–Hispano–Européenne: Quelle Cooperation?', in *A New Security Agenda for Future Regional Co--operation in the Mediterranean Region* (Rome: Fourth Mediterranean Dialogue International Research Seminar, Nato Defense College) November 2001.

Bell, D. (1975) 'Ethnicity and social change', in N. Glazer and D. P. Moynihan (eds), *Ethnicity – Theory and Experience* (Cambridge, MA: Harvard University Press).

Bensaâd, A. (2002) 'Voyage au bout de la peur avec les clandestins du Sahel', in *Le Monde Diplomatique Histoires d'Immigration*, 15–20.

Benyon, J. (1986) 'Spiral of decline: race and policing', in Z. Layton-Henry and P. B. Rich (eds), *Race, Government and Politics in Britain* (London: Macmillan).

Bernstein, A., Schlemmer, L. and Simkins, C. (1999) 'A proposed policy framework for controlling cross-border migration to South Africa,', in A. Bernstein and M. Weiner (eds) *Migration and Refugee Policies: An Overview* (London: Pinter).

Bernstein, A. and Weiner, M. (eds) (1999) *Migration and Refugee Policies: An Overview* (London: Pinter).

Binur, Y. (1990) *My Enemy, Myself* (New York: Penguin).

Birks, J. S., Sinclair, C. A. and Seccombe, I. J. (1986) 'Migrant Workers in the Arab Gulf: The Impact of Declining Oil Revenues', *International Migration Review*, 20: 4.

Birrell, R. (2001) 'Immigration on the Rise: The 2001–2002 Immigration Program', *People and Place*, 9(2).

Birrell, R. and Khoo, S.-E. (1995) *The Second Generation in Australia: Educational and Occupational Characteristics*, Statistical Report No. 14 (Canberra: AGPS).

Björgo, T. and Witte, R. (eds) (1993) *Racist Violence in Europe* (London: Macmillan).

Black, R. (1998) *Refugees, Environment and Development* (London: Longman).

Blackburn, R. (1988) *The Overthrow of Colonial Slavery 1776–1848* (London and New York: Verso).

Blustein, P. (2002) 'Bush shift on foreign aid strengthens US position at summit', *Washington Post*, 16 March 2002.

Böhning, W. R. (1984) *Studies in International Labour Migration* (London: Macmillan; New York: St Martin's Press).

Böhning, W. R. (1991a) 'Integration and immigration pressures in Western Europe', *International Labour Review*, 130: 4.

Böhning, W. R. (1991b) 'International migration to Western Europe: what to do?', paper presented to the Seminar on International Security (Geneva: Graduate Institute of International Studies, 15–20 July 1991).

Borjas, G. J. (1989) 'Economic theory and international migration', *International Migration Review*, Special Silver Anniversary Issue, 23: 3.

Borjas, G. J. (1990) *Friends or Strangers: The Impact of Immigration on the US Economy* (New York: Basic Books).

Borjas, G. J. (1999) *Heaven's Door* (Princeton: Princeton University Press).

Borjas, G. J. (ed.) (2000) *Issues in the Economics of Immigration* (Chicago: University of Chicago Press).

Boudahrain, A. (1985) *Nouvel Ordre Social International et Migrations* (Paris: L'Harmattan/CIEMI).

Boudahrain, A., (1991) 'The new international convention: a Moroccan perspective', *International Migration Review*, 25: 4.

Boulanger, P. (2000) 'Un regard français sur l'immigration Kurde en Europe', *Migrations Société*, 12: 72.

Bourdieu, P. and Wacquant, L. (1992) *An Invitation to Reflexive Sociology* (Chicago: University of Chicago Press).

Boyd, M. (1989) 'Family and personal networks in migration', *International Migration Review*, Special Silver Anniversary Issue, 23: 3.

Boyle, P., Halfacree, K. and Robinson, V. (1998) *Exploring Contemporary Migration* (Harlow, Essex: Longman).

Breton, R., Isajiw, W. W., Kalbach, W. E. and Reitz, J. G. (1990) *Ethnic Identity and Equality* (Toronto: University of Toronto Press).

Brettell, C. B. and Hollifield, J. F. (eds) (2000) *Migration Theory: Talking Across Disciplines* (New York and London: Routledge).

Briggs, V. M., Jr (1984) *Immigration Policy and the American Labor Force* (Baltimore, MD, and London: Johns Hopkins University Press).

Brubaker, W. R. (ed.) (1989) *Immigration and the Politics of Citizenship in Europe and North America* (Lanham, MD: University Press of America).

Bundesgrenzschutzamt (2001) *Fulfillment of the Border Police Tasks in the Federal Republic of Germany* (Report, personally communicated by Isabelle Vosswinkel, Polizeihauptkommissarin in BGS).

Cahill, D. (1990) *Intermarriages in International Contexts* (Quezon City: Scalabrini Migration Center).

Calavita, K. (2003) 'Italy: Immigration, Economic Flexibility and Policy Responses', in W. Cornelius, P. L. Martin and J. F. Hollifield (eds), *Controlling Immigration: A Global Perspective*, Second Edition (Stanford, CA: Stanford University Press).

Camarota, S. (2002) *The Open Door* (Washington, DC: Center for Immigration Studies).

Carnoy, M. (1994) *Faded Dreams: The Politics and Economics of Race in America* (Cambridge: Cambridge University Press).

Castells, M. (1996) *The Rise of the Network Society* (Oxford: Blackwells).

Castells, M. (1997) *The Power of Identity* (Oxford: Blackwells).

Castells, M. (1998) *End of Millennium* (Oxford: Blackwells).

Castles, S. (1986) 'The guest-worker in western Europe: an obituary', *International Migration Review*, 20: 4.

Castles, S. (1989) *Migrant Workers and the Transformation of Western Societies* (Ithaca, NY: Cornell University).

Castles, S. (1995) 'How nation-states respond to immigration and ethnic diversity', *New Community*, 21: 3.

Castles, S. (2000) 'The impacts of emigration on countries of origin', in S. Yusuf, W. Wu and S. Evenett (eds), *Local Dynamics in an Era of Globalization* (New York: Oxford University Press for the World Bank).

Castles, S. (2001) 'International migration and the nation-state in Asia', in M. A. B. Siddique (ed.), *International Migration in the 21st Century* (Cheltenham/Northampton, MA: Edward Elgar).

Castles, S., Alcorso, C., Rando, G. and Vasta, E. (eds) (1992b) *Australia's Italians: Culture and Community in a Changing Society* (Sydney: Allen & Unwin).

Castles, S., Booth, H. and Wallace, T. (1984) *Here for Good: Western Europe's New Ethnic Minorities* (London: Pluto Press).

Castles, S., Cope, B., Kalantzis, M. and Morrissey, M. (1992c) *Mistaken Identity – Multiculturalism and the Demise of Nationalism in Australia*, Third Edition (Sydney: Pluto Press).

Castles, S. and Davidson, A. (2000) *Citizenship and Migration: Globalisation and the Politics of Belonging* (London: Macmillan).

Castles, S., Foster, W., Iredale, R. and Withers, G. (1998) *Immigration and Australia: Myths and Realities* (St Leonards, NSW: Allen & Unwin).

Castles, S. and Kosack, G. (1973 and 1985) *Immigrant Workers and Class Structure in Western Europe* (Oxford: Oxford University Press).

Castles, S., Rando, G. and Vasta, E. (1992a) 'Italo-Australians and Politics', in S. Castles, C. Alcorso, G. Rando and E. Vasta (eds), *Australia's Italians: Culture and Community in a Changing Society* (Sydney: Allen & Unwin).

Castles, S. and Vasta, E. (2003) 'Australia: new conflicts around old dilemmas', in W. Cornelius, P. L. Martin and J. F. Hollifield (eds) *Controlling Immigration: A Global Perspective*, Second Edition (Stanford, CA: Stanford University Press).

CCCS (Centre for Contemporary Cultural Studies) (1982) *The Empire Strikes Back* (London: Hutchinson).

Cernea, M. M. and McDowell, C. (eds) (2000) *Risks and Reconstruction: Experiences of Resettlers and Refugees* (Washington, DC: World Bank).

Chazan, N. (1994) 'Engaging the state: Associational life in Sub-Saharan Africa', *State Power and Social Forces* (Cambridge: Cambridge University Press).

Chimni, B.S. (1998) 'The geo-politics of refugee studies: A view from the South', *Journal of Refugee Studies*, 11: 4.

Chin, K. (1999) *Smuggled Chinese: Clandestine Immigration to the United States* (Philadelphia: Temple University Press).

Chiswick, B. R. (2000) 'Are immigrants favorably self-selected? An economic analysis', in C B. Brettell and J. F. Hollifield (eds), *Migration Theory: Talking Across Disciplines* (New York and London: Routledge).

Choo, A. L. (1992) 'Asian-American political clout grows stronger', *Wall Street Journal*, 21 February 1992.

CIC (Citizenship and Immigration Canada) (2002) *Canada's Recent Immigrants: A Comparative Portrait Based on the 1996 Census* (Ottawa: CIC).

Cinanni, P. (1968) *Emigrazione e Imperialismo* (Rome: Riuniti).

Çinar, D. (1994) 'From aliens to citizens: a comparative analysis of rules of transition', in R. Bauböck (ed.), *From Aliens to Citizens* (Aldershot: Avebury).

Cohen, P. and Bains, H. S. (eds) (1988) *Multi-Racist Britain* (London: Macmillan).

Cohen, R. (1987) *The New Helots: Migrants in the International Division of Labour* (Aldershot: Avebury).

Cohen, R. (1991) 'East-West and European migration in a global context', *New Community*, 18: 1.

Cohen, R. (1995) 'Asian indentured and colonial migration', in R. Cohen (ed.), *The Cambridge Survey of World Migration* (Cambridge: Cambridge University Press).

Cohen. R. (ed.) (1995) *The Cambridge Survey of World Migration* (Cambridge: Cambridge University Press).

Cohen, R. (1997) *Global Diasporas: An Introduction* (London: UCL Press).

Cohen, R. and Kennedy, P. (2000) *Global Sociology (*Basingstoke: Palgrave).

Cohen, R. and Layton-Henry, Z. (eds) (1997) *The Politics of Migration* (Cheltenham/ Northampton, MA: Edward Elgar)

Cohn-Bendit, D. and Schmid, T. (1993) *Heimat Babylon: Das Wagnis der multikulturellen Demokratie* (Hamburg: Hoffmann & Campe).

Collins, J. (1978) 'Fragmentation of the working class', in E. L. Wheelwright and K. Buckley (eds), *Essays in the Political Economy of Australian Capitalism*, Vol. 3 (Sydney: ANZ Books).

Collins, J. (1991) *Migrant Hands in a Distant Land: Australia's Post-War Immigration*, Second Edition (Sydney: Pluto Press).

Collins, J. and Castles, S. (1991) 'Restructuring, migrant labour markets and small business in Australia', *Migration*, 10.

Collins, J., Gibson, K., Alcorso, C., Castles, S. and Tait, D. (1995) *A Shop Full of Dreams: Ethnic Small Business in Australia* (Sydney: Pluto Press).

Commission of the European Communities (1989) *Eurobarometer: Public Opinion in the European Community*. Special Issue: *Racism and Xenophobia* (Brussels: Commission of the European Community).

Commission of the European Communities (1990) *Policies on Immigration and the Social Integration of Migrants in the European Community* (Brussels: Commission of the European Community).

Committee for Economic Development (2001) *Reforming Immigration* (New York: Committee for Economic Development).

Cornelius, W. A. (2001) 'Death at the border: efficacy and unintended consequences of US immigration control policy', *Population and Development Review*, 27(4).

Cornelius, W., Martin, P. and Hollifield, J. (1994) *Controlling Immigration: A Global Perspective* (Stanford, CA: Stanford University Press).

Cornelius, W., Martin, P. and Hollifield, J. (eds) (2003) *Controlling Immigration: A Global Perspective*, Second Edition (Stanford, CA: Stanford University Press).

Costa-Lascoux, J. (1989) *De l'Immigré au Citoyen* (Paris: La Documentation Française).

Crock, M. and Saul, B. (2002) *Future Seekers: Refugees and the Law in Australia* (Sydney: Federation Press).

Croissandeau J. M. (1984) 'La formation alternative au chomage?', *Le Monde de l'Education*, February.

Cross, G. S. (1983) *Immigrant Workers in Industrial France: The Making of a New Laboring Class* (Philadelphia: Temple University Press).

Cross, M. (1995) 'Race, class formation and political interests: a comparison of Amsterdam and London', in A. Hargreaves and J. Leaman, *Racism, Ethnicity and Politics in Contemporary Europe* (Aldershot: Edward Elgar).

Cross, M. and Keith, M. (eds) (1993) *Racism, the City and the State* (London: Routledge).

Crossette, B. (2000) 'UN warns that trafficking in human beings is growing', *New York Times*, 25 June 2000.

Crossette, B. (2002a) 'Annan says terrorism roots are broader than poverty', *New York Times*, 7 March 2002.

Crossette, B. (2002b) 'UN coaxes out the wheres and whys of global immigration' *The New York Times*, 7 July.

CSIMCED (1990) *Unauthorized Migration: An Economic Development Response* (Washington, DC: US Government Printing Office).

Curtin, P. (1997) 'Africa and global patterns of migration', in W. Gungwu (ed), *Global History and Migrations* (Boulder: Westview).

Davidson, A. (1997) *From Subject to Citizen: Australian Citizenship in the Twentieth Century*, (Cambridge: Cambridge University Press).

Dávila, R. (1998) *The Case of Venezuela* (The Hague: UN Technical Symposium on International Migration and Development paper).

Davis, M. (1990) *City of Quartz: Excavating the Future in Los Angeles* (London: Verso).

de Lattes, A. and de Lattes, Z. (1991) 'International migration in Latin America: Patterns, implications and policies', Informal Expert Group Meeting on International Migration (Geneva: UN Economic Commission for Europe/UNPF paper).

de Lepervanche, M. (1975) 'Australian immigrants 1788–1940', in E. L. Wheelwright and K. Buckley (eds), *Essays in the Political Economy of Australian Capitalism*, Vol. 1 (Sydney: ANZ Books).

de Montvalon, J. B. (2002) 'Le gouvernement Raffarin 2 s'ouvre davantage à la société civile', *Le Monde*, 19 June 2002.

Decloîtres, R. (1967) *The Foreign Worker* (Paris: OECD).

Deng, F. (2001) *Report of the Seminar on Internal Displacement in Indonesia: Towards an Integrated Approach* (New York: UN Commission on Human Rights).

Derisbourg, J.P. (2002) 'L'Amérique latine entre Etats-Unis et Union européenne', *Politique Etrangère*, 67: 2.

DeSipio, L. (2001) 'Building America, one person at a time: Naturalization and political behavior of the naturalized in contemporary American politics', in G. Gerstle and J. Mollenkopf (eds), *E Pluribus Unum?* (New York: Russell Sage Foundation).

DeWind, J., Hirschman, C. and Kasinitz, P. (eds) (1997) *Immigrant Adaptation and Native-born Responses in the Making of Americans, International Migration Review* (Special Issue) Vol. 31 (New York: Center for Migration Studies).

Dieux, H. (2002) 'Transferts de pauvreté au Portugal' *Le Monde Diplomatique*, 5 July.
DIMA (Department of Immigration and Multicultural Affairs) (1999) *A New Agenda for Multicultural Australia* (Canberra: DIMA).
DIMIA (Department of Immigration and Multicultural Affairs and Indigenous Affairs) (2001) *Key Facts in Immigration* (Canberra: DIMIA).
Dohse, K. (1981) *Ausländische Arbeiter and bürgerliche Staat* (Konistein/Taunus: Hain).
Drbohlav, D. (2001) 'The Czech Republic', in C. Wallace and D. Stola (eds), *Patterns of Migration in Central Europe* (Basingstoke/New York: Palgrave).
Dubet, F. and Lapeyronnie, D. (1992) *Les Quartiers d'Exil* (Paris: Seuil).
El-Sohl, C. (1994) 'Calculating the risks of resettlement: Egyptian peasant families in Iraq', in S. Shami (ed.), *Population Displacement and Resettlement: Development and Conflict in the Middle East* (New York: Center for Migration Studies).
Engels, F. (1962) 'The condition of the working class in England', in *Marx, Engels on Britain* (Moscow: Foreign Languages Publishing House).
Entzinger, H. B. (1985) 'The Netherlands', in T. Hammar (ed.), *European Immigration Policy: A Comparative Study* (Cambridge: Cambridge University Press).
Entzinger, H. B. (2002) 'The rise and fall of multiculturalism: The case of the Netherlands', in C. Joppke and E. Morawaska (eds), *Towards Assimilation and Citizenship: Immigration in Liberal Nation-States* (London: Palgrave).
Esman, M. J. (1994) *Ethnic Politics* (Ithaca, NY, and London: Cornell University Press).
Essed, P. (1991) *Understanding Everyday Racism* (London and Newbury Park, New Delhi: Sage).
European Parliament (1985) *Committee of Inquiry into the Rise of Fascism and Racism in Europe: Report on the Findings of the Inquiry* (Strasbourg: European Parliament).
Evans, L. and Papps, I. (1999) 'Migration dynamics in the GCC countries', in R. Appleyard (ed.), *Emigration Dynamics in Developing Countries*, Vol. IV: *The Arab Region* (Aldershot: Ashgate).
Fadil, M. A. (1985) 'Les effets de l'émigration de main d'oeuvre sur la distribution des revenus et les modèles de consommation dans l'économie égyptienne', *Revue Tiers Monde*, 26: 103.
Faist, T. (2000) *The Volume and Dynamics of International Migration and Transnational Social Spaces* (Oxford: Oxford University Press).
Fakiolas, R. (2002) 'Greek migration and foreign immigration in Greece', in R. Rotte and P. Stein (eds), *Migration Policy and the Economy: International Experiences* (Munich: Hans Seidel Stiftung).
Farrag, M. (1999) 'Emigration dynamics in Egypt', in R. Appleyard (ed.), *Emigration Dynamics in Developing Countries*, Vol. IV: *The Arab Region* (Aldershot: Ashgate).
Fassman, H. and Münz, R. (1994) *European Migration in the Late Twentieth Century* (Laxenburg, Austria: International Institute for Applied System Analysis).

Favell, A. (1998) *Philosophies of Integration: Immigration and the Idea of Citizenship in France and Britain* (London: Macmillan).
Fawcett, J. T. and Arnold, F. (1987) 'Explaining diversity: Asian and Pacific immigration systems', in J. T. Fawcett and B. V. Cariño (eds), *Pacific Bridges: The New Immigration from Asia and the Pacific Islands* (New York: Center for Migration Studies).
Fawcett, J. T. and Cariño, B. V. (eds) (1987) *Pacific Bridges: The New Immigration from Asia and the Pacific Islands* (New York: Center for Migration Studies).
Feagin, J. R. (1989) *Racial and Ethnic Relations* (Englewood Cliffs, NJ: Prentice-Hall).
Feldblum, M. (1999) *Reconstructing Citizenship* (Albany, NY: State University of New York Press).
Fergany, N. (1985) 'Migrations inter-arabes et développement', *Revue Tiers Monde*, 26: 103.
Ferguson, E. (1997) 'Drowned in a sea of apathy', *Guardian Weekly*, 26 January 1997.
Findlay, S. (2001) 'Compelled to move: the rise of forced migration in Sub-Saharan Africa', in M. Siddique (ed.), *International Migration into the 21st Century* (Cheltenham/Northampton, MA: Edward Elgar).
Fishman, J. A. (1985) *The Rise and Fall of the Ethnic Revival: Perspectives on Language and Ethnicity* (Berlin, New York and Amsterdam: Mouton).
Fix, M. and Passel, J. S. (1994) *Immigration and Immigrants: Setting the Record Straight* (Washington, DC: The Urban Institute).
Foot, P. (1965) *Immigration and Race in British Politics* (Harmondsworth: Penguin).
Foster, W. (1996) *Immigration and the Australian Economy* (Canberra: DIMA).
Fox-Genovese, E. and Genovese, E. D. (1983) *Fruits of Merchant Capital: Slavery and Bourgeois Property in the Rise and Expansion of Capitalism* (New York and Oxford: Oxford University Press).
Fregosi, R. (2002) 'Au-delà de la crise financière et institutionnelle, l'Argentine en quête d'un véritable projet', *Politique Etrangère*, 67: 2.
French, H. W. (1990) 'Sugar harvest's bitter side: some call it slavery', *New York Times*, 27 April 1990.
French, H. W. (1991) 'Haitians expelled by Santo Domingo', *New York Times*, 11 August 1991.
Funcke, L. (1991) *Bericht der Beauftragten der Bundesregierung für die Integration der ausländischen Arbeitnehmer und ihrer familienangehörigen* (Bonn: German Government).
Gallagher, A. (2002) 'Trafficking, smuggling and human rights: tricks and treaties', *Forced Migration Review*, 12.
Gallagher, D. and Diller, J. M. (1990) *At the Crossroads between Uprooted People and Development in Central America* (Washington, DC: Commission for the Study of International Migration and Cooperative Economic Development, Working Paper No. 27).
Garrard, J. A. (1971) *The English and Immigration: A Comparative Study of the Jewish Influx 1880–1910* (Oxford: Oxford University Press).
Gaspard, F. (1990) *Une petite Ville en France* (Paris: Gallimard).

Geddes, A. (2000) *Immigration and European Integration: Towards Fortress Europe?* (Manchester and NY: Manchester University Press).

Geddes, A. and Favell, A. (1999) *The Politics of Belonging: Migrants and Minorities in Contemporary Europe* (Aldershot: Ashgate).

Geertz, C. (1963) *Old Societies and New States – The Quest for Modernity in Asia and Africa* (Glencoe, ILL: Free Press).

Gellner, E. (1983) *Nations and Nationalism* (Oxford: Blackwell).

Gerstle, G. and Mollenkopf, J. (eds) (2001) *E Pluribus Unum?* (New York: Russell Sage Foundation).

Gibney, M. J. (2000) *Outside the Protection of the Law: The Situation of Irregular Migrants in Europe* (Oxford: Refugee Studies Centre).

Gilroy, P. (1987) *There Ain't no Black in the Union Jack* (London: Hutchinson).

Gilroy, P. and Lawrence, E. (1988) 'Two-tone Britain: White and black youth and the politics of anti-racism', in P. Cohen and H. S. Bains (eds), *Multi-Racist Britain* (London: Macmillan).

Glazer, N. and Moynihan, D. P. (1975) 'Introduction', in N. Glazer and D. P. Moynihan (eds), *Ethnicity: Theory and Experience* (Cambridge, MA: Harvard University Press).

Glick-Schiller, N. (1999) 'Citizens in transnational nation-states: the Asian experience', in K. Olds, P. Dicken, P. F. Kelly, L. Kong, and H. W.-C. Yeung (eds), *Globalisation and the Asia-Pacific: Contested Territories* (London: Routledge).

Glover, S., Gott, C., Loizillon, A., Portes, J., Price, R., Spencer, S., Srinivasan, V. and Willis, C. (2001) *Migration: An Economic and Social Analysis* (London: Home Office).

Go, S. P. (1998) 'The Philippines: a look into the migration scenario in the nineties', in OECD, *Migration and Regional Economic Integration in Asia* (Paris: OECD).

Go, S. P. (2002) 'Detailed case study of the Philippines', in R. Iredale, C. Hawksley, and K. Lyon (eds), *Migration Research and Policy Landscape: Case Studies of Australia, the Philippines and Thailand* (Wollongong: Asia-Pacific Migration Research Network).

Goldberg, D. T. and Solomos, J. (eds) (2002) *A Companion to Racial and Ethnic Studies* (Malden, MA and Oxford: Blackwell).

Gonzalez, J. L. I. (1998) *Philippine Labour Migration: Critical Dimensions of Public Policy* (Singapore: Institute of South-East Asian Studies).

Guimezanes, N. (1995) 'Acquisition of nationality in OECD countries', in OECD, *Trends in International Migration: Annual Report* (Paris: OECD).

Gurr, T. R. and Harf, B. (1994) *Ethnic Conflict in World Politics* (Boulder, CO: Westview).

Gutmann, A. (ed.) (1994) *Multiculturalism: Examining the Politics of Recognition* (Princeton, NJ: Princeton University Press).

Habermas, J. (2001) *The Postnational Constellation* (Boston: MIT Press).

Halliday, F. (1985) 'Migrations de main d'oeuvre dans le monde arabe: l'envers du nouvel ordre économique', *Revue Tiers Monde*, 26: 103.

Hammar, T. (ed.) (1985a) *European Immigration Policy: A Comparative Study* (Cambridge: Cambridge University Press).

Hammar, T. (1985b) 'Sweden', in T. Hammar (ed.), *European Immigration Policy: A Comparative Study* (Cambridge: Cambridge University Press).

Hammar, T. (1990) *Democracy and the Nation-State: Aliens, Denizens and Citizens in a World of International Migration* (Aldershot: Avebury).

Hardt, M. and Negri, A. (2000) *Empire* (Cambridge, MA: Harvard University Press).

Hargreaves, A. and Leaman, J. (1995) *Racism, Ethnicity and Politics in Contemporary Europe* (Aldershot: Edward Elgar).

Harris, N. (1995) *The New Untouchables: Immigration and the New World Worker* (London: I.B. Tauris).

Harris, N. (1996) *The New Untouchables: Immigration and the New World Worker* (Harmondsworth: Penguin).

Hárs, A., Sik, E. and Toth, J. (2001) 'Hungary', in C. Wallace and D. Stola (eds) *Patterns of Migration in Central Europe* (Basingstoke/New York: Palgrave).

Häussermann, H. and Kazepov, Y. (1996) 'Urban poverty in Germany', in E. Mingione (ed.), *Urban Poverty and the Underclass* (Oxford: Blackwell).

HCI (1991) *Journal Official, Assemblée Nationale* (Paris: French Government, 7 June 1991).

Held, D., McGrew, A., Goldblatt, D. and Perraton, J. (1999) *Global Transformations: Politics, Economics and Culture* (Cambridge: Polity).

Hobsbawn, E. (1994) *Age of Extremes: The Short Twentieth Century, 1914–1991* (London: Michael Joseph)

Hoffmann, L. (1990) *Die unvollendete Republik* (Cologne: Pappy Rossa Verlag).

Hollifield, J. F. (2000) 'The politics of international migration: how can we "bring the state back in"?', in C. B. Brettell and J. F. Hollifield (eds), *Migration Theory: Talking Across Disciplines* (New York and London: Routledge).

Hollifield, J. F. (2003) 'France', in W. Cornelius, P. L. Martin and J. F. Hollifield (eds), *Controlling Immigration: A Global Perspective* (Stanford, CA: Stanford University Press).

Home Office (1989) *The Response to Racial Attacks and Harassment: Guidance for the Statutory Authorities – Report of the Inter-Departmental Racial Attacks Group* (London: Home Office).

Home Office (2000) *Statistics on Race and the Criminal Justice System* (London: Home Office)

Homze, E. L. (1967) *Foreign Labor in Nazi Germany* (Englewood Cliffs, NJ: Princeton University Press).

Hönekopp, E. (1999) *Central and East Europeans in the Member Countries of the European Union since 1990: Development and Structure of Migration, Population and Employment* (Munich: Institute for Employment Research).

Horowitz, D. and Noiriel, G. (1992) *Immigrants in Two Democracies: French and American Experience* (New York: New York University Press).

Houstoun, M. F., Kramer, R. G. and Barrett, J. M. (1984) 'Female predominance in immigration to the United States since 1930: A first look', *International Migration Review*, 18: 4.

HREOC (Human Rights and Equal Opportunity Commission) (1991) *Racist Violence: Report of the National Inquiry into Racist Violence in Australia* (Canberra: Australian Government Publishing Service).

Hugo, G. (1993) *The Economic Implications of Emigration from Australia* (Canberra: Australian Government Publishing Service).
Hugo, G. (1994) 'Migration and the family', *Occasional Papers Series for the International Year of the Family: 12* (Vienna: United Nations).
Hulse, C. V. (2002) 'Gephardt is preparing a measure to legalize illegal immigrants', *New York Times*, 23 July 2002.
Hunter, S. (2002) *Islam, Europe's Second Religion* (New York: Praeger).
Ignatieff, M. (1994) *Blood and Belonging: Journeys into the New Nationalism* (New York: Vintage).
INS (2002a) *Statistical Yearbook of the Immigration and Naturalization Service, 1999* (Washington, DC: US Government Printing Office).
INS (2002b) *Statistical Yearbook of the Immigration and Naturalization Service, 2000* (Washington, DC: US Government Printing Office) (http://www.ins.usdoj/gov/graphics/aboutins/statistics/yearbook2000.pdf).
International Migration Review (1985) Special Issue on Civil Rights and the Sociopolitical Participation of Migrants, 19: 3.
International Migration Review (1989) Special Silver Anniversary Issue, 23: 3
International Migration Review (1992) Special Issue on the New Europe and International Migration, 26: 2.
IOM (1999) *Trafficking in Migrants* (Geneva: IOM Policy and Responses).
IOM (2000a) *Migrant Trafficking and Human Smuggling in Europe* (Geneva: IOM).
IOM (2000b) *World Migration Report 2000* (Geneva: IOM).
Ireland, P. (1994) *The Policy Challenge of Ethnic Diversity* (Cambridge, MA: Harvard University Press).
ISMU-Cariplo (2001) *Sesto Rapporto sulle Migrazioni in Italia* (Milan: Franco Angeli).
Istat (2001) *Annuario Statistico Italiano* (Rome: Istituto Poligrafico dello Stato).
Jackson, J. A. (1963) *The Irish in Britain* (London: Routledge and Kegan Paul).
Jakubowicz, A. (1989) 'The state and the welfare of immigrants in Australia', *Ethnic and Racial Studies*, 12: 1.
Johnson, I. (2002) 'Muslim extremism perplexes Germany on eve of elections', *The Wall Street Journal*, 20 September 2002.
Johnson, I. and Gugath, B. (2002) 'Turkish voters are transforming political landscape in Germany', *The Wall Street Journal*, 29 September 2002.
Joly, D. (1988) 'Les musulmans à Birmingham', in R. Leveau and G. Kepel (eds), *Les Musulmans dans la Société Française* (Paris: Presse de la Fondation Nationale des Sciences Politiques).
Jones, S. (2000) *Making Money Off Migrants: The Indonesian Exodus to Malaysia* (Hong Kong and Wollongong: Asia 2000 Foundation and Centre for Asia-Pacific Social Transformation Studies).
Joppke, C. (1999) *Immigration and the Nation-State* (Oxford: Oxford University Press).
Jupp, J. (ed.) (2001) *The Australian People: An Encyclopedia of the Nation, its People and their Origins* (Cambridge: Cambridge University Press).
Jupp, J., York, B. and McRobbie, A. (1989) *The Political Participation of Ethnic Minorities in Australia* (Canberra: Australian Government Publishing Service).

Kaldor, M. (2001) *New and Old Wars: Organized Violence in a Global Era*, Second Edition (Cambridge: Polity).

Kassim, A. (1998) 'The case of a new receiving country in the developing world: Malaysia', in *United Nations Technical Symposium on International Migration and Development* (The Hague: Unpublished paper).

Kastoryano, R. (1996) *La France, l'Allemagne et leurs immigrés: négocier l'identité* (Paris: Armand Colin).

Kay, D. and Miles, R. (1992) *Refugees or Migrant Workers? European Volunteer Workers in Britain 1946–1951* (London: Routledge).

Keeley, C.B. (2001) 'The international refugee regimes(s): the end of the Cold War matters', *International Migration Review*, 35: 1.

Kepel, G. and Leveau, R. (1987) *Les Banlieues d'Islam* (Paris: Seuil).

Kerr, M. and Yassin, E. S. (1982) *Rich and Poor States in the Middle East: Egypt and the New Arab Order* (Boulder, CO: Westview Press).

Kharoufi, M. (1994) 'Forced migration in the Senegalese Mauritanian conflict: The consequences for the Senegal River Valley', in S. Shami (ed.), *Population Displacement and Resettlement* (New York: Center of Migration Studies).

Kindleberger, C. P. (1967) *Europe's Postwar Growth: The Role of Labor Supply* (Cambridge, MA: Harvard University Press).

King, R. (2000) 'Southern Europe in the changing global map of migration', in R. King, G. Lazaridis and C. Tsardanidis (eds), *Eldorado or Fortress? Migration in Southern Europe* (London: Macmillan).

King, R. (ed.) (2001) *The Mediterranean Passage: Migration and New Cultural Encounters in Southern Europe* (Liverpool: Liverpool University Press).

King, R., Lazaridis, G. and Tsardanidis, C. (eds) (2000) *Eldorado or Fortress? Migration in Southern Europe* (London: Macmillan).

Kiser, G. and Kiser, M. (eds) (1979) *Mexican Workers in the United States* (Albuquerque: University of New Mexico Press).

Klekowski Von Koppenfels, A. (2001) *The Role of Regional Consultative Processes in Managing International Migration* (Geneva: IOM).

Klug, F. (1989) ' "Oh to be in England": the British case study', in N. Yuval-Davis and F. Anthias (eds), *Woman–Nation–State* (London: Macmillan).

Komai, H. (1995) *Migrant Workers in Japan* (London: Kegan Paul International).

Kondo, A. (2001) 'Citizenship rights for aliens in Japan', in A. Kondo (ed.), *Citizenship in a Global World* (Basingstoke: Palgrave).

Koopmans, R. and Statham, P. (2000) *Challenging Immigration and Ethnic Relations Politics* (Oxford: Oxford University Press).

Kop, Y. and Litan, R. E. (2002) *Sticking Together: The Israeli Experiment in Pluralism* (Washington, DC: The Brookings Institute).

Koslowski, R. (2000) *Migrants and Citizens* (Ithaca, NY: Cornell University Press).

Kramer, R. (1999) *Developments in International Migration to the United States* (Washington, DC: Department of Labor).

Kratochwil, H. K. (1995) 'Cross-border population movements and regional economic integration in Latin America', *IOM Latin America Migration Journal*, 13: 2.

Kritz, M. (2001) 'Population growth and international migration: is there a link?', in A. Zolberg and P. Benda (eds) *Global Migrants, Global Refugees* (New York/ Oxford: Berghahn Books).

Kritz, M. M., Keely, C. B. and Tomasi, S. M. (eds) (1983) *Global Trends in Migration* (New York: Center for Migration Studies).
Kritz, M. M., Lin, L. L. and Zlotnik, H. (eds) (1992) *International Migration Systems: A Global Approach* (Oxford: Clarendon Press).
Kubat, D. (1987) 'Asian immigrants to Canada', in J. T. Fawcett and B. V. Cariño (eds), *Pacific Bridges: The New Immigration from Asia and the Pacific Islands* (New York: Center for Migration Studies).
Kyle, D. and Koslowski, R. (2001) *Global Human Smuggling* (Baltimore and London: Johns Hopkins University Press).
Kymlicka, W. (1995) *Multicultural Citizenship* (Oxford: Clarendon Press).
La lettre de la citoyenneté (2001) Number 53, September–October 2001.
Laacher, S. (2002) 'Comment les "papiers" peuvent changer la vie', in *Le Monde Diplomatique, Histoires d'Immigration.*
Lapeyronnie, D., Frybes, M., Couper, K. and Joly, D. (1990) *L'Intégration des Minorités Immigrées, Étude Comparative: France – Grande Bretagne* (Paris: Agence pour le Développement des Relations Interculturelles).
Laqueur, W. (1972) *A History of Zionism* (New York: Holt, Rinehart & Winston).
Larsson, S. (1991) 'Swedish racism: the democratic way', *Race and Class*, 32: 3.
Layton-Henry, Z. (1981) *A Report on British Immigration Policy since 1945* (Coventry: University of Warwick).
Layton-Henry, Z. (1986) 'Race and the Thatcher Government', in Z. Layton-Henry and P. B. Rich (eds), *Race, Government and Politics in Britain* (London: Macmillan).
Layton-Henry, Z. (ed.) (1990) *The Political Rights of Migrant Workers in Western Europe* (London: Sage).
Layton-Henry, Z. (2003) 'Britain: from immigration control to migration management', in W. Cornelius, P. L. Martin and J. F. Hollifield (eds), *Controlling Immigration: A Global Perspective*, Second Edition (Stanford, CA: Stanford University Press).
Layton-Henry, Z. and Rich, P. B. (eds) (1986) *Race, Government and Politics in Britain* (London: Macmillan).
Lazcko, F. (2001) 'Irregular migration in Central and Eastern Europe: An overview of return policies and procedures', in C. Wallace and D. Stola (eds), *Patterns of Migration in Central Europe* (Basingstoke: Palgrave).
Lebon, A. (2000) *Immigration et presence étrangère en France en 1999* (Paris: La documentation française).
Lee, J. S. and Wang, S.-W. (1996) 'Recruiting and managing of foreign workers in Taiwan', *Asian and Pacific Migration Journal*, 5: 2–3.
Leggewie, C. (1990) *Multi Kulti: Spielregeln für die Vielvölkerrepublik* (Berlin: Rotbuch).
Lequin, Y. (ed.), (1988) *La Mosaïque France* (Paris: Larousse).
Leveau, R., Wihtol de Wenden, C. and Mohsen-Finan, K. (eds) (2001) *Nouvelles citoyennetés: Réfugiés et sans-papiers dans l'espace européen* (Paris: Institut français des relations internationales).
Lever-Tracy, C. and Quinlan, M. (1988) *A Divided Working Class* (London: Routledge).

Lidgard, J. M. (1996) 'East Asian migration to Aotearoa/New Zealand: Perspectives of some new arrivals', *Population Studies Centre Discussion Papers: 12* (Hamilton: University of Waikato).
Light, I. and Bonacich, E. (1988) *Immigrant Entrepreneurs* (Berkeley, CA: University of California Press).
Lim, L. L. (1996) 'The migration transition in Malaysia', *Asian and Pacific Migration Journal*, 5: 2–3.
Lim, L. L. and Oishi, N. (1996) 'International labor migration of Asian women', *Asian and Pacific Migration Journal*, 5: 1.
Lloyd, C. (1991) 'Concepts, models and anti-racist strategies in Britain and France', *New Community*, 18: 1.
Lluch, V. (2002) 'Apartheid sous plastique à El Ejido', *Le Monde Diplomatique, Histoires d'Immigration*, 85–89.
Loescher, G. (2001) *The UNHCR and World Politics: A Perilous Path* (Oxford: Oxford University Press).
Lohrmann, R. (1987) 'Irregular migration: A rising issue in developing countries', *International Migration*, 25: 3.
Lopez-Garcia, B. (2001) 'La régularisation des Maghrébins sans papiers en Maroc', in R. Leveau, C. Wihtol de Wenden and K. Mohsen-Finan (eds), *Nouvelles cityoyennetés: Réfugées et sans-papiers dans l'espace européen* (Paris: IFRI).
Loutete-Dangui, N. (1988) 'L'immigration étrangère au Congo', in Association Internationale des Démographes de Langue Française, *Les Migrations Internationales* (Paris: Edition de PINED).
Lucassen, J. (1995) 'Emigration to the Dutch colonies and the USA', in R. Cohen (ed.), *The Cambridge Survey of World Migration* (Cambridge: Cambridge University Press).
Luso-American Development Foundation (1999) *Metropolis International Workshop Preceedings* (Lisbon: Luso-America Development Foundation).
Lutz, H., Phoenix, A. and Yuval-Davis, N. (1995) 'Introduction: nationalism, racism and gender', in H. Lutz, A. Phoenix and N. Yuval-Davis (eds), *Crossfires: Nationalism, Racism and Gender in Europe* (London: Pluto Press).
MacMaster, N. (1991) 'The "seuil de tolérance": the uses of a "scientific" racist concept', in M. Silverman (ed.), *Race, Discourse and Power in France* (Aldershot: Avebury).
Maguid, A. (1993) 'The importance of systematizing migration information for making policy: recent initiatives and possibilities for Latin America and the Caribbean', *IOM Latin America Migration Journal*, 11: 3.
Mann, J. A. (1979) 'For Millions in Colombia, Venezuela is El Dorado', *New York Times*, 23 December 1979.
Marcus, J. (1995) *The National Front and French Politics* (New York: New York University Press).
Mares, P. (2001) *Borderline: Australia's Treatment of Refugees and Asylum Seekers* (Sydney: UNSW Press).
Marie, Claude-Valentin (2000) 'Measures taken to combat the employment of undocumented foreign workers in France', in OECD, *Combating the Illegal Employment of Foreign Workers* (Paris: OECD).

Marshall, T. H. (1964) 'Citizenship and social class', in *Class, Citizenship and Social Development: Essays by T. H. Marshall* (New York: Anchor Books).
Martin, P. L. (1991) *The Unfinished Story. Turkish Labour Migration to Western Europe* (Geneva: ILO).
Martin, P. L. (1993) *Trade and Migration: NAFTA and Agriculture* (Washington, DC: Institute for International Economics).
Martin, P. L. (1996) 'Labor contractors: a conceptual overview', *Asian and Pacific Migration Journal*, 5: 2–3.
Martin, P. L. (2002) *Immigration and the Changing Face of Rural and Agricultural America* (Washington, DC: Urban Institute, unpublished paper).
Martin, P. L., Mason, A. and Nagayama, T. (1996) 'Introduction to special issue on the dynamics of labor migration in Asia', *Asian and Pacific Migration Journal*, 5: 2–3.
Martin, P. L. and Miller, M. J. (2000a) 'Smuggling and trafficking: A conference report', *International Migration Review*, 34: 3.
Martin, P. L. and Miller, M. J. (2000b) *Employer Sanctions: French, German and US Experiences* (Geneva: ILO).
Martin, P. L. and Taylor, J. E. (2001) 'Managing migration: the role of economic policies', in A. Zolberg and P. Benda, *Global Migrants, Global Refugees* (New York: Berghahn Books).
Martin, P. L. and Widgren, J. (1996) 'International migration: a global challenge', *Population Bulletin*, 51: 1.
Martinez, J. N. (1989) 'Social effects of labour migration: the Colombian experience', *International Migration*, 27: 2.
Martiniello, M. (1994) 'Citizenship of the European Union: a critical view', in R. Bauböck (ed.), *From Aliens to Citizens* (Aldershot: Avebury).
Massey, D. S., Alarcón, R., Durand, J. and González (1987) *Return to Aztlan – The Social Process of International Migration from Western Mexico* (Berkeley: University of California Press).
Massey, D. S., Arango, J., Hugo, G. and Taylor, J. E. (1993) 'Theories of international migration: a review and appraisal', *Population and Development Review*, 19.
Massey, D. S., Arango, J., Hugo, G. and Taylor, J. E. (1994) 'An evaluation of international migration theory: the North American case', *Population and Development Review*, 20.
Massey, D. S., Arango, J., Hugo, G., Kouaouci, A., Pellegrino, A. and Taylor, J. E. (1998) *Worlds in Motion, Understanding International Migration at the End of the Millennium* (Oxford: Clarendon Press).
McAllister, I. (1988) 'Political attitudes and electoral behaviour', in J. Jupp (ed.), *The Australian People: An Encyclopedia of the Nation, its People and their Origins* (Sydney: Angus & Robertson).
McCarthy, J. (1995) *Death and Exile: The Ethnic Cleansing of Ottoman Muslims 1821–1922* (Princeton: Darwin Press).
Meissner, D., Papademetriou, D. and North, D. (1987) *Legalization of Undocumented Aliens: Lessons from Other Countries* (Washington, DC: Carnegie Endowment for International Peace).
Messina, A. (1989) 'Anti-immigrant illiberalism and the "new" ethnic and racial minorities in Western Europe', *Patterns of Prejudice*, 23: 3.

Messina, A. (1996) 'The not so silent revolution: postwar migration to Western Europe,' *World Politics*, 49.

Messina, A. (ed.) (2002) *West European Immigration and Immigrant Policy in the New Century* (Westport, CT, and London: Praeger)

Miles, R. (1989) *Racism* (London: Routledge).

Miller, A. (2002) *Leaving the Land of the Setting Sun: Transnationalization of Algerian Terrorism and Transatlantic Security* (Madison, WI: University of Wisconsin, Senior Honors thesis).

Miller, J. (1985) 'Wave of Arab migration ending with oil boom', *New York Times*, 6 October 1985.

Miller, J. (1991) 'Egyptians now replace other Arabs in Saudi jobs', *New York Times*, 4 February 1991.

Miller, M. J. (1978) *The Problem of Foreign Worker Participation and Representation in France, Switzerland and the Federal Republic of Germany* (Madison, WI: University of Wisconsin)

Miller, M. J. (1981) *Foreign Workers in Western Europe: An Emerging Political Force* (New York: Praeger).

Miller, M. J. (1984) 'Industrial policy and the rights of labor: the case of foreign workers in the French automobile assembly industry', *Michigan Yearbook of International Legal Studies*, vi.

Miller, M. J. (1986) 'Policy ad-hocracy: the paucity of coordinated perspectives and policies', *The Annals*, 485.

Miller, M. J. (1989) 'Continuities and Discontinuities in Immigration Reform in Industrial Democracies', in H. Entzinger and J. Carter (eds), *International Review of Comparative Public Policy*, Vol. 1 (Greenwich, CT and London: JAI Press).

Miller, M. J. (1991) 'La nouvelle loi américaine sur l'immigration: vers un modèle d'après-guerre froide', *Revue Européenne des Migrations Internationales*, 7:3.

Miller, M. J. (1994) 'Towards understanding state capacity to prevent unwanted migration: employer *sancions enforcement* in France, 1975–1990', *Western European Politics*, 17:2.

Miller, M. J. (1998) 'International migration in post-Cold War international relations', in Pontifical Council for the Pastoral Care of Migrants and Itinerant People (ed.), *Migration at the Threshold of the Third Millenium* (The Vatican: Pontifical Council for the Pastoral Care of Migrants and Itinerant People).

Miller, M. J. (1999) 'Prevention of unauthorized migration', in A. Bernstein and M. Weiner (eds), *Migration and Refugee Policies: An Overview* (London and New York: Pinter).

Miller, M. J. (2002) 'Continuity and change in postwar French legalization policy', in A. Messina (ed.), *West European Immigration and Immigrant Policy in the New Century* (Westport, CT and London: Praeger).

Miller, M. J. and Martin, P. (1996) 'Prospects for cooperative management of international migration in the 21st century', *Asian and Pacific Migration Journal*, 5:2–3.

Milward, A. (1992) *The European Rescue of the Nation State* (London: Routledge).

Mingione, E. (1996) *Urban Poverty and the Underclass* (Oxford: Blackwell).

Mingione, E. and Qassoli, F. (2000) 'The participation of immigrants in the underground economy in italy', in R. King, G. Lazaridis and C. Tsardanidis (eds), *Eldorado or Fortress? Migration in Southern Europe* (Basingstoke: Macmillan).

Mitchell, C. (1989) 'International migration, international relations and foreign policy', *International Migration Review*, Special Silver Anniversary Issue, 23: 3.

Mitchell, C. (1992) *Western Hemisphere Immigration and United States Foreign Policy* (University Park, PA: The Penn State University Press).

Moch, L. P. (1992) *Moving Europeans: Migration in Western Europe Since 1650* (Bloomington, IN: Indiana University Press).

Moch, L. P. (1995) 'Moving Europeans: historical migration practices in Western Europe', in R. Cohen (ed.), *The Cambridge Survey of World Migration* (Cambridge: Cambridge University Press).

Mohsen-Finan, K. (ed.) (2002) *L'Algerie: Une improbable sortie de crise?* (Paris: Institut Français des Relations internationales).

Montwieler, N. H. (1987) *The Immigration Reform Law of 1986* (Washington, DC: The Bureau of National Affairs Inc.).

Mori, H. (1997) *Immigration Policy and Foreign Workers in Japan* (London: Macmillan).

Morokvasic, M. (1984) 'Birds of passage are also women', *International Migration Review*, 18: 4.

Morrison, J. (1998) *The Cost of Survival: The Trafficking of Refugees to the UK* (London: British Refugee Council).

Mosse, G. C. (1985) *Towards the Final Solution* (Madison: University of Wisconsin Press).

Münz, R. (1996) 'A continent of migration: European mass migration in the twentieth century', *New Community*, 22: 2.

Muus, P. J. (1991) *Migration, Minorities and Policy in the Netherlands. Recent Trends and Developments – Report for SOPEMI* (Amsterdam: University of Amsterdam, Department of Human Geography).

Muus, P. J. (1995) *Migration, Immigrants and Policy in the Netherlands – Report for SOPEMI 1995* (Amsterdam: Centre for Migration Research).

Myers, N. and Kent, J. (1995) *Environmental Exodus: An Emergent Crisis in the Global Arena* (Washington, DC: Climate Institute).

Nieves, E. (2002) 'Illegal immigrant death rates rises sharply in barren areas', *The New York Times*, 6 August 2002.

Nirumand, B. (ed.) (1992) *Angst vor den Deutschen: Terror gegen Ausländer und der Zerfall des Rechtstaates* (Reinbek bei Hamburg: Rowohlt).

NMAC (National Multicultural Advisory Council, Commonwealth of Australia) (1999) *Australian Multiculturalism for a New Century: Towards Inclusiveness* (Canberra: NMAC).

Noble, K. B. (1991) 'Congo expelling Zairian citizens', *New York Times*, 11 December 1991.

Noiriel, G. (1988) *Le creuset français: Histoire de l'immigration XIXe-XXe siècles* (Paris: Seuil).

Nygård, A.-C. and Stacher, I., 'Towards a harmonised migration and asylum regime in Europe', in C. Wallace and D. Stola (eds), *Patterns of Migration in Central Europe* (Basingstoke: Palgrave).

OECD (1987) *The Future of Migration* (Paris: OECD).
OECD (1992) *Trends in International Migration: Annual Report 1991* (Paris: OECD).
OECD (1994) *Trends in International Migration: Annual Report 1993* (Paris: OECD).
OECD (1995) *Trends in International Migration: Annual Report 1994* (Paris: OECD).
OECD (1997) *Trends in International Migration: Annual Report 1996* (Paris: OECD).
OECD (1998a) *Migration, Free Trade and Regional Integration in North America* (Paris: OECD).
OECD (1998b) *Trends in International Migration: Annual Report 1998*, Paris: OECD.
OECD (2000) *Combating the Illegal Employment of Foreign Workers* (Paris: OECD).
OECD (2001) *Trends in International Migration: Annual Report 2000* (Paris: OECD).
Oezcan, V. (2002) *Germany: Immigration in Transition* (Migration Policy Institute) (http://www.migrationinformation.org/ Vol.2002).
Oishi, N. (1995) 'Training or employment? Japanese immigration policy in dilemma', *Asian and Pacific Migration Journal*, 4: 2–3.
Okólski, M. (2001) 'Incomplete migration: a new form of mobility in Central and Eastern Europe. The Case of Polish and Ukrainian Migrants', in C. Wallace and D. Stola (eds), *Patterns of Migration in Central Europe* (Basingstoke: Palgrave).
Okuda, M. (2000) 'Asian newcomers in Shinjuku and Ikebukuro area, 1988–98: reflections on a decade of research', *Asian and Pacific Migration Journal*, 9: 3.
Okunishi, Y. (1996) 'Labor contracting in international migration: The Japanese case and implications for Asia', *Asian and Pacific Migration Journal*, 5: 2–3.
OMA (Office of Multicultural Affairs) (1989) *National Agenda for a Multicultural Australia* (Canberra: AGPS).
ONS (Office of National Statistics) (2001) *Population Trends. Autumn: 105* (London: ONS)
Oriol, P. (2001) 'Des commissions consultatives au droit de vote', *Migrations Société*, 13: 73.
Ouedraogo, D. (1994) 'Population, migrations et développement', *Revue Européenne des Migrations Internationales*, 10: 3.
Papantoniou-Frangouli, M. and Leventi, K. (2000) 'The legalization of aliens in Greece', *International Migration Review*, 34: 3.
Parisot, T. (1998) 'Quand l'immigration tourne à l'esclavage', *Le Monde Diplomatique*, 20–21 June 1998.
Pascoe, R. (1992) 'Place and community: the construction of Italo-Australian space', in S. Castles, C. Alcorso, G. Rando and E. Vasta (eds), *Australia's Italians: Culture and Community in a Changing Society* (Sydney: Allen & Unwin).
Péan, L. (1982) 'L'alliance hégémonique insulaire', *Le Monde Diplomatique*, August 1982.
Pelligrino, A. (1984) 'Illegal immigration from Colombia', *International Migration Review*, 18: 3.

Pe-Pua, R., Mitchell, C., Iredale, R. and Castles, S. (1996) *Astronaut Families and Parachute Children: The Cycle of Migration from Hong Kong* (Canberra: AGPS).

Però, D. (2001) 'Inclusionary rhetoric/exclusionary practice: an ethnographic critique of the Italian Left in the context of migration', in R. King (ed.), *The Mediterranean Passage: Migration and New Cultural Encounters in Southern Europe* (Liverpool: Liverpool University Press).

Perotti, A. and Thepaut, F. (1991) 'Les répercussions de la guerre du golfe sur les arabes et les juifs de France', *Migrations Société*, 3: 14.

Pfahlmann, H. (1968) *Fremdarbeiter and Kriegsgefangene in der deutschen Kriegswirtschaft 1939–45* (Darmstadt: Wehr and Wissen).

Phizacklea, A. (ed.) (1983) *One Way Ticket? Migration and Female Labour* (London: Routledge and Kegan Paul).

Phizacklea, A. (1990) *Unpacking the Fashion Industry: Gender, Racism and Class in Production* (London: Routledge).

Picquet, M., Pelligrino, A. and Papail, J. (1986) 'L'immigration au Venezuela', *Revue Européenne des Migrations Internationales*, 2: 2.

Pillai, P. (1999) 'The Malaysian state's response to migration', *Sojourn*, 14(1).

Piore, M. J. (1979) *Birds of Passage: Migrant Labor and Industrial Societies* (Cambridge: Cambridge University Press).

Pipes, D. and Durán, K. (2002) *Muslim Immigrants in the United States* (Washington, DC: Center for Immigration Studies).

Pool, I. and Bedford, R. (1996) *Macro Social Change in New Zealand: Historical and International Contexts*, Hamilton: University of Waikato Population Studies Centre Discussion Papers no. 18.

Portes, A. (ed.) (1995) *The Economic Sociology of Immigration* (New York: Russell Sage Foundation).

Portes, A. (1999) 'Conclusion: towards a new world: the origins and effects of transnational activities', *Ethnic and Racial Studies*, 22: 2.

Portes, A. and Bach, R. L. (1985) *Latin Journey: Cuban and Mexican Immigrants in the United States* (Berkeley: University of California Press).

Portes, A. and Böröcz, J. (1989) 'Contemporary immigration: theoretical perspectives on its determinants and modes of incorporation', *International Migration Review*, 23: 83.

Portes, A., Guarnizo, L. E. and Landolt, P. (1999) 'The study of transnationalism: pitfalls and promise of an emergent research field', *Ethnic and Racial Studies*, 22: 2.

Portes, A. and Rumbaut, R. E. (1996) *Immigrant America: A Portrait*, Second Edition (Berkeley, Los Angeles, London: University of California Press).

Potts, L. (1990) *The World Labour Market: A History of Migration* (London: Zed Books).

Price, C. (1963) *Southern Europeans in Australia* (Melbourne: Oxford University Press).

Prost, A. (1966) 'L'immigration en France depuis cent ans', *Esprit*, 34: 348.

Purdum, S. (2002) 'Several US allies criticized in Powell Report on slave trading', *New York Times*, 6 June 2002.

Rath, J. (1988) 'La participation des immigrés aux élections locales aux Pays-Bas', *Revue Européenne des Migrations Internationales*, 4: 3.

Ravenstein, E. G. (1885) 'The laws of migration', *Journal of the Statistical Society*, 48.

Ravenstein, E. G. (1889) 'The laws of migration', *Journal of the Statistical Society*, 52.

Rawls, J. (1985) 'Justice as fairness: political not metaphysical', *Philosophy and Public Affairs*, 14: 3.

Reed, R. (1977) 'National Front: British threat', *New York Times*, 18 August 1977.

Reitz, J. G. (1998) *Warmth of the Welcome: The Social Causes of Economic Success for Immigrants in Different Nations and Cities* (Boulder, CO: Westview Press).

Research Perspectives on Migration (1997) 1: 2 (Washington, DC: The Urban Institute and the Carnegie Endowment for International Peace).

Rex. J. (1986) *Race and Ethnicity* (Milton Keynes: Open University Press).

Rex, J. and Mason, D. (eds) (1986) *Theories of Race and Ethnic Relations* (Cambridge: Cambridge University Press).

Reyneri, E. (2001) *Migrants' Involvement in Irregular Employment in the Mediterranean Countries of the European Union* (Geneva: International Labour Organization).

Reynolds, H. (1987) *Frontier* (Sydney: Allen & Unwin).

Ricca, S. (1990) *Migrations internationales en Afrique* (Paris: L'Harmattan).

Richard, A. O. (1999) *International Trafficking in Women to the United States: A Contemporary Manifestation of Slavery and Organized Crime* (Washington, DC: Center for the Study of Intelligence).

Rocha-Trindade, M. (ed.) (1993) *Recent Migration Trends in Europe* (Lisbon: Universidade Aberta).

Romero, F. (1993) 'Migration as an issue in European interdependence and integration: the case of Italy', in A. Milward, F. Lynch, R. Ranieri, F. Romero and V. Sørensen (eds), *The Frontier of National Sovereignty* (London: Routledge).

Rosenau, J. N. (1997) *Along the Domestic Foreign Frontier* (Cambridge: Cambridge University Press).

Rotte, R. and Stein, P. (eds) (2002) *Migration Policy and the Economy: International Experiences* (Munich: Haus Seidel Stiftung).

Roussillon, A. (1985) 'Migrations de main-d'oeuvre et unité arabe: les enjeux du modèle irakien', *Revue Tiers Monde*, no. 10.

Rubio-Marin, R. (2000) *Immigration as a Democratic Challenge* (Cambridge: Cambridge University Press).

Rudolph, H. (1996) 'The new *Gastarbeiter* system in Germany', *New Community*, 22: 2.

Safir, N. (1999) 'Emigration Dynamics in the Maghreb', in R. Appleyard (ed.) *Emigration Dynamics in Developing Countries*, Vol. IV: *The Arab Region* (Aldershot: Ashgate).

Sanz, L. C. (1989) 'The impact of Chilean migration on employment in Patagonia', *International Migration*, 27: 2.

Sassen, S. (1988) *The Mobility of Labour and Capital* (Cambridge: Cambridge University Press).

Scalabrini Migration Center (2001) *Scalabrini Migration Atlas 2001* (Quezon City, Philippines: Scalabrini Migration Center) (http://www.scalabrini.asn.au/atlas/Vol.2001).

Schierup, C.-U. and Ålund, A. (1987) *Will they still be Dancing? Integration and Ethnic Transformation among Yugoslav Immigrants in Scandinavia* (Stockholm: Almquist & Wiksell International).
Schnapper, D. (1991) 'A host country of immigrants that does not know itself', *Diaspora*, 1: 3.
Schnapper, D. (1994) *La Communauté des Citoyens* (Paris: Gallimard).
Scott, J. (2002) 'Foreign born in US at record high', *New York Times*, 7 February 2002.
Seccombe, I. J. (1986) 'Immigrant workers in an emigrant economy', *International Migration* 24: 2.
Seccombe, I. J. and Lawless, R. I. (1986) 'Foreign worker dependence in the Gulf and international oil companies', *International Migration Review*, 20: 3.
Segura, C. (2002) *Migration and Development in the Dominican Republic*, Unpublished paper.
Select Commission on Immigration and Refugee Policy (1981) *Staff Report* (Washington, DC)
Semyonov, M. and Lewin-Epstein, N. (1987) *Hewers of Wood and Drawers of Water* (Ithaca, NY: ILR Press).
Seol, D.-H. and Skrentny, J. D. (2003) 'South Korea: Importing undocumented workers', in W. Cornelius, P. L. Martin and J. F. Hollifield (eds) *Controlling Immigration: A Global Perspective* (Stanford CA: Stanford University Press).
Seton-Watson, H. (1977) *Nations and States* (London: Methuen).
Shain, Y. (1989) *Frontiers of Loyalty: Political Exiles in the Age of the Nation State* (Middletown, CT: Wesleyan University Press).
Shami, S. (ed.) (1994) *Population Displacement and Resettlement: Development and Conflict in the Middle East* (New York: Center for Migration Studies).
Shami, S. (1999) 'Emigration Dynamics in Jordan, Palestine and Lebanon', in R. Appleyard (ed.), *Emigration Dynamics in Developing Countries*, Vol. IV: *The Arab Region* (Aldershot: Ashgate).
Shaw, M. (2000) *Theory of Global State* (Cambridge: Cambridge University Press).
Shimpo, M. (1995) 'Indentured migrants from Japan', in R. Cohen (ed.), *The Cambridge Survey of World Migration* (Cambridge: Cambridge University Press).
Shu, J., Khoo, S. E., Struik, A. and McKenzie, F. (1994) *Australia's Population Trends and Prospects 1993* (Canberra: AGPS).
Siddique, M. (ed.) (2001) *International Migration into the 21st Century* (Cheltenham/Northampton: Edward Elgar).
Silvestri, S. (1999) 'Libya and transatlantic relations: An Italian view', in R. Haass (ed.), *Transatlantic Tension* (Washington, DC: Brookings).
Simon, G. (ed.) (1990) *Les effets des migrations internationales sur les pays d'origine: le cas du Maghreb* (Paris: SEDES).
Singaby, T. E. (1985) 'Migrations et capitalisation de la campagne en Egypte: La reconversion de la famille paysanne', *Revue Tiers Monde*, 26: 103.
Sinn, E. (ed.) (1998) *The Last Half Century of Chinese Overseas* (Hong Kong: Hong Kong University Press).
Sivanandan, A. (1982) *A Different Hunger* (London: Pluto Press).
Skeldon, R. (1992) 'International migration within and from the East and South-east Asian region: a review essay', *Asian and Pacific Migration Journal*, 1: 1.

Skeldon, R. (ed.) (1994) *Reluctant Exiles? Migration from Hong Kong and the New Overseas Chinese* (Hong Kong: Hong Kong University Press).

Smith, A. D. (1986) *The Ethnic Origins of Nations* (Oxford: Blackwell).

Smith, A. D. (1991) *National Identity* (Harmondsworth: Penguin).

Smith, J. P. and Edmonston, B. (1997) *The New Americans: Economic, Demographic and Fiscal Effects of Immigration* (Washington, DC: National Academy Press).

Smith, R. (2001) 'Current dilemmas and future prospects of the Inter-American Migration System', in A. Zolberg and P. Benda (eds) *Global Migrants, Global Refugees* (New York: Berghahn Books).

Solomos, J. (1993) *Race and Racism in Contemporary Britain*, Second Edition (London: Macmillan).

Solomos, J. and Back, L. (1995) *Race, Politics and Social Change* (London: Routledge).

Solomos, J. and Wrench, J. (1993) *Racism and Migration in Europe* (Oxford and New York: Berg).

Soysal, Y. N. (1994) *Limits* of *Citizenship: Migrants and Postnational Membership in Europe* (Chicago and London: University of Chicago Press).

Spencer, W. (2002) *The Middle East* (Guilford, CT: McGraw-Hill/Dushkin).

Stahl, C. (ed.) (1988) *International Migration Today: Emerging Issues* (Paris: UNESCO).

Stahl, C. (1993) 'Explaining international migration', in C. Stahl, R. Ball, C. Inglis and P. Gutman (eds), *Global Population Movements and their Implications for Australia* (Canberra: Australian Government Publishing Service).

Stalker, P. (1994) *The Work of Strangers* (Geneva: ILO).

Stalker, P. (2000) *Workers Without Frontiers: The Impact of Globalization on International Migration* (Boulder: Lynne Rienner).

Stark, O. (1991) *The Migration of Labour* (Oxford: Blackwell).

Stasiulis, D. K. (1988) 'The symbolic mosaic reaffirmed: multiculturalism policy', in K. A. Graham (ed.), *How Ottawa Spends, 1988/89* (Ottawa: Carleton University Press).

Stasiulis, D. K. and Jhappan, R. (1995) 'The fractious politics of a settler society: Canada', in D. K. Stasiulis and N. Yuval-Davis (eds), *Unsettling Settler Societies* (London: Sage).

Stasiulis, D. K. and Yuval-Davis, N. (eds) (1995) *Unsettling Settler Societies* (London: Sage).

Statistics Canada (2002) *Immigrant Population by Place of Birth and Period of Immigration, 1996 Census, Canada*, Vol. 2002 (Statistics Canada)

Steinberg, S. (1981) *The Ethnic Myth: Race, Ethnicity and Class in America* (Boston, MA: Beacon Press).

Stichter, S. (1985) *Migrant Labourers* (Cambridge: Cambridge University Press).

Stirn, H. (1964) *Ausländische Arbeiter im Betrieb* (Frechen/Cologne: Bartmann).

Stola, D. (2001) 'Poland', in C. Wallace and D. Stola (eds), *Patterns of Migration in Central Europe* (Basingstoke: Palgrave).

Strozza, S. and Venturini, A. (2002) 'Italy is no longer a country of emigration. Foreigners in Italy: how many, where they come from', in R. Rotte and P. Stein (eds), *Migration Policy and the Economy: International Experiences* (Munich: Hans Seidel Stiftung).

Studlar, D. T. and Layton-Henry, Z. (1990) 'Non-white minority access to the political agenda in Britain', *Policy Studies Review*, 9: 2 (Winter).

Suhrke, A. and Klink, F. (1987) 'Contrasting patterns of Asian refugee movements: the Vietnamese and Afghan syndromes', in J. T. Fawcett and B. V. Cariño (eds), *Pacific Bridges: The New Immigration from Asia and the Pacific Islands* (New York: Center for Migration Studies).

Süssmuth, R. (2001) *Zuwanderung gestalten, Integration fördern: Bericht der unabhängigen Kommission 'Zuwanderung'* (Berlin: Bundsesminister des Innern).

Suzuki, H. (1988) 'A new policy for foreign workers in Japan? Current debate and perspective', *Waseda Business and Economic Studies*, 24.

Tapinos, G. (1984) 'Seasonal workers in French agriculture', in P. Martin (ed.), *Migrant Labor in Agriculture* (Davis: Gianni Foundation of Agricultural Economics).

Taylor, E. J. (1987) 'Undocumented Mexico–US migration and the returns to households in rural Mexico', *American Journal of Agricultural Economics*, 69.

Taylor, J. E. (1999) 'The new economics of labour migration and the role of remittances in the migration process', *International Migration*, 37(1).

Taylor, J. E., Martin, P. L. and Fix, F. (1997) *Poverty amid Prosperity* (Washington, DC: The Urban Institute).

Teitelbaum, M. S. and Weiner, M. (1995) *Threatened Peoples, Threatened Borders: World Migration and US Policy* (New York and London: Norton).

Tekeli, I. (1994) 'Involuntary displacement and the problem of resettlement in Turkey from the Ottoman Empire to the present', in S. Shami (ed.), *Population Displacement and Resettlement: Development and Conflict in the Middle East* (New York: Center for Migration Studies).

Tiersky, R. (1994) *France in the new Europe: Changing yet Steadfast* (Belmont, CA: Woodsworth).

Tirtosudarmo, R. (2001) 'Demography and security: transmigration policy in Indonesia', in M. Weiner and S. S. Russell (eds), *Demography and National Security* (New York and Oxford: Berghahn Books).

Togman, J. (2002) *The Ramparts of Nations* (Westport and London: Praeger).

Tribalat, M. (1995) *Faire France: Une Enquête sur les Immigrés et leurs Enfants* (Paris: La Découverte).

Trlin, A. D. (1987) 'New Zealand's admission of Asians and Pacific Islanders', in J. T. Fawcett and B. V. Cariño (eds), *Pacific Bridges: The New Immigration from Asia and the Pacific Islands* (New York: Center for Migration Studies).

Tsuzuki, K. (2000) '*Nikkei* Brazilians and local residents: a study of the H Housing Complex in Toyota City', *Asian and Pacific Migration Journal*, 9(3).

UNHCR (1995) *The State of the World's Refugees: In Search of Solutions* (Oxford: Oxford University Press).

UNHCR (2000a) *Global Report 2000: Achievements and Impact* (Geneva: United Nations High Commissioner for Refugees).

UNHCR (2000b) *The State of the World's Refugees: Fifty Years of Humanitarian Action* (Oxford: Oxford University Press).

UNHCR (2002) *Afghan Humanitarian Update, 63* (Geneva: UNHCR).

UNPD (2000) *Replacement Migration: Is it a Solution to Declining and Ageing Populations?* (New York: UNPD).

USCR (US Committee for Refugees) (1996) *World Refugee Survey 1996* (Washington, DC: Immigration and Refugee Services of America).
USCR (US Committee for Refugees) (2001) *World Refugee Survey 2001* (Washington, DC: USCR, Immigration and Refugee Services of America).
USCR (US Committee for Refugees) (2002) *World Refugee Survey 2002* (Washington DC: USCR, Immigration and Refugee Services of America).
US Department of Labor (1989) *The Effects of Immigration on the US Economy and Labor Market* (Washington, DC: US Government Document).
US General Accounting Office (2000) *Alien Smuggling* (Washington, DC: US General Accounting Office).
Van Hear, N. (1998) *New Diasporas: the Mass Exodus, Dispersal and Regrouping of Migrant Communities* (London: UCL Press).
Vasta, E. (1990) 'Gender, class and ethnic relations: the domestic and work experiences of Italian migrant women in Australia', *Migration*, 7.
Vasta, E. (1992) 'The second generation', in S. Castles, C. Alcorso, G. Rando and E. Vasta (eds), *Australia's Italians: Culture and Community in a Changing Society* (Sydney: Allen & Unwin).
Vasta, E. (1996) 'Dialectics of domination: racism and multiculturalism', in E. Vasta and S. Castles (eds), *The Teeth are Smiling: The Persistence of Racism in Multicultural Australia* (Sydney: Allen & Unwin).
Vasta, E. and Castles, S. (eds) (1996) *The Teeth are Smiling: The Persistence of Racism in Multicultural Australia* (Sydney: Allen & Unwin).
Vasta, E., Rando, G., Castles, S. and Alcorso, C. (1992) 'The Italo-Australian community on the Pacific rim', in S. Castles, C. Alcorso, G. Rando and E. Vasta (eds), *Australia's Italians: Culture and Community in a Changing Society* (Sydney: Allen & Unwin).
Verbunt, G. (1985) 'France', in T. Hammar (ed.), *European Immigration Policy: A Comparative Study* (Cambridge: Cambridge University Press).
Vertovec, S. (1999) 'Conceiving and researching transnationalism', *Ethnic and Racial Studies*, 22(2).
Vuddamalay, V. (1990) 'Tendances nouvelles dans le commerce étranger en France', *Migrations et Société*, 2: 11.
Waldinger, R., Aldrich, H., Ward, R. and Associates (1990) *Ethnic Entrepreneurs Immigrant Business in Industrial Societies* (Newbury Park, London, New Delhi: Sage).
Wallace, C. and Stola, D. (eds) (2001) *Patterns of Migration in Central Europe* (Basingstoke/New York: Palgrave).
Wallman, S. (1986) 'Ethnicity and boundary processes', in J. Rex and D. Mason, (eds), *Theories of Race and Ethnic Relations* (Cambridge: Cambridge University Press).
Weber, M. (1968) *Economy and Society: An Outline of Interpretive Sociology*, G. Roth, and C. Wittich (eds) (New York: Bedminster Press).
Weil, P. (1991a) 'Immigration and the rise of racism in France: the contradictions of Mitterrand's policies', *French Society and Politics*, 9: 3–4.
Weil, P. (1991b) *La France et ses Étrangers* (Paris: Calmann-Levy).
Weiner, M. (ed.) (1993) *International Migration and Security* (Boulder, CO: Westview Press).

Weiner, M. and Hanami, T. (eds) (1998) *Temporary Workers or Future Citizens? Japanese and US Migration Policies* (New York: New York University Press).

Werner, H. (1973) *Freizügigkeit der Arbeitskräfte und die Wanderungsbewegungen in den Ländern der Europäischen Gemeinschaft* (Nuremburg: Institut für Arbeitsmarkt-und Berufsforschung).

Westin, C. (2000) *Settlement and Integration Policies towards Immigrants and their Descendants in Sweden* (Geneva: ILO).

White, G. (1999) 'Encouraging unwanted immigration: a political economy of Europe's efforts to discourage North African immigration', *Third World Quarterly*, 20: 4.

Widgren, J. (1994) *The Key to Europe: A Comparative Analysis of Entry and Asylum Policies in Western Countries* (Vienna: ICMPD).

Wieviorka, M. (1991) *L'Espace du Racisme* (Paris: Seuil).

Wieviorka, M. (1992) *La France Raciste* (Paris: Seuil).

Wieviorka, M. (1995) *The Arena of Racism* (London: Sage).

Wihtol de Wenden, C. (1987) *Citoyenneté, Nationalité et Immigration* (Paris: Arcantère Éditions).

Wihtol de Wenden, C. (1988) *Les Immigrés et la Politique: cent-cinquante ans d'évolution* (Paris: Presses de la Fondation Nationale des Sciences Politiques).

Wihtol de Wenden, C. (1995) 'Generational change and political participation in French suburbs', *New Community*, 21: 1.

Wihtol de Wenden, C. and Leveau, R. (2001) *La Beurgeoisie: les trois ages de la vie associative issue de l'immigration* (Paris: CNRS Editions).

Willard, J. C. (1984) 'Conditions d'emploi et salaires de la main d'oeuvre étrangère', *Economie et Statistiques*, January 1984.

Wong, D. (1996) 'Foreign domestic workers in Singapore', *Asian and Pacific Migration Journal*, 5: 1.

Wooden, M. 1994. 'The economic impact of immigration', in M. Wooden, R. Holton, G. Hugo and J. Sloan (eds), *Australian Immigration: A Survey of the Issues* (Canberra: AGPS).

Wüst, A. (2002) *Wie Wählen Neubürger?* (Opladen: Leske & Budrich).

Zlotnik, H. (1999) 'Trends of international migration since 1965: what existing data reveal', *International Migration*, 37(1).

Zolberg, A. R. and Benda, P. M. (eds) (2001) *Global Migrants, Global Refugees: Problems and Solutions* (New York and Oxford: Berghahn Books).

Zolberg, A. R., Suhrke, A. and Aguao, S. (1989) *Escape from Violence* (New York: Oxford University Press).

Author Index

Subject Index

Note: page numbers in **bold** indicate extended discussion.